Dictionary of Physical Geography

Dictionary of Physical Geography

H.M. Saxena
Rahul Saxena
Pooja Saxena

Rawat Publications
Jaipur • New Delhi • Bangalore • Guwahati • Kolkata

ISBN 978-81-316-0283-6 (HB)
ISBN 978-81-316-0285-0 (PB)

Reprinted, 2016

Published by
Prem Rawat for **Rawat Publications**
Satyam Apts, Sector 3, Jawahar Nagar, Jaipur 302 004 (India)
Phone: 0141 265 1748 / 7006 Fax: 0141 265 1748
E-mail: info@rawatbooks.com
Website: www.rawatbooks.com

New Delhi Office
4858/24, Ansari Road, Daryaganj, New Delhi 110 002
Phone: 011 2326 3290

Also at *Bangalore, Guwahati* and *Kolkata*

Typeset by Rawat Computers, Jaipur
Printed at Nice Printing Press, New Delhi

Preface

Physical geography encompasses a wide range of themes covering geomorphology, climatology, oceanography, ecology and environmental geography. This dictionary provides, in one volume a comprehensive coverage of the major terms used in physical geography. Although, physical geography is a vast subject and it is very difficult to include all the terms and definitions in one volume. But we have tried to include all important facts and concepts which are necessary for its basic knowledge.

Our main objective for the preparation of this dictionary is to provide definitions and explanations of technical terms used in physical geography. Basically, this dictionary is designed for the students of geography as well as for those who are interested in physical geography. Not only this, the book will also be useful for those who are preparing for the competitive examinations.

In this dictionary, all the terms, concepts and principles of physical geography have been explained in brief and in simple language. In all, we have included more than 2,500 terms in the present volume. Another distinctive feature of this dictionary is that it contains as many as 233 illustrations, which will be helpful in proper understanding of the technical terms.

The authors acknowledge their gratitude and thanks all the authors, both foreign and Indian, whose works have been consulted to clarify much of what we wanted to say. We are thankful to the members of our family and friends for their moral support. We are also thankful to the entire team of Rawat Publications, Jaipur for publication of the book.

H.M. Saxena
Rahul Saxena
Pooja Saxena

A

aa flow A lava flow with a surface typified by angular, jagged blocks. The word **aa** is Hawaiian. It contrasts with **pahoehoe flow**. Examples are found in Mauna Loa in the Hawaiian Island, Mt. Etna in Sicily and Mt. Hekla in Iceland.

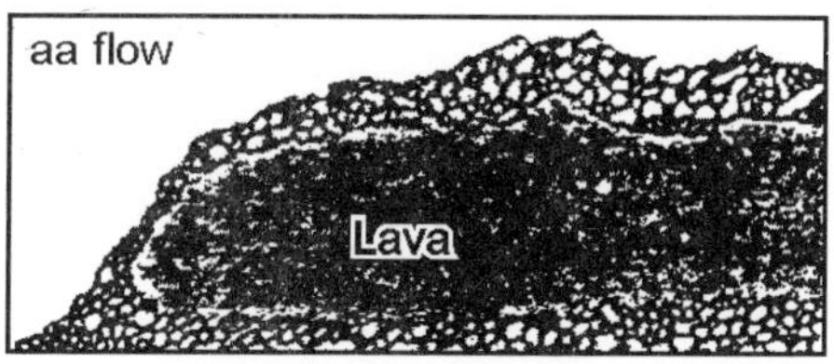

abandoned channel A former stream channel through which water no longer flows.

abandonment facies association The facies associations are formed under conditions of rising sea level, when clastic deposition has ceased and sediment is deposited very slowly.

abime It is a vertical shaft in karstic limestone terrain.

abiogenesis The concept which states the development of living organism from non-living matter. This concept was earlier accepted for the origin of life on the earth but with modern research this is not acceptable.

abiotic The non-biological factors in ecosystems, such as solar radiation and water.

ablation This is a process through which reduction of a glacier takes place by melting, evaporation, iceberg calving or deflation. Melting is the most important process in temperate and sub-polar regions, while in Antarctic the most important ablation process is the calving of ice shelves.

ablation moraine The debris of rock left at the side of a glacier after ablation has occurred.

ablation zone The part of glacier or ice sheet lying below the equilibrium line, where ice surface is lowered by melting during summer.

abrasion The mechanical wearing away of a rock by friction, rubbing, scraping or grinding. It is a processes by which solid rock is eroded by rock fragments transported by running water, glacier ice, wind and breaking waves. Abrasion is most-effective when the impact of the particles on bed rock is vigourous, and the particles themselves are coarse, hard and angular.

absolute age (true age) Age of a mineral, rock, fossil or geologic event measured in terms of actual number of years.

absolute humidity The amount of water vapour contained within a unit volume of air, commonly expressed in gram per cubic metre (g cm^{-3}) of water vapour in the air.

absolute instability The condition of the atmosphere in which the environment lapse-rate exceeds the dry adiabatic lapse-rate. If air pockets begin to rise, as a result of initial heating and convection, they will lose heat adiabatically owing to expansion but remain warmer than the surrounding air; they will therefore, continue to rise to great heights.

absolute plate motion The motion of a Lithospheric plate with respect to a fixed frame of reference. Various frames of reference have been used, including those defined by hot spots, no net torque of all the plates, and palaeomagnetic Euler poles.

absolute stability The condition of the atmosphere in which the environmental lapse-rate is less than the saturated adiabatic lapse-rate.

absolute temperature The Kelvin temperature scale based on absolute zero (0°K), the point at which thermal molecular motion ceases. It is used in meteorology to express upper air temperature 0°K corresponds to –273.15°C or –460°F.

abstraction (1) A term used in geomorphology to denote river capture or piracy of streams. Although not in common usage, the action involves the widening and/or lowering of a stream or river valley more rapidly than those of its neighbours, thereby capturing or abstracting the adjoining drainage. (2) A term used to describe the process by which ice-sheets grow by 'laking-up' water within the hydrological cycle. Water abstracted in this way is temporarily prevented from returning to the oceans, thus leading to a glaciocustatic fall of sea level.

abyss From the Latin abyssus/ Greek abusos (literally 'bottomless'). It refers to an extremely deep ocean trough.

abyssal A term applied to the deepest part of the ocean floor (mainly between 2,200 and 5,500 metres), on which fine textured deposits (ooze) of calcareous or siliceous composition have accumulated to considerable thicknesses over long periods of geological time.

abyssal hills The part of the ocean floor consisting of hills rising as much as 1,000 m above the surrounding floor.

abyssal plains These are flat areas of the ocean floor, having a slope of less than 1:1000. Most abyssal plains lie at the base of a continental rise and are simply areas where abyssal hills are completely covered with sediment.

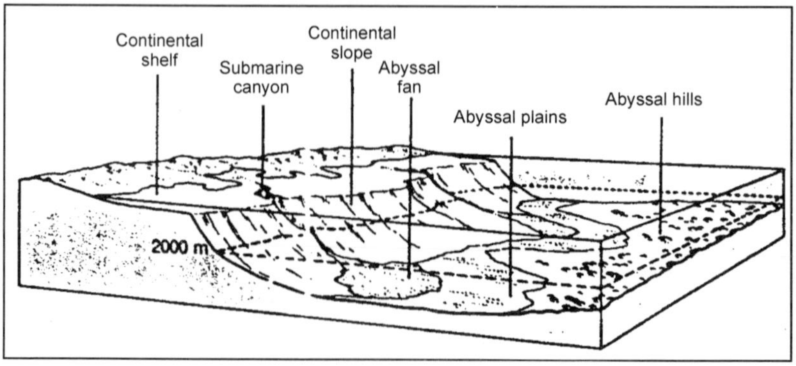

Abyssal hills and Abyssal plains

abyssal storm (Benthic storm) A large pulse of energy—possibly transferred from the surface that accelerates contour currents on the ocean floor to about 40 cm per second raising large amounts of fine sediment.

abyssal zone This is a zone of greatest ocean depth, i.e., below a depth of 2000 m. This zone lies seaward of and deeper than the bethyl zone (200–2,000 m) and covers approximately 75 per cent of the total ocean floor. It is the most extensive earth environment, cold, dark, with slow moving currents (less than a few centimetres per second), supporting fauna that typically are black or grey, delicately structured, and not streamlined.

accelerated erosion An increase in the rate of 'natural' erosive processes (such as rain wash or rain splash) owing to the activities of people (for example, in the clearing of vegetation, building construction, ploughing of fields, and overgrazing by domestic animals). The occurrence of accelerated erosion is shown by such obvious features as gullies on hill slopes in areas subject to soil erosion and the rapid erosion of previously stable stream banks.

accessory mineral The mineral components of a rock which do not occur in sufficient quantities to merit their inclusion in the definition or classification of the rock, that is, not an essential mineral.

accidented relief A descriptive term for an irregular or highly dissected terrain. It may have been derived from a French phrase, 'relief accident', meaning broken ground.

acclimatization The process of adjustment by an animal, normally with significant physiological effects, to a marked change in climate, related to a large change of altitude or temperature extremes.

accordant drainage A drainage pattern in which the streams are mainly guided by the underlying geological structure; for example some streams will follow the dip of the rocks, others will develop along lines of geological weakness (such as soft rock outcrops and fault-lines), and none will cut across structural features such as anticlinal folds.

accordant summits Those summits of hill or mountain groups which rise to a uniform surface at approximately the same elevation above sea-level. The uniform surface need not be horizontal (it may be gently curved, tilted or even undulating) but the important criterion is the summits of a mountain mass if rise and fall together, regardless of the underlying geology and structure.

accretion A process of accumulation. (1) In meteorology, the incremental growth of an ice particle each time it collides with a supercooled water droplet. (2) In geomorphology, the accumulation of mineral material in a particular locality or environment, e.g., lacustrine mud in a lake basin. (3) In astronomy and comogeny, a theory which suggests that the planets grew by the condensation of small cosmic dust particles into larger bodies.

accumulated temperature Normally the total number of days (or hours) since a given date, during which the mean temperature has been above or below a given threshold.

accumulation The addition of ice and snow to a glacier, as snow is added from snowfall and avalanches, it is compressed to form ice; this process usually takes place at the start of a glacier where temperatures are lower.

accumulation slope A type of slope in which material is being gained, leading to net increase in ground. The form of the slope is controlled by the rate of accumulation.

accumulation zone (i) In geomorphology, the term refers to that part of a slope which over a period of time, undergoes a net gain of material, leading to a progressive raising of the ground surface in that zone, (ii) that part of a glacier or ice sheet lying above the equilibrium line on which the dominant process is the addition of snow and ice. Winter snowfall is the most important component of accumulation; the snow which is not removed by ablation during the following summer remains as net accumulation and is gradually transformed by compaction and recrystallisation into glacier ice. The accumulation zone thus comprises a layered structure with a series of accumulation layers, separated by ablation surface.

acid lava This is a type of volcanic lava which is rich in silica, has a high melting point, (85°C) and flows slowly owing to its high viscosity. Acid lava forms steep-sided, dome-like volcanoes.

acid profile The acid concentration in ice core layers as a function of depth as determined from electrical measurements. The magnitude of some volcanic eruptions in the northern hemisphere has been estimated from the acidity of annual layers in ice cores taken in Greenland. This methodology is sometimes referred to as acidity signal or acidity record.

acid rain Rain contaminated by chemicals (notably sulphur dioxide, producing dilute sulphuric acid), which have been released from industrial chimneys and in particular coal burning power stations. It is increasingly believed that acid rain is responsible for the acidification of

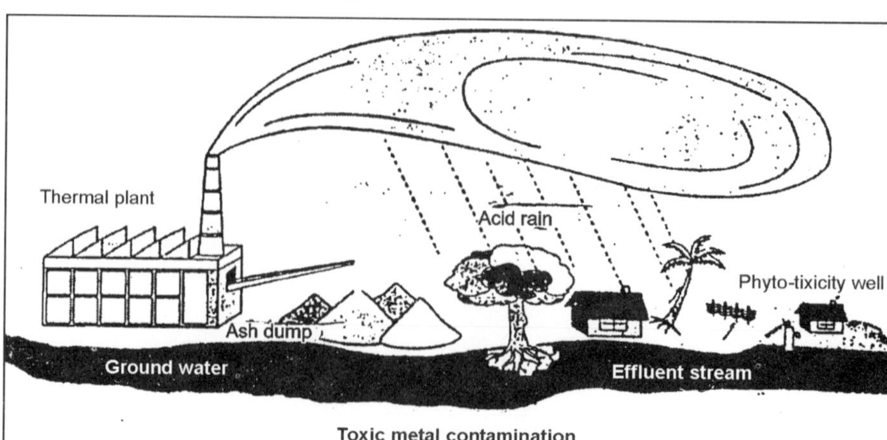

acid rain

rivers and lakes in uplands, the widespread destruction of fish and other wild life. Acid rain is now widely regarded as a serious environmental hazard, and campaigns to curtail the emission of sulphur dioxide are being mounted.

acid rock Any of the igneous rocks which contain a high proportion of silica (more than 66%). The main silica mineral is quartz, but some feldspars are also quite rich in silica. Acid igneous rocks are usually light in colour. The most common is granite. When they are weathered to form soil, they are generally infertile.

acid soil A bare, deficient soil with a pH below 7, possibly due to the parent material or due to leaching of the soluble bases (especially sodium and calcium) acidification, the development of soil acidity that is related to increasing hydrogen ion concentration.

aclinic line The magnetic equator, an irregularly curved line near the equator along which the compass needle does not dip from the horizontal.

actinometer An instrument used for measuring the chemical and heating influences of the solar radiation.

active fault A fault along which movement occurs fairly regularly, thereby causing an earth tremor or an earthquake.

active glacier A glacier which experiences relatively rapid flow, owing to a combination of a steep gradient and large inputs of snow and ice and outputs of melt water.

active layer In periglacial conditions, where permafrost exists, only the upper layer of ground thaws in summer; this upper layer, which is affected by summer thawing and winter freezing, is the active layer. Its lower limit is the permafrost table which causes the active layer to be poorly drained.

actual isotherm An isotherm for which the plotted values relate to actual mean temperatures, not those reduced to sea level by altitudinal correction.

adaptive radiation The evolutionary diversification of a group of organisms in response to the ecological pressures of different habitats. When new groups of organisms evolve or occupy a newly accessible environment they will tend to fill, over time, all available niches; they will 'radiate' along evolutionary lines in a genetic response to the stimulus of environmental diversity.

adiabatic It refers to the changes of temperature which occur in a mass of gas (air) when it is compressed (heated) or expanded (cooled) without the aid of any external sources of heating or cooling. The rate at which this occurs is termed the lapse rate. Generally, when a parcel of air rises into zones of lower atmospheric pressure, it expands at the expense of its internal energy and becomes warmer. For a saturated rising parcel, the fall of temperature is checked by the release of latent heat of condensation.

adjusted stream A stream which flows parallel to the strike of underlying rocks.

adret The side of a hill or valley that receives the most sunlight and which may, therefore, in high altitude areas have the most intensive land use and settlement.

adsorption The physical or chemical bonding of molecules, gaseous, liquid or dissolved, to the surface of solid bodies or their interiors, if the surfaces are porous or permeable.

advection The movement of property in the fluid natural environment (air and water) due solely to the velocity field of the fluid. Thus, heat is transferred through the atmosphere by winds. Advection may be resolved into two components; horizontal and vertical. In meteorology, advection refers frequently only to horizontal motion whereas vertical advection, particularly on the scale of the individual cloud, is often called convection.

advection fog A type of fog formed when a relatively warm, moist and stable mass of air moves laterally over a cooler surface, thereby reducing the temperature of the lower layers of the air mass until it reaches the dew point, when condensation takes place.

adventitive cone A small volcanic cone or crater which appears on the flanks of a major volcano. Sometimes referred to as a 'parasitic' cone.

aeolian A term pertaining to the wind; hence wind-borne, wind-blown or wind-deposited materials are often referred to as aeolian.

aeon (eon) A term which was once used to describe a very long but indefinite period of time but is now being replaced by a precise usage for a period of 10 years. Thus, the age of the earth is about 4.7 aeons.

aeration zone In the context of the hydrological cycle, the zone between the soil moisture zone and the capillary zone immediately above the water table. In this zone, vadose water is moving downwards under the influence of gravity and the zone may vary in thickness from zero to several hundred metres in arid regions.

aerial photograph A photograph taken from the air showing the characteristics of an area particularly useful in the making of maps. It may be taken at an oblique (i.e., slanting) angle or a vertical (i.e., straight down—a bird's eye view) angle.

aerobic A term denoting organisms which live in a location with free oxygen.

aerodynamic roughness Uneven flow of air caused by irregularities in the surface (which may be of solid, or of air of different density) over which the flow takes place.

aerography A term introduced in 1973 by E.H. Rapoport to describe studies of the geographical ranges of texa, e.g., subspecies, species, genera, families, etc.

aerology The scientific study of all the atmosphere above the surface layers.

aeromagnetic survey Survey of the earth's magnetic field based on data from magnetometers towed behind aircraft or suspended below helicopters to measure the total intensity of the geomagnetic field, or occasionally, components of this field. The resulting measurement can then be compared with theoretical models for the value of the field (magnetic anomalies), can be interpreted in terms of changes in the magnetic properties of the rocks below the survey line or grid.

aeronomy The scientific study of the upper atmosphere (above 50 km) where dissociation and ionisation of gas molecules occur.

aerosols An intimate mixture of two substances, one of which is in the liquid or solid state dispersed uniformly within a gas. The term is normally used to describe smoke, condensation nuclei, freezing nuclei or fog contained within the atmosphere, or other pollutants such as droplets containing sulphur-dioxide or nitrogen dioxide, aerosols tend to obscure visibility by scattering light. They tend to vary in size between about a millimicron (10^{-9} m) and one micron (10^{-6} m). Clouds are not normally considered to be aerosols because the droplets are too large and tend to fall due to gravity. Aerosols can remain in the atmosphere for long periods, collisions with air molecules keep them aloft.

aerosphere The entire gaseous envelope surrounding planet earth.

aestivation The dormancy of certain animals during the summer season, dry season or prolonged droughts. It is an important means of adaptation for desert animals, and can be contrasted with hibernation—a state of dormancy in the winter months.

affluent A stream or river flowing into another, a tributary.

afterglow A joint flow is the western sky seen occasionally just after sunset when the sun is 3° to 4° below the horizon, probably the result of scattering of white light by dust particles in the atmosphere.

aftershock A relatively minor vibration of the earth's crust following the main series of earthquake shock waves. Generally, it originates at or near the focus of the main earthquake and represents the subsequent readjustment of rocks that have been overstressed during the main rupture, according to the magnitude of the main earthquake the aftershocks may continue for hours, days or months, often causing the collapse of buildings or structures, damaged by the initial earthquake.

agate (mocha stone) A semiprecious stone, consisting of a type of silica (chalcedony), which forms concentric bands of variegated colours when seen in section. It occurs in cavity in volcanic and some other rocks.

age An interval of geological time in the chronometric standard scale of chrono-stratigraphic classification. Ages are subdivisions of epochs and may themselves be subdivided into chrons.

agestrophic This is a type of atmospheric motion in the troposphere in which the horizontal pressure gradient is not balanced with the deviating force (coriolis force) owing to the wind velocity. It is associated with vertical motion and hence the formation of cloud and weather.

agglomerate A type of rock which has been formed by volcanic activity. It consists of coarse blocks of rock or lava which have been thrown out by a volcanic eruption and then cemented together in volcanic ash.

aggradation A term used to describe the building up of sediments by rivers and wave action, hence, an aggraded river valley, and aggraded beach profile, more strictly, the term refers to deposition carried out to restore or maintain the condition of grade, e.g., an influx of sediment, may cause a stream to become overloaded; it will therefore deposit sediment and steepen its course (aggrade), thus increasing its energy and capacity to transport the increased load.

aggredational ice Ice which forms in horizontal layers underground at the base of the active layer in a permafrost environment. It develops as the upper surface of the permafrost layer rises.

aggregate In pedology, a single mass or cluster of soil particles which adhere together in such a way that they behave mechanically as a unit.

aggregation ratio It is a ratio of percentage weight of clay minerals, determined by mineralogical analysis, to the percentage weight of clay particles, determined by sedimentation methods. The ratio is meant to account for problems with the activity of a soil in which clay mineral particles and act as clay, but have a size corresponding to a value in the silt size range.

aggressivity In the context of limestone solution, the propensity of water to dissolve calcium carbonate. When water comes in contact with air, it dissolves an amount of carbon dioxide into the water. The resultant carbonic acid can dissolve calcium carbonate (via the theoretical compound $CaHCO_3$) until the aggressiveness of the water diminishes (when the CO_2 is used up). It is then said to be saturated with respect to calcium carbonate. When two saturated solutions of calcium carbonate are mixed, it is possible for the resulting water to be aggressive. This phenomenon was first identified by A. Bogli (1971).

agoclimatology The study of the interaction between climatological and hydrological factors and agriculture, including animal husbandry and forestry.

agonic line A shifting, irregular, imaginary line running through the earth's north and south magnetic poles along which the compass needle points to true north, hence the line of no magnetic variation (zero magnetic delineation).

agro forestry A system where trees are planted or encouraged on land where crops are grown. In other words, tree plantation is done along with agriculture.

agrometerology The application of the meteorological concepts in agriculture, specially in the measurement and analysis of the physical environment in agricultural systems.

agulhas current This is a part of the large-scale circulation of the southern Indian Ocean. It is a surface water current that flows along the east coast of southern Africa between latitudes 25° S and 40° S in south westerly direction. Flow velocity varies seasonally between 0.2 and 0.6 m/s.

A-horizon The top zone in a soil profile from which leaching removes soluble salts and colloids but where an admixture of mineral and organic matter is retained. Resting on the top of a horizon there may be an O-layer, which is where organic material accumulates.

aiguille A French term which has been widely adopted to describe narrow needle-shaped rocks. They are particularly well developed in the Chamonix region of the European Alphs.

air mass An almost homogenous mass of air of great lateral extent (generally hundreds or thousands of kilometre), which marked horizontal temperature and humidity uniformity acquired from prolonged contact in its place of origin or source region. It is separated from an adjacent air mass by well-defined fronts, which move as the air mass travels long distances, thereby modifying its original characteristics to a greater or lesser degree. According to its source region, an air mass will be classified as: Arctic (A), Antarctic (AA), Polar (P), Tropical (T) or Equatorial (E) based on its temperature characteristics.

air parcel An imaginary body of air to which may be assigned any or all of the dynamic and thermodynamic properties of the atmosphere. Investigation of the stability of the atmosphere is made most simply by the 'parcel method' in which it is hypothesised that a test parcel of air moves vertically with respect to its environment as represented by an assent curve on a Tephigram.

air pollution Air pollution is defined as the presence of natural or man-made or man caused contaminants in a given quantity, and for a given time in the atmosphere that are damaging to human plant or animal life; or to property; or sensually interfere with the comfortable enjoyment of life. Common air pollutants include smoke, sulphur dioxide, carbon monoxide, carbon dioxide, nitric oxide, nitrogen dioxide, ozone and lead.

air stream An air flow which has a distinctive source or place of origin but which is not necessarily homogenous. Thus, an air stream is distinguished by its direction of approach rather than any inherent temperature or moisture characteristics.

air wave A sound wave which travels through the air from a seismic shot. The speed of such a wave is approximately 330 m/s.

air-quality standard The prescribed level of a pollutant in the outside air that should not be exceeded during a specified time in a given area.

air-sea interaction It occurs because the atmosphere and the oceans constitute a single mechanical and thermodynamic system of two coupled fluids, and transfers of heat momentum, solids and gases take place across the air-sea boundary. The interaction occurs in a manner which is so complex that cause and effect cannot always be distinguished.

airy model A model to account for isostasy which is the lithosphere assumes a constant density (PC = 2670 kg/m^3) but in which topographic elevations (h) are compensated by the presence of 'roots' replacing high density metal rocks (pm = 3300 kg/m^3) by lower density lithosopheric rocks. The depth of the root (d) is equal to hpc/(p-pc).

airy phase When a high frequency seismic wave is superimposed on a low frequency ground wave, the two frequencies gradually approach one another until they merge, at which point they form a single wave with a relatively large amplitude, called the 'airy phase'.

aklè A type of parallel wavy dune pattern in which the inter dune areas are enclosed by crescentic elements of the dune ridges. It is sometimes called a fish scale dune pattern.

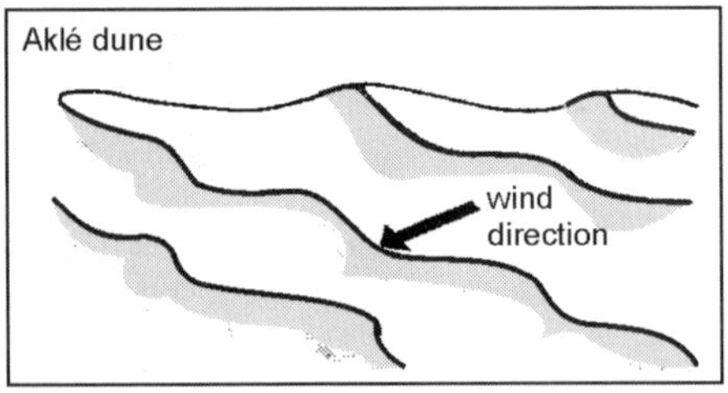

alas A large thermokarst depression with steep bounding walls and a flat floor, sometimes occupied by a shallow lake. Alases are well developed in Siberia, where they are upto 40 m deep and 15 km in diameter and result from the localised melting of the permafrost following destruction of the forest cover, ice wedges melt to give narrow linear depressions and hummocky terrain with mounts that eventually collapse to create a larger depression of melt water lakes, alases grow laterally, and may merge to give alas valleys.

Alaska current Oceanic water current produced by the deflection of the N. Pacific current by the N. American continent. It flows in a north-westward direction along the south-eastern margins of Alaska. The Alaska current is also called the Aleutian current in some texts.

albedo The proportion of the total solar radiation which is reflected by the earth surface, expressed as a decimal or percentage. The earth's mean albedo is 0.4 (40%), but the actual albedo from place to place varies greatly, depending on the precise nature of the surface (ice, snow, different types of vegetation, soils of differing hues, urban areas, etc.). The albedo for fresh clean snow exceeds 0.8, but is much reduced for coarsely crystalline glacier ice (0.5). For grassland, albedo ranges between 0.3 and 0.2 and for a dark peaty soil is less than 0.1.

alberta low These are storms common in Alberta, Canada and associated with heavy rain and snow. The storms form as a result of cyclone regeneration after passage over the Canadian Rockies; as they move

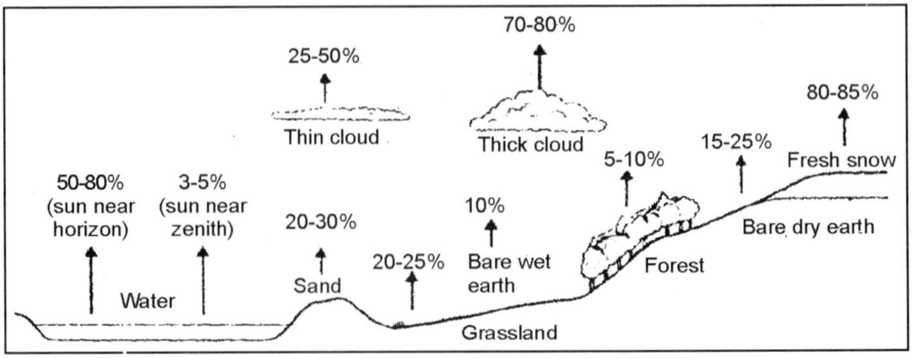

albedo

eastwards they bring very cold conditions with blizzards.

alcove An accurate steep-sided cavity on the side of a rock outcrop which has been produced by erosion by water, specially spring rapping or solutional processes.

alcrete Aluminium rich duricrust, often in the form of indurated bauxites. Generally, the products of the accumulation of aluminium rescuioxides within the zone of weathering.

Aleutian current (sub-arctic current) The oceanic current that flows westwards south of the Aleutian islands and parallel to but north of the N. Pacific current. The water mass is a mixture of water from the Kuroshio and Oyashio currents.

Aleutian low Region of the N. Pacific, near the Aleutian Islands where the average value of atmospheric pressure is low, owing to the frequency of low pressure systems (cyclones) moving into and occupying the region. Any one of these systems, when present, may be called an 'Aleutian low'.

Aleutian trench The oceanic trench which marks the boundary between the N. American plate and the Pacific plate. The subduction of the Pacific plate changes from normal to oblique from west to east along the trench, with the boundary becoming a transform pull before subduction continues in the Kuril Trench towards the eastern end of the Aleutian Trench there is an increasingly wide accretionary wedge and an absence of andestic volcanoes.

alfalfa A leguminous plant, medicago sativa, which is similar to clover. It has long roots and can survive spells of dry weather, alfalfa is a good fodder crop for cattle and is widely grown in the USA and Argentina with no leaves, flowers or vascular system.

alfisol These are relatively young, acid soils characterised by a clay-enriched 'B' horizon, commonly occurring beneath deciduous forests in humid, subhumid, temperate and subtropical climates.

algae A large group of simple plants containing chlorophyll. They are either aquatic or live in damp places and range in form from minute, i.e., Plankton to enormous sea weeds, algae bloom. A sudden increase in number of aquatic algae resulting from changes in the chemical content or temperature of water bodies.

aligned sequence This is a series of glacial overspill channels, successively crossing the divides between river valleys and showing a distinct alignment. It has been assumed in the past that the channels were developed along or close to a former ice sheet margin which dammed the proglacial streams to give a series of lakes, nourished by glacial melt water, on the site

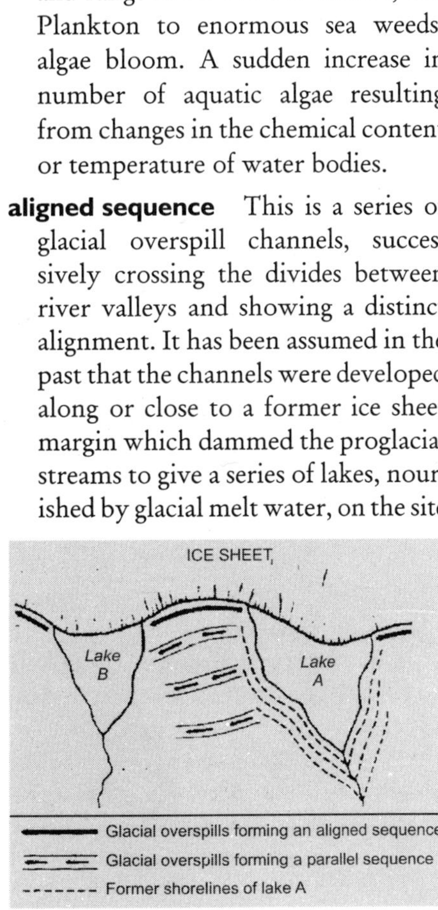

of pre-existing valleys. The channels were eroded as the lakes overspilled from one to another.

alimentation The accumulation in quantity of ice, through snowfall or avalanching in a firm field contributing to a glacier.

alkali flat A dried-out lake floor where intense evaporation has left behind a level plain of alkaline sediments, e.g., in the Jordan valley of Israel.

alkali soils (1) Soils having a high degree of alkalinity (pH > 8.5) and/or having a high exchangeable sodium content (15% or more of the exchange capacity). (2) Soils which contain enough alkali (sodium) to interfere with the growth of most plants.

alkaline soil Soil with a pH of 7 or more, which usually has a high concentration of carbonates.

alkalisation The process by which the exchangeable sodium content of a soil is increased.

Allen's rule A rule which states that warm blooded animals have a tendency for shorter legs, tails and ears in colder climates than in warmer climates. This correlates with the increased need to conserve body heat.

allerϕd A brief period of Late-Glacial time, dating from approximately 11000 BP, during which the climate became relatively warm and trees such as birches spread in S. Britain. The allerϕd was succeeded by a short phase (some 500 years in duration) when the climate again became colder and subject to higher winter snowfalls; at this time, periglacial erosion became very active, and many dry valleys were either formed or greatly modified.

allochthonous It refers to the material forming rocks which has been transported to the site of deposition whereas an allochthonous sediment is one in which the main constituents have been formed *in situ*.

allogenic The word referred as, 'Having an origin elsewhere', e.g., a river which is able to maintain its flow across an arid region (often a lowland) because it derives its discharge mainly from an adjacent humid region (often an upland), such as the Tan river in Kenya. Many limestone rivers are also allogenic, e.g., the Tarn in S. France, which rises on granite uplands and then flows westwards through the limestone plateau region of the Grands causes to Millau and beyond.

allogenic stream A stream which derives its discharge from outside the local area. The term is particularly used when local conditions do not generate much stream flow, for example in arid areas or ones with permeable rocks. Here stream flow may be derived from distant parts of the topographic catchment where precipitation and run-off are effective. Appearances can sometimes be misleading because stream flow may be augmented by contributions from local groundwater that are not readily appreciated.

allogenic succession This process is caused by an external environmental factor rather than by the organisms themselves. Instances are the change in vegetation induced by the inflow and accumulation of sediment in a pond or a change in regional climate.

allometric growth The growth of any organic system in dynamic equilibrium such that the ratio remains constant between each part of the system and each other part and also between each part of the system and the whole, e.g., as the number of stream segments in a network increases, the proportion of the segments falling into each stream order remain approximately unchanged.

allopatric Adjective applied to the condition of two species or subspecies being distributed in areas which do not overlap.

allophane An amorphous hydrated alumino silicate gel. The chemical composition is highly variable. The name is applied to any amorphous substance in clay.

alluvial cone This is a type of alluvial fan in which the slope angles are steeper and the deposited material is generally coarser and thicker.

alluvial fan A fan-shaped mass of alluvium formed where a rapidly flowing stream leaves a steep and narrow valley and enters a lowland or broad valley. At such points a reduction in gradient, and thus stream velocity occurs causing overloading and deposition. Alluvial fans form at different sites, e.g., where a small tributary valley joins a major glacial trough, as in the Rhon Valley, Switzerland; where a stream incised into a fault scarp flows into the scarp-pot zone; and where are desert stream passes through the mountain front to the Piedmont zone.

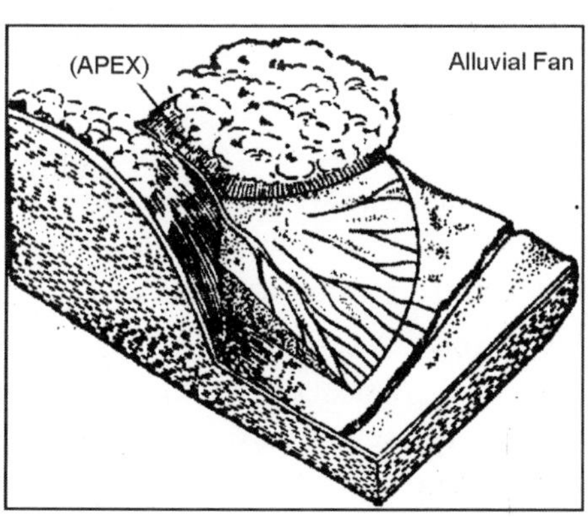

alluvial soils A soil group of the azonal soils, comprising those immature soils which form on recent alluvial deposits.

alluvial terrace It is similar to that of river terrace and was formed in the same way, i.e., by down cutting action of a river following rejuvenation, leaving at higher levels portions of its former flood plain.

alluviation The process of infilling by alluvium, deposited by rivers anywhere along their courses.

alluvium The sediments led down by the streams, alluvium is unconsolidated material forming features such as alluvial fans, flood plains, river terraces and deltas. The most common constituents are clay and silt (from the suspended sediment load of the stream) and sand and grave (from the bed load of the stream).

almwind A local wind of the Fohn variety which blows south across the Tatra mountains. and descends as a warmer wind onto the plains of South Poland in spring, thereby causing avalanches.

alp A topographical term to describe the high altitude bench or shoulder standing above the glaciated trough in a mountainous terrain. The relatively gentle slope, though snow covered in winter, provides valuable summer pasture for grazing animals.

alpha diversity The diversity that results when competition between species produces greater specialisation within the different niches that they occupy and reduces the variation within specific species, also known as niche diversification.

alpine The zone of a mountain above the tree line and below the level of permanent snow.

alpine glacier A long tongue-like glacier occupying a clearly defined mountain valley. The glacier is nourished by an accumulation zone, which may comprise a number of coalescent cirques or a high level ice field. The glacier may descend steeply from its source by way of an ice fall, or possess a relatively smooth unbroken long profile. Alpine glaciers are usually active, owing to the large winter snowfalls associated with high mountains and the resultant considerable inputs of ice, which pass quite rapidly through the glacier to the melt zone (ablation zone) at lower altitudes. Ice velocities are usually within the range 5–20 cm day^{-1}, but may be upto 5 m day^{-1} on steep ice falls.

alpine glow A series of phenomena observed over mountains around sunrise and sunset. At sunset, it begins when the sun is 2° above the horizon and when snow covered peaks assume a yellow-pint-purple tint.

alpine orogeny (alpine folding) A mountain building movement which culminated in mid-tertiary times having created the alpine ranges which stretch from the Pyrenees, Atlas and Alps in W. Europe and N. Africa through the Caucasus to the Himalayas of Asia.

altaids Structural trends, created by an orogeny in late Paleozoic times (carboniferous-Permian), which have given the predominant grain to the highland massifs of central Europe and northern Asia. Running in great arcuate lines, generally from W to E. These worn down mountain ranges comprise the Armorican range in central Europe and many of the ranges of S. Russia and N. China.

alteration A geological term to describe the change in the mineralogical composition of a rock, typically brought about by the action of hydrothermal solutions during metamorphism.

altimeter An instrument used to plot altitude, based on the average decrease in atmospheric pressure with increasing height. While *altimerty* is the measurement of heights and an **altigraph** is a self-recording altimeter.

altimetric frequency curve A frequency curve constructed by dividing an area into squares, determining the maximum altitude in each square and plotting the frequency of these determinations. A useful technique for rapid determination of generalised altitudes in an area, and much used by denudation chronologists for identifying erosion surface remnants.

altiphano An elevated plateau in the Andes, particularly in Bolivia. It is generally surrounded by mountains, which generate streams (and formerly glaciers) contributing to the inland drainage system of altiplano. Lake Titicaca and Lake Poopo occupied centres of these high plateaux are examples.

altiplanation A process of levelling by weathering and mass movement in a periglacial environment during which benches or terraces are cut in solid rock. The planation is achieved by a combination of the freeze-thaw process, solifluction and congeliturbation in which rock material is broken by frost action before being transported downslope. Thus, the altiplanation terraces partly rock cut usually include aggradational features of rock debris. The term is being replaced by cryoplanation.

altithermal A US term for the postglacial period from approximately 6,000 BC to 1,000 BC (8,000 +3,000 BP) when mid-summer mean temperature in western N. America went upto 4°F warmer than at present. This led to widespread drought and aggradation of the trunk river valleys owing to headward erosion in the tributaries.

altitude (1) The vertical height of a point above mean sea level. (2) The vertical height of a mountain or hill measured from base to summit.

altocumulus The word derived from the Latin altum (height) and cumulus (heap). A type of cloud (symbol-Ac), white and/or grey in colour, occurring in a patch sheet, or layer and consisting of bands, rolls or waves of globular masses, often separated by blue sky, although they may be close enough to merge, thereby obliterating the blue sky.

altostratus From the Latin altum (height) and stratus (spread out). A type of cloud (symbol As), grayish in colour, occurring in sheets or layers, and of fibrous or uniform appearance. It has areas thin enough to reveal a hazy indication of a watery looking sun as viewed through ground glass. Its base can occur between 2,000 and 7,000 m (6,500–23,000 ft), i.e., at middle latitudes. It forms from a thickening of cirrostratus, is associated with a warm front and is often a precursor of rainy weather.

alveolate relief A terrain of dome-shaped hills (Coupola), particularly common in granite areas of the humid tropics. The domes are thought to be the protsidiong tops of buried inselberg which would be uncovered by the striping of regolith should the drainage pattern become rejuvenated.

amber Fossil conifer resin which is brittle and hard, translucent to transparent, and yellow to brown in colour. It is found in sediments or on the shore and takes a fine polish.

ambient temperature The temperature of any part of the atmosphere which immediately surrounds an entity such as a cloud.

amenity A feature of the environment, which is perceived as being pleasant and attractive, e.g., fine scenery, an equable climate, open space, etc.

amensalism An interaction of species populations, in which one population is inhibited while the other is unaffected.

ammonium fixation The absorption or adsorption of ammonium ions by the soil in such a way that they become relatively insoluble in water and largely unexchangeable by the usual processes of caution.

amorphous cloud Continuous cover of low, featureless cloud (e.g., nimbostratus), often producing rain.

amphiboli (amphibole) One of the ferromagnesian silicate mineral group, e.g., hornblende.

amphibolite A medium-grained, dark-coloured regional metamorphic rock composed of hornblende and plagioclase with minor epidote, sphene, biotite and quartz.

amphidromic system A tidal system in which the high water rotates around a central point (the nodal or amphidromic point), the rotation being caused by the coriolis force of the earth's rotation. The rotationary movement is anticlockwise in the N. hemisphere and clockwise in the S. hemisphere. It is one of the consequences of the modifying influence that the earth's rotation has on a standing oscillation created by tide producing forces. At the amphidromic point itself the tidal range is almost zero but the tidal ranges increase as they are terraced outwards from this point along catidal lines. The time of low and high water progress as they are traced around the amphidromic point.

amphitheatre Flat top, steep-walled depression, shaped like a horse shoe and resembling an ancient Greek theatre. It may be result of glacial erosion, forming a cirque, or the collapse of a volcano.

amplitude (1) The amount of elevation of the crest of a wave or ripple above the base of the adjacent trough. (2) In ecology, the term refers to the range of tolerance of a species.

amygdale It is spheroidal, ellipsoidal or (literally) almond-shaped cavity or vescile within a lava, filled with secondary minerals, e.g., calcite, quartz or zealites.

anabatic wind An upslope wind formed when air on hillsides is heated by insolation conduction to a greater extent than air at the same horizontal level but vertically above the valley floor. This causes convectional rising of the heated air, which is replaced by cooler air from the valley floor.

anabranching channel A type of distributory river channel that separates from its trunk stream and may flow parallel to it for several kilometres before rejoining it. The anabranching channel remains undivided, and so differs from an anastomoring channel which has major distributaries that branch and then rejoin it.

anaclinal A rarely used term to describe any topographical features, which is, opposed to the general geological dip. Thus, anaclinal streams run counter to the dip of the rocks.

anacrobic In the biological sense, organisms living in the absence of free oxygen. An anacrobic soil is thus, in an airless state, notably when waterlogged.

ana-front A front which has ascending air at one side, particularly one experiencing the unusual phenomenon of rising cold air.

analemna A scale drawn on a globe to show the daily declination of the sun enabling the determination of those parallels where the sun is directly over head at any specific time of year.

analog image In remote sensing, an image in which continuous variation in the scene being sensed is represented by continuous variation in image tone, such as the image produced from photosensitive chemicals in a photographic film.

anamolistic cycle The tidal cycle related to the varying distance between the moon and the earth. It is normally taken as 27.5 days.

anascism The vertical component of the waves moving upwards from the focus of an earthquake.

anastomoris The term related to the braiding of rivers, i.e., the tendency of some streams to divide and reunite producing a complex pattern of channels.

anchialine The term applied to flooded caves lying deep below the ground surface. They form in limestone environments and from volcanic tubes.

anchor ice Submerged ice which is attached to the bottom. It occurs in polar scar where the upper water layer, some metres in thickness is overturned by convection during freezing. Under these circumstances, ice may attach itself to the bottom or around seaweeds, if the water depth is sufficiently shallow.

Andean orogenic belt Part of the active cordillera mountain belt on the eastern margin of the Pacific Ocean, the Andes extend approximately 10,000 km from the Caribbean Sea to the Scotia Sea. They developed during the Mesozoic and Cenozoic eras, largely as a result of the subduction of normal oceanic lithosphere. Unlike the N. American cordillera, very few allochthonouf terrenes have been recognised, and there has been no collision of major continental masses as in the Himalayas. These differences have led to the concept of a distinctive Andeao type orogeny.

andesite A fine-grained volcanic rock composed of andesine, similar in mineralogy to a diorite.

Andesite Line A boundary line drawn on a map of the Pacific ocean delimiting the basic rocks of the true ocean basin of the central Pacific from those of greater acidity which lie on the continental side of the line. Thus, no igneous intermediate rocks (andesine, desite, rhyolite) are to be found *in situ* inside the andesite line.

andhis A local term, used in the northwest of Indian subcontinent for dust storms accompanying violent sequels, caused by strong convection.

andosols An order of volcanic soils, deep and light in texture, that contain compounds of iron and aluminium.

andromy The migration of some fish species from salt to fresh water for breeding.

anemograph A self-recording instrument for measuring the speed and sometimes the direction of the wind.

anemometer An instrument used for measuring the speed of wind. Most do so either from the rotation of a wind-driven turbine (rotating cups anemometer) or from the wind pressure through a tube aligned by a vane to point into the wind (pressure tub anemometer).

aneroid barometer An instrument for measuring atmospheric pressure (invented in 1843). It consists of a shallow, thin corrugated metal box, evacuated of air, the flexible sites of which contract, and expand respectively with rising and falling atmospheric pressure. Movements due to pressure changes are conveyed and minified through a train of levers to a chain, which actuates a pointer on a dial, or a pen on chart paper (barograph).

angara A continental mass of Asia, China and the far east that existed during Paleozoic. In the latest Paleozoic, it became joined to Euramerica along the line of the present Ural Mountains.

angara land A name given by the Austrian geologist Eduard Syess (1831–1914) to a small shield in North Central Siberia, where Pre-Cambrian rocks are exposed and which was considered to be the nucleus for subsequently developed structural features in Asia.

angle of declination The angle between true north (geographical north) and the direction of the magnetic meridian (magnetic declination).

angle of incidence A term used in optics and meteorology for the angle between the incident ray of light arriving at a surface and the normal to that surface.

angle of reflection The angle between a ray of light reflected from a surface and the normal to that surface.

angle of refraction The angle between a ray of light that has been refracted, e.g., through a body of water and the normal from the surface at which the ray is refracted.

angle of repose A term which has been applied to the angle at which granular material comes to rest (also called the angle of rest) and approximates to the angle of scree slope. Strictly natural slopes for one material may have a variable angle of repose according to whether the material has just come to rest or about to move. Hence it is also related to the angle of initial yield or the angle of residual sheer.

anglian The glaciation or the glacial stage of the Pleistocene which immediately proceeded the great interglacial (Hexamine) in Britain, the antepenultimate cold stage, i.e., proceeding the Wolstonian which in turn proceeded the Devensian. It is thought to be equivalent in age to the elaster glaciation of N. Europe and the mindel glaciation of the Alps.

angular momentum The product of the linear velocity of a point on the surface of the earth (as it moves from W to E owing to the earth's rotation) and the perpendicular distance of that point from the earth's axis of rotation, a line drawn through the N and S poles. Its value is highest at equator (where the perpendicular distances are greatest) and decreases systemati-

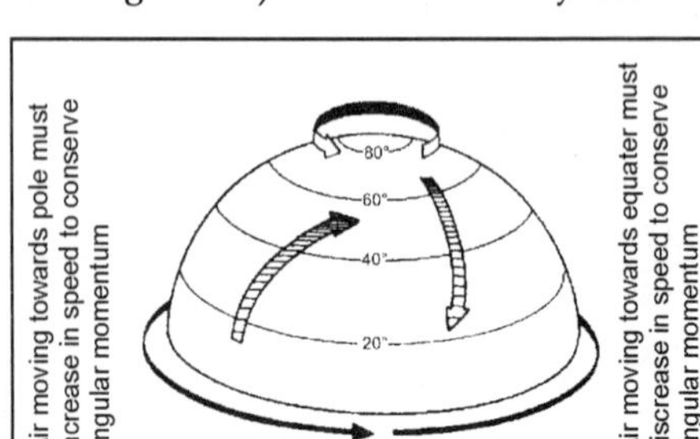

angular momentum

cally with latitude until it becomes zero at the poles.

angular unconformity A stratigraphic unconformity represented by younger strata overlying older strata which dip at a different angle, usually a steeper one than the younger strata.

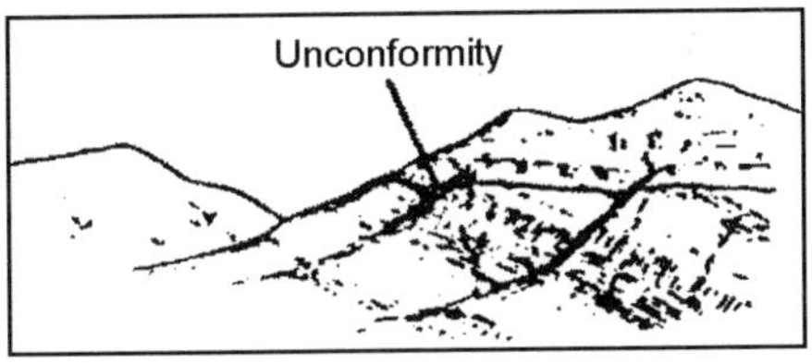

Angular unconformity

angular velocity The rate at which a rotating body turns, expressed in revolutions per unit of time, degrees or radians. The angular velocity of the earth is 15° longitude per hour or 7.29 $\times 10^{-5}$ radians per second. It is used in meteorology to indicate the horizontal rate of air rotating around a depression.

angulate drainage pattern A type of trellis drainage in which the tributary stream, instead of entering the trunk stream at (or nearly at) right angles, joins it at very acute or obtuse angles either because of the underlying fault pattern or because of the dominant joint orientations of the rocks in a region of homogenous strata. Such influences are known as structural guidance.

anhedral (allotrimorophic) A morphological term referring to grains in igneous rocks which have no regular crystalline shape.

anhydride Calcium sulphate ($CaSO_4$). One of the evaporate minerals which occur in some sedimentary rocks, often associated with gypsum.

anisotropy The condition of a mineral or geological stratum having different optical or physical properties in different directions.

annular drainage pattern A drainage pattern in which the tributary streams follow accurate forces determined by lines of weakness in the underlying rock structure. Annular drainage is thus characteristic of dissected domes comprising alternating hard and soft rocks. The latter are etched out by headward erosion effected by tributaries of the main radial consequent streams developed initially on the flanks of the dome.

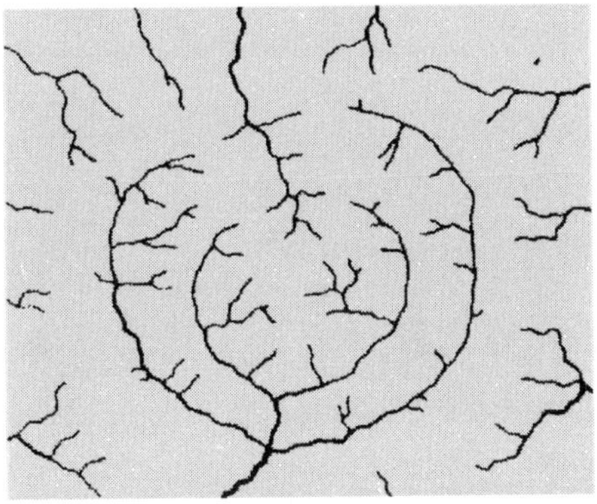

Annular drainage pattern

annulus (planetry) A ring-like structure used, for example, to describe a discrete ring of ejecta around a martian rampart crater.

anomly The departure of any element or feature from uniformity or from a normal state, used particularly in meteorology in connection with temperature and in oceanography in connection with salinity. In geomagnetism, a gravity anomly is the difference between the observed gravity reading and that computed for an idealised globe.

Antarctic An adjectival term relating to the S. Polar region as opposed to the N. Polar region (Arctic), e.g., the Antarctic ocean and the Antarctic ice cap. It is also used as a noun to describe the whole of the earth's surface which lies within the Antarctic circle, a parallel of latitude at 66°32′ S. The land mass of Antarctica is almost entirely covered by the world's largest ice sheet, 5 million sq miles (13 million sq km) in area and with a maximum thickness of approximately 2,300 m (7,500 ft).

Antarctic air Very cold, continental polar air mass, which originates over the frozen surface of Antarctica and the surrounding pack ice, and over the coldest waters of Antarctic Ocean. As this air moves northwards over warmer water, it becomes convectively unstable.

Antarctic bottom water A dense bottom water mass, formed in the Weddell and Ross Seas, which moves in an easterly direction around Antarctica under the influence of the deep-reaching, wind-driven surface Antarctic circum-polar and current (west wind drift). It is typified by a salinity of 34.66 parts per thousand and a low temperature (–2°C to –0.4°C), the high salinity and dense nature is caused by the removal of pure water as sea ice.

Antarctic circum-polar current (West wind drift) The largest and most important ocean current in the southern hemisphere. It flows in an eastward direction around Antarctica and occupies a wide tract of water in the S. Pacific, S. Atlantic and Indian Oceans. There is very little separation between surface and bottom water circulation within this area. The current is remarkably constant and is characterised by low salinity (less than 34.7 parts per thousand) and cold waters (1°C to 5°C). It is the only current which flows right around the world.

Antarctic convergence A convergence line in the seas that circles Antarctica between latitudes 50° S and 60°S. It is where the cold waters from the Antarctic region meet and sink beneath the warm water from the middle latitudes so forming the Antarctic intermediate water. This conveyance line is now called Antarctic Polar front.

Antarctic Front Frontal boundary between cold Antarctic air and warmer air, usually line to the north of it.

Antarctic intermediate water A water mass formed at the surface near the Antarctic conveyance, at about 50°S. It is typified by a low salinity (33.8 part per thousand) and low temperature (2.2°C). As it spreads northwards, it sinks to depths of 900 m and can be traced even in the North Atlantic at 25°N.

Antarctic meteorites Meteorite finds from ablation zones on Antarctic glaciers, which represent about 25 per cent of all the meteorites so far discovered. They vary in composition but are usually quite fresh and may have fallen from one million years ago to the present date. They are, therefore, of great importance in the study of meteorites. Antarctic meteorites include the first meteorites known to have been derived from the moon and mars.

Antarctic meteorology Deals with the circulation of the atmosphere and the climate south of the Antarctic circle.

Antarctic Ocean Term used to describe the oceanic waters surrounding Antarctica. It extends northwards to about 40° S latitude. The limit of the northward drift of ice from the Antarctic region where there is a marked change in water temperature and salinity. The Antarctic Ocean is typified by low water temperatures (–1.8°C to 10°C).

Antarctic plate A present day major lithosphere plate that extends beyond the Antarctic continent to the surrounding constructive plate margins.

Antarctic polar current A surface current which flows in a westward direction around Antarctic under the influence of easterly winds blowing off the ice cap.

antecadent moisture The soil moisture condition in an area before a rainfall, which moderates the area's run-off response to the rainfall. Antecedent moisture condition is usually expressed as an index and may be estimated by weighing past rainfall events to derive an antecedent precipitation index (API).

antecedent drainage A process of drainage development in which an ancient river is able to maintain its course across more youthful fold and fault structures as these develop. A prerequisite is that the river's capacity for downcutting must equal or exceed the rate at which the structure grows upwards across its path. Where river courses are markedly discordant to structure, antecedence is sometimes suspected. However, the process is often extremely difficult to prove, and can usually be discounted on various grounds.

antecedent platform theory The theory, propounded by Sir John Murray that coral reefs and atolls developed through upward growth from submarine platforms.

antecedent precipitation index An index of moisture conditions in a catchment area used to assess the amount of effective rainfall that will form direct surface run-off. If there has been no rain for several weeks, less rainfall will get into the streams than if the ground surface is already saturated. The index is calculated on a daily basis and assumes that soil moisture declines exponentially when there is no rainfall.

anteconsequent stream A stream which flows consequent on an early uplift but antecedent to later stages of the same tectonic uplift.

antemeridian Any meridian which is 180° of longitude from another meridian.

anthorpogen A primarily Russian term for the period during which man has been an inhabitant of cave (i.e., the past two to 3 million years).

anthracite A type of coal relatively hard, jet black, with a metallic lustre, subconchoidal fracture, unbended with less than 10 per cent volatiles and more than 90 per cent carbon. It burns with intense heat and a known luminous flame.

anthropo geomorphology The study of the influence on human activities on the physical landscape and the processes which shape it, e.g., the effect of agriculture, dam construction and deforestation on processes of fluvial erosion, transport and sedimentation, the impact of groynes and seawall construction on beach sediment transportation and coastal erosion; the acceleration of slope processes and increases of slope instability due to cultivation and constructional works of various kinds, and the creation of thermocast as a result of road and house building and the stripping of surface gravel from permafrost.

anticendre The point opposite to the epicentre, above the focus of an earthquake.

anticlinal trap A fold structure with an arch of non-porous rock overlying porous strata (reservoir rock), providing a trap in which oil, gas, or water may accumulate. In Middle East, oil fields, large, upright folds occur in thick, competent limestones which extend for many kilometres; fracturing along the crests of the anticlines increases permeability in the reservoir rocks.

anticlinal valley A valley developed along the axis of an anticline. Inward facing escarpments are developed where the upper beds of the anticline are relatively resistant, while softer lower materials occupy the valley floor. It is a common landform in gently folded strata of varying resistance such as, occur in Southern Britain.

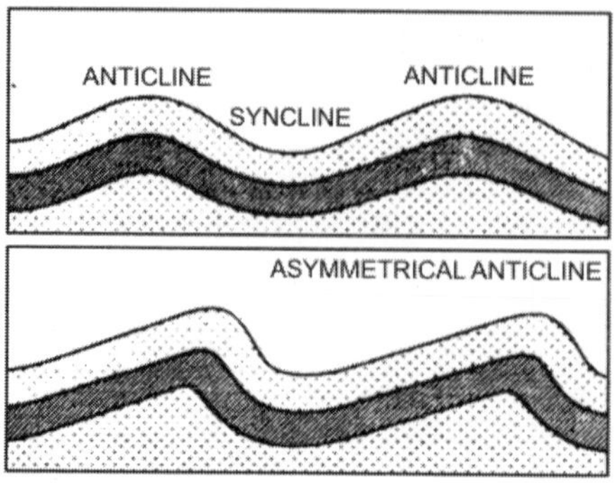

Anticlinal

anticline An upfold in the rocks resulting from compressive stresses in the earth's crust. The strata dip is opposite directions from the central line, or axis, of the anticline. Anticlines may be symmetrical or asymmetrical. After erosion, the oldest rocks are exposed in the central core of the fold.

anticlinerium A regional antiformal structure composed of a series of smaller, higher order anticlines and synclines, some of which may be small enough to be viewed in outcrop.

anticyclogenesis Process whereby an anticyclone or a ridge of high pressure is formed and developed.

anticyclolysis Process whereby an anticyclone or ridge of high pressure is dissipated or weakened.

anticyclone A large area of high atmospheric pressure, usually stationary for slow moving. It is normally associated with widely-spaced isobars, resulting in light and variable breezes of calm conditions, anticyclones occur where there is convergence of air at high altitudes in the atmosphere, consequent large scale subsidence, and slow divergence (by way of clockwise peripheral winds) near the earth's surface. Anticyclones comprise air of low relative humidity (reflecting the dryness of their from the upper atmosphere), and low environmental lapse rates; the later resulting in absolute stability or very limited instability so that weather conditions are usually fair.

anticyclonic gloom The condition of low visibility associated with anticyclones accompanied, in the colder month of the year, by well-developed temperature inversions that can trap dust and other pollutants, and often have radiation fog in the lower layers. The stability of the high pressure systems can make the resultant reduced visibility very persistent and may establish smoke conditions.

antidune A type of small scale cross bedding feature formed from a sand deposit on a river bed. It develops as a sand wave from a so called 'normal' dune as the current velocity increases in the highly loaded river during the antidune and ridge-like structures are created by fast flowing stream and these are characterised by erosion on their gentler downstream stops and deposition on their steeper upstream slopes. Consequently, as the individual sand grains move down current so the antidune moves up current. It is, therefore, a transitory feature, forming, being eroded and reforming.

antiforms A general term denoting an upfold of strata in the earth's crust, but one in which the precise statigraphic relationship of the rock are not known. Thus, the rocks exposed at the core of the upfold need not be, as in an articline, the oldest of an area of complex overfolding (Nappe) is worned down by denudation. It is possible for the stratigraphical succession to be inverted, even though the fold exposed at the surface is arched (convex upwards).

antipleion An area or a specific meteorological station where the mean annual temperature is lower than the average for the region.

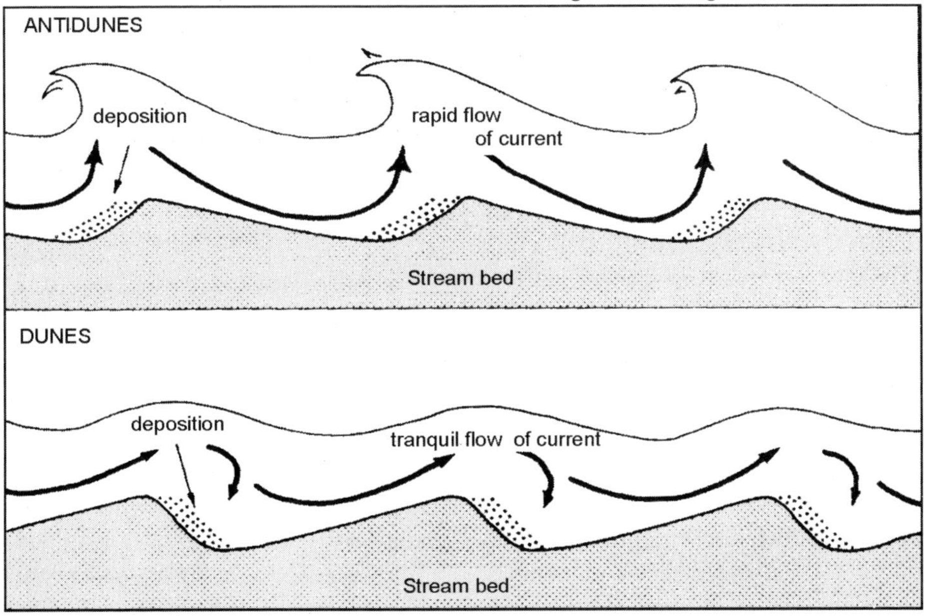

Formation of antidunes

antipodal blunge The tidal effect occurring at the point on the earth's surface opposite that where the pull of the moon's gravity is strongest. Hence, it is the tidal effect at the point where lunar attraction is weakest.

antipodes A place on the earth's surface which is directly opposite to another when measured through the diameter of the earth. This means that they are both 180° of longitude apart and one of them is many degrees from the S. Pole as the other is from the N. Pole. The nearest land antipodal to the British Isles is antipodes island, which lies SE of New Zealand at 49°42′ S and 178°50′ E. It is antipodal to the Channel Island.

antitrades A deep layer of westerly winds in the troposphere above the surface trade winds. In a simple way, they represent the upper limits of the Hadley cell within which occurs the poleward transfer of heat and momentum and water vapour.

anvil cloud A very high cumulonimbus cloud, in which the topmost parts spread out at the base of the stratosphere in the direction of the high level winds. The upper parts of cumulonimbus clouds display a fibrous texture, which is due to the presence of snow flakes and small ice crystals. Cumulonimbus clouds result from highly instable atmosphere conditions, which produce rapid convectional uplift and cooling and large scale condensation. They give rise to heavy showers of rain, hail, bleets or snow and are often associated with thunderstorms.

aphanitic Microcrystalline and cryptocrystalline rock textures. Pertaining to a texture of which the crystalline components are not visible with the naked eye.

aphelion The furthest point of a heavenly body from the sun during its orbit. In the case of the earth, this is 152 million km (94.5 million miles) on 4th July, when the sun-earth distance is 1.5 per cent greater than the annual mean distance.

aphotic zone A zone below 300 m depth in very deep lakes and oceans into which light does not penetrate and where photosynthesis is therefore impossible.

aphytic zone The portion of the floor of lakes, seas and oceans which owing to their depth are not colonised by plants.

apogie The term used for: (1) The meridional altitude of the sun at midday on the longest day of the year. (2) The point on the orbit of a planet where it is furthest away from the earth, with particular reference to the moon. (3) The farthest point of a satellite, used in remote sensory from the point about which it is orbiting. (4) The highest point reached by a rocket, used in remote sensing, when fired upwards from the earth's surface.

Apollo (1) A solar system asteroid (No. 1862), diameter 1.6 km; approximate mass 2×10^{12} kg; rotational period 3.063 hours, orbital period 1.81 years. Its orbit crosses that of earth. (2) The name of the NASA manual lunar programme that ran from 1963 to 1972.

apophysis An irregular or sheet like vin or dyke which originates from a larger igneous rock body.

Appalachian orogenic belt A 3200 km long, Paleozoic, orogenic belt extending from New Found Land to Alabama. Extension of the belt along strike includes the Caledonian orogenic belt in N.W. Europe, now separated by the Atlantic Ocean.

apparent dip The angle between the horizontal and the tilt of the stratum or bedding plane measured vertically in any direction except that which is at right angles (90°) to the strike. The angular value of the apparent dip will always be less than that of the time dip.

appleton layer A zone in the upper atmosphere (thermosphere) (300 km/190 miles above the Earth's surface) which reflects short wave radio waves back to earth. Also known as the F2 Layer.

applied geomorphology Studies which involve the application of geomorphological methods of survey and analysis towards the solution of problems occurring within man's physical environment. These include such problems as flood control, irrigation, stabilisation of slopes, land reclamation and coastal erosion migration.

applied meteorology The use of archived and real time atmosphere data to solve practical problems in a wide range of economic, social and environmental fields.

apposed glacier A glacier formed when two separate glaciers coalesce. These are outwash spread in front of an alluvial fan.

apse line A theoretical line joining the points of aphelion and perihelion, the times of which become progressively later (C.1.255 per annum), because the line moves very slowly around the orbit of the earth in the same direction as its orbital movement

aquiclude A rock which is porous and thus able to hold but in which the pore water is so strongly held by surface tension that water movement thought the rock is virtually impossible. Clay and unjointed chalk and aquieludes; within these rocks there may be a high content of water (upto 50%) of the total volume), but this can be expelled only by the application of extreme pressure.

aquifer A rock layer which will absorb water and allow it to pass freely through. The term is also applied to any water-saturated stratum of earth or gravel that has sufficient porosity and permeability to yield ample supplies of ground water in the forms of wells or springs. If it is underlined by a layer of impermeable rock it will prevent water passing downwards that by directing it laterally. If the stratus is bounded on its upper surface by an impermeable rock layer it will form a confined aquifer.

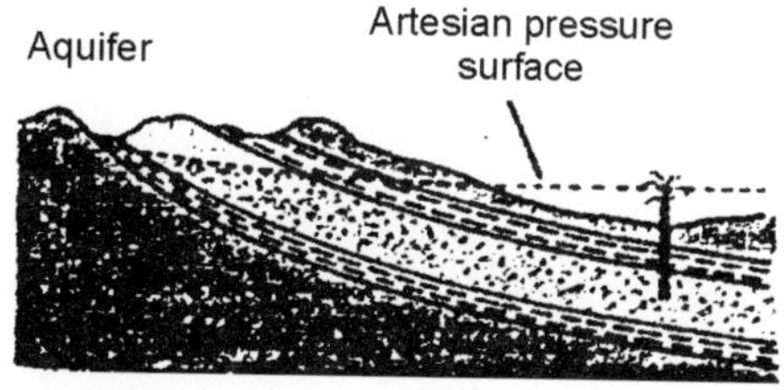

aquifuge A US term (rarely used in Britain) denoting a rock layer or rock type which is completely impermeable, i.e., it will not hold water nor will it allow water to pass though, e.g., obsidian, quartzite. Sometimes the term aquiclude is used as a synonym for aquifuge although the terms have somewhat different meanings.

aquitard A rock with low values of hydrantic conductivity which allows some movement of water through it, but at the rates of flow lower than those of adjacent aquifers.

aratan plate One of the present day minor lithospheric plates, separated from the African plate by the spreading Red Sea, the continuation of the Carlsberg Ridge in the Gulf of Aden and the Dead Sea transport system, while the boundary with the Indo-Australian plate is the owen fracture zone; the plate is colliding with the iron plate.

arc (1) In geometry, any part of a curve. (2) In topology, any line which need not be part of a curve going two nodes. (3) In geology, a curvy linear line of islands.

arc of meridian A measurement made along a meridian in order to determine the exact size and shape of the earth.

arch A natural opening through a mass of rock or boulder clay. It is most commonly used on the sea coast where waves have cut through a promontory, e.g., Durdle Door Dorset, the Green bridge of Wales, the Bow Fiddle, Banff, Scotland; or where sea-caves have coalesced and partly collapsed. When the keystone of the marine arch collapses, the feature will become a stack.

Archaean (Archaeozoic, Azoic) One of the three subdivisions of the pre-Cambrian lasting from about 4000 to about 2500 Ma ago.

archipelago A closely grouped cluster of islands, e.g., the Tuamotu archipelago of the S. Pacific, the cycleades of the Aegean Sea.

Arctic An adjectival term relating to the N. polar region, as opposed to the S. polar region (Antarctic), e.g., the Arctic Ocean. It is also used adjectivally to describe animals and plants characteristic of these northern regions and also in relation to weather and climatic types. The whole of the earth's surface which lies within the Arctic circle, a parallel of latitude at 66°32′ N, is frequently referred to as the Arctic (i.e., using Arctic as a noun).

arctic air mass An extremely cold air mass originating in the snow and ice covered arctic, and travelling almost directly S. polar air mass.

arctic bottom water (ABW) Cold, dense water that sinks to a maximum depth of about 6000 m, from the sub-polar gyre in the Greenland and Norwegian seas fills the basin of those seas, then spills as an intermittent southward flow through narrow channels in the ridge between Scotland, Iceland and Greenland.

arctic front Frontal boundary between cold arctic air and warmer air masses, usually lying to the south of it. Many depressions originate on it. In north-western Canada in winter, for example, the frontal zone incorporates cold, dry continental polar air and modified maritime arctic air from the Gulf of Alaska to the north of continental tropical air.

arctic haze A reddish-brown atmospheric haze, which is often observed in the Arctic, especially in the winter and spring when atmospheric conditions are calm. It consists primarily of atmospheric pollutants derived from industrial sources in Europe and northern Asia. These include sooty and acidic particles.

arctic meterology Meteorological conditions in the arctic reflect the combined influence of the earth's orbital geometry on the day length and angle of the sun, the surface properties especially the seasonal snow and sea ice cover, and the global atmospheric circulation. The Arctic Circle delimits the polar cap which experiences an annual cycle of winter night and summer day, but the timing of the actual seasons is substantially out of phase.

arctic smoke A phenomenon sometimes observed in very high latitudes where the sea surface appears to be giving off smoke (actually steam). It develops where the sea temperature is much colder. As water vapour is evaporated from the sea surface into the atmosphere it is immediately condensed into tiny water droplets, forming a low-level mist. Arctic smoke is thus a local form of steam fog.

arc-trench gap (fore-arc) The region between an oceanic trench and the adjacent volcanic island arc. The arc-trench gap is at least 100 km wide in nearly all cases, and upto 570 km at the eastern end of the Aleutian arc. The width of the arc-trench gap increases with time through the growth of an accretionary wedge and the development of a fore-arc-basin.

arculate delta A river delta that extends into the sea of a lake in a fan shape with the outer edge of the delta having a rounded outline, for example, the Nile delta in Egypt. Small arcuate deltas may also be developed where heavily loaded streams enter lakes. In the Swiss Alps, a melt water stream from the Mount Mine glacier has formed an arcuate delta in an adjoining lake in only 10 years.

area The extent of a surface, measured in square units, e.g., 640 acres = 1 sq mile. To convert units per square mile to units per km, one has

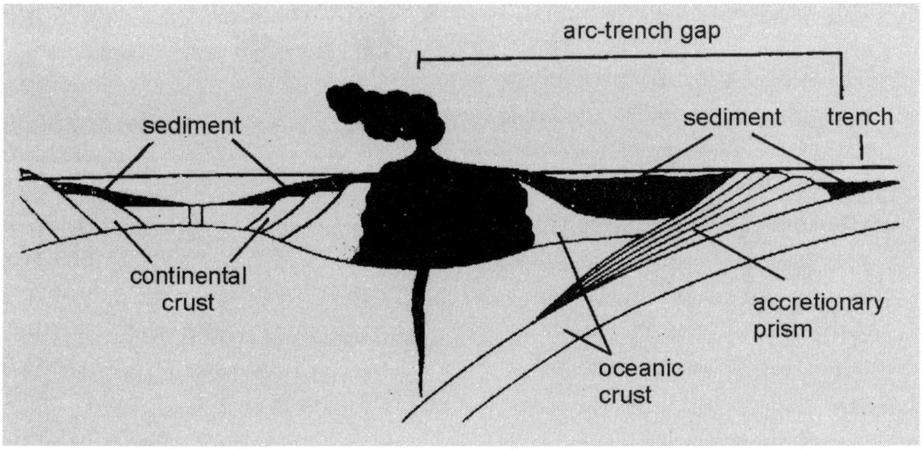

Arc-trench gap

to multiply by a factor of 2.58999; to convert units per km^2 to units per square mile one has to multiply by a factor of 0.3861.

Area of Outstanding Natural Beauty (AONB) Areas in England and Wales which for reasons of scenery, interest and amenity enjoy special protection under the terms of the National Parks and Access to the Countryside Act (1949). Today there are 39 AONBs in England and Wales and they cover around 15,500 km^2 or about 10 per cent of the total land area.

area-height diagram A group showing the relationships between area and altitude. The vertical axis indicates elevation above a given datum (usually sea level) while the horizontal axis denotes either the area between each pair of selected contours, or the percentage of the total area occupied by the area between each pair of selected contours.

areal data Data which refer to the physical extent of a surface, measured in square units.

areal differentiation During the 1940s and 1950s, following the lead given by Hartshorne, this was regarded as the main objective of geographical study. It placed great stress on the study of variations from place to place in the character.

areal erosion Erosion by an ice sheet of an area too large to be visible on an erosion feature and identifiable only through mapping at a continental scale.

arena A shallow, broadly circular basin lammed in by a rim of higher land.

arenaceous Pertaining to sandy or sand-like appearance. The term is applied to sedimentary rocks with a grain size 0.0625 to 2.00 mm.

arête A French term, which has been widely accepted to describe a narrow, rocky and often jagged ridge, which divides the steep walls of two adjacent cirques. It may be on the actual crestline of a mountain or merely on a subsidiary ridge. In the USA, it is termed a combe-ridge and in German, a Grat.

aretic drainage A drainage pattern which is confined to an inland basin, with no outlet to the sea; e.g., the Tarim Basin drainage of Central Asia, Cooper's Creek in Central Australia. Also termed internal drainage.

argillaceous Pertaining to containing or composed of clay. Applied to locks which contain clay-sized material and clay minerals.

argillic brown earths One of the subdivisions of the brown earths, which are themselves part of the soil classification of England and Wales. They are characterised brown or reddish in colour and have a loamy horizon overlying a marked subsurface horizon of clay accumulation.

arheic A term derived from the Greek verb (rhein) to flow, which means literally 'without flow'. It is used by geomorphologists to describe those arid areas where no stream valleys have developed.

arid cycle of erosion The cycle of erosion occurred in deserts. It is an idealised cycle which applies only in desert areas, when landforms used to undergo the stages of youthfulness, maturity and old and are responsible for the creation of arid topography.

arid, aridity A climate having insufficient precipitation to support vegetation is defined as arid, while a more statistical definition denotes on arid area as having less than 250 mm (10 in) annual rainfall. A further definition regards the arid zone—the portion of the earth between latitude 15° and 30° north of the equator and the same latitudes south which contains most of the world's warm deserts. Any area of the earth that receives less than a specific amount of rainfall annually, usually about 250 mm is known as arid.

aridisol One of the ten orders defined in the Seventh Approximation soil classification. It includes infertile, alkaline and saline soils of desert areas and is characterised by a thick accumulation of basic mineral salts at or near the surface owing to capillarity (e.g., calcium and sodium). This capillary action may cause underground water to concentrate sodium to such a degree that the soil becomes toxic. Except for a few tolerant species, the vegetation is extremely sparse on aridisols, which renders them liable to soil erosion.

aridity index Indication of moisture deficit. All climatic classification include arid categories defined either by quantitative or more usually by mainly subjective criteria. C.W. Thormthwaite first used the term aridity index and calculated it as 100 times the water deficit, divided by the potential evaporation.

arkose Arenaceous rock containing quartz and 25 per cent or more of feldspar. The feldspar is easily destroyed during chemical change or transportation, implying that arkoses were deposited rapidly under fairly arid environmental conditions.

Armonican orogeny The mountain building episode of carbo-permen times. It is the W. European equivalent of the variscan orogeny of central Europe, named by E. Suess in 1888 after the old name for Brittany (Armorica). It is the part of the Hercynian orogeny of Eurasia which created mountain chains from Ireland to China (Altaides). Since it folded and faulted the carboniferous coal measures, a great deal is known about Armonican structures.

armoured mud balls (also called clay balls, pudding balls, mud pebbles and mud balls) These roughly spherical lumps are cohesive sediment, which generally have diameters of few centimetres though much larger examples have been reported. Many examples are lumps of clay or cohesive mud that have been gouged from streambeds or banks by vigorous currents. They often occur in bed lands and along ephemeral streams, but can also be found on beaches in tidal channels, etc.

armouring A term used of heterogeneous river bed material when coarse grains are concentrated sufficiently at the bed surface to stabilise the bed and inhibit transportation of underlying finer material. An amour layer is coarser and better sorted than the substrate, and is typically only 1–2 grains thick.

arroyo A Spanish term commonly used in Latin America to describe a stream bed which is periodically dry.

A trench with a roughly rectangular cross, section excavated in valley bottom alluvium with a throw-flowing stream channel on the floor of the trench (Graf, 1983). Although the term gully is used for similar features, a gully is V-shaped in cross section instead of rectangular and is excavated in colluvium instead of alluvial fill. Dodge (1902) first used the term in geomorphological research, identifying arroyos as the product of stream channel and entrenchment resulting from the changes in climate and landuse that causes increased run-off.

artesian basin A large synclinal structure comprising an aquifer (or series of aquifers) sandwiched between overlying and underlying impermeable strata. Rain water percolates into the ground at the margins of the basin, where the permeable rocks are exposed, and migrates down-dip towards the access of the artesian basin. The ground water lying at depth here is under great hydrostatic pressure, and when wells are bored through the overlying impermeable stratum the ground water will rise spontaneously to the surface.

artesian well A well that flows at the surface without pumping, because it is sunk into a confined aquifer whose hydraulic head (sometimes calls the potentiometer or piezometrise head) lies above ground level.

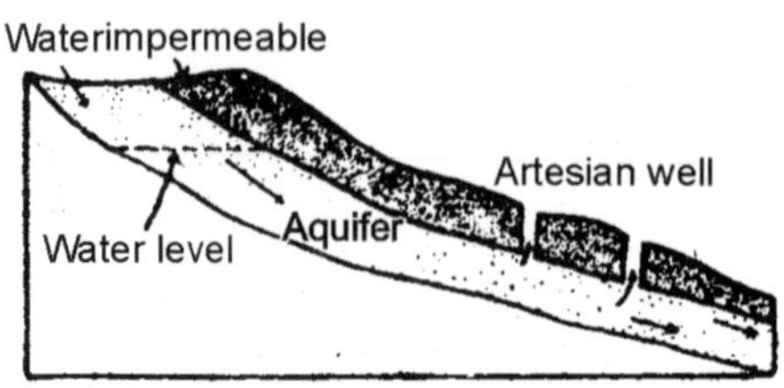

Artesian wells

artificial freezing Method of controlling groundwater and improving strength of ground by pumping a refrigerate, e.g., calcium chloride or liquid nitrogen, through tubes in the ground; as the ground freezes around closely spaced tubes, a continuous frozen zone may be formed. This is an expensive technique but suitable for mixed strata as the results are more predictable than dewatering or grouting.

artificial rain Rain, or, increased rain, produced by seeding clouds artificially with dry ice (frozen carbon dioxide), silver iodide, or other appropriate particles, which act as condensation nuclei.

artificial recharge The techniques of increasing groundwater supplies by the spreading of river water widely over flat areas, encouraging infiltration and percolation, or the pumping of such water into (1) broad shallow basins formed by excavation, or (2) deep pits or wells. In alluvium-floored Wadis in desert areas such as Saudi Arabia, transverse low earth-dams are constructed to impound water during infrequent flash floods. This water sinks rapidly into the permeable alluvium where it is stored and withdrawn from shallow wells as and when required.

asbestos A term for the fibrous verities of several distinct mineral species, all of which are silicates. They include such mineral as chrysotile (fibrous serpentine), and are important because like other fibres, they can be woven and have the added advantage of being fireproof.

aseismic plates Any of the earth's crustal plates within which there are relatively few earthquakes.

aseismic ridge Long, linear, inactive, volcanic, topographic feature found in many deep ocean basins. An example is the Walvis ridge in the south-eastern Atlantic, which extends for 3000 km and in places reaches a height of 2 km above the deep ocean floor.

ash Fine powdery material emitted during a volcanic eruption, and deserting from the break up of solid lava. The ash is often so fine (the particles may be less than 0.25 mm in diameter) that it can be carried by the wind over vast distances.

ash cone A volcanic cone made entirely, or almost entirely of ash, in general, ash cones are small in stature, although there are a few of major proportions, e.g. Volcan de Fugo, Guaternala, which reached an elevation of 11,000 ft (3,350 m). Such cones are usually concave in profile and relatively gentle-sided.

ash fall A rain of air-borne volcanic ash falling from an eruption cloud.

aspect The direction in which a slope faces. This has effects on the climate of the slope, in terms of total insolation received, exposure to rain-bearing winds, amount and duration of frost and snow cover etc. These, in turn, may significantly affect the operation of geomorphological processes, and the degree of development of certain landform types.

asphalt Brown or black, solid or semisolid, bituminous substance made almost entirely of carbon and hydrogen. It melts between 65° and 95°C and is soluble in carbon disulphide. It is formed by the evaporation of volatile hydrocarbons and occurs in oil-bearing rocks.

assemblage zone A biostratigraphic unit defined and identified by a group of associated fossils rather than by a single index fossil.

association In biological usage, the term is given to an assemblage of plants living in close interdependence in a similar ecological niche. They exhibit similar habitat and growth requirements, although there are usually one or more dominant species. A precise botanical definition regards it as a climax community that is the largest subdivision of a climax biome or formation.

association analysis A method which subdivides a classification of vegetation into a statistical hierarchy: the normal association analysis subdivides a plant population so that all association disappears. The population is sampled and first divided into two broad groups related to one species and exhibits the highest degree of association—those in which this species occurs and areas without it. The procedure is then repeated for each group, i.e., in discovering which species now has the highest degree of association which when determined forms the base for further subdivision.

asteroids A belt of some 1500 heavenly bodies which revolve in the solar system between Jupiter and Mars, formed possibly from the disintegration of a former planet. None is larger in diameter than 800 km (500 miles).

Asthenosphere The uppermost zone of the earth's mantle, lying at a depth usually within the range 60–200 km.

Within the asthenosphere, the rocks are probably close to melting point as a result of the concentration of heat from radioactive decay; this is indicated by the weakening and slow passage of earthquake waves through the zone. The asthenosphere, therefore, has a plastic quality, allowing slow flowage to occur under high pressure. It is believed that lateral movements within the overlying lithosphere, as envisaged in the theory of plate tectonics, are facilitated by the presence of the asthenosphere.

astrobleme A term put forward by Dictz (1961) meaning 'Star Wound' and referring to the erosional remnant or scar of a structure of extra-terrestrial origin produced before the Pliocene by the impact of a meteorite on the earth surface.

asymmetrical fold A fold (either an anticline or a syncline) in which the axial plane is not vertical and which results in one of the limbs dipping more steeply than the other.

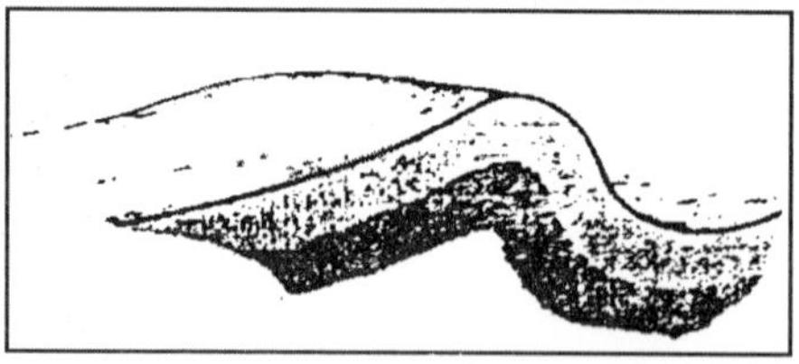

Asymmetric fold

asymmetrical valley A valley whose slopes on one side are steeper than those on the other. The formation of such a valley may be due to geological structure (i.e., the dip of the strata may cause a stream to experience uniclinal shifting and thus to undercut and steeper the down-dip side) or to climatic influences. Many asymmetrical valleys in Britain and Europe are attributed to periglacial conditions, when differential exposure of the valley slopes to solar radiation and/or snow-bearing winds resulted in differential denudation by frost action and solifluction. The active slopes (in many instances, those facing to the south-west) were modified more rapidly than the inactive slopes (often facing north-eastwards and remaining frozen and snow covered), resulting in asymmetry. However, the development of asymmetry is a complex process and yet not fully understood, many factors may be involved, including vegetation cover and the action of wind, not only in transporting fine sediments from the exposed slope but in evaporating soil moisture and thus impeding mass movements.

Atlantic conveyor A system of ocean currents that plays a major role in the transport of heat from low to high latitudes and therefore, in global climates. It is driven by convective overturning of water near the edge of Northern Sea ice, where cold, saline water sinks to the ocean floor and travels South as the North Atlantic deep water eventually to the Southern ocean. Its place is taken at the surface by warmer water flowing northward.

Atlantic Ocean One of the main oceanic areas of the world between North and South America at western side and Europe and Africa at eastern side. It is relatively shallow having an average depth of 3310 m; and it is the warmest (average temperature 3.73°C) and most saline (average salinity 34.9 parts per thousand) of the major oceans.

Atlantic period A term used to describe the warm and moist climatic phase which occurred in north west Europe between 7500 BP and 5200 BP in post-glacial (Flandrian) times.

atlantic type coast (Transverse type coast)—A strongly embayed type of coastline which is formed where the trend of mountain ridges and major lowlands is at right angles to the coastline as a whole. Typically, such a coast borders are relatively young ocean that is widening because of sea-floor spreading. This type of coast was first recognised by Aduard Suess.

atmometer An instrument for measuring the rate of evaporation. It is normally in the form of an open-ended glass tube from which water can evaporate.

atmosphere A relatively thin layer of odourless, tasteless and colourless gases (air) and dust surrounding the earth (aerosphere). It is subdivided into concentric shells according to characteristics of temperature lapse rates: The lowest is the troposphere, separated by the tropopause from the

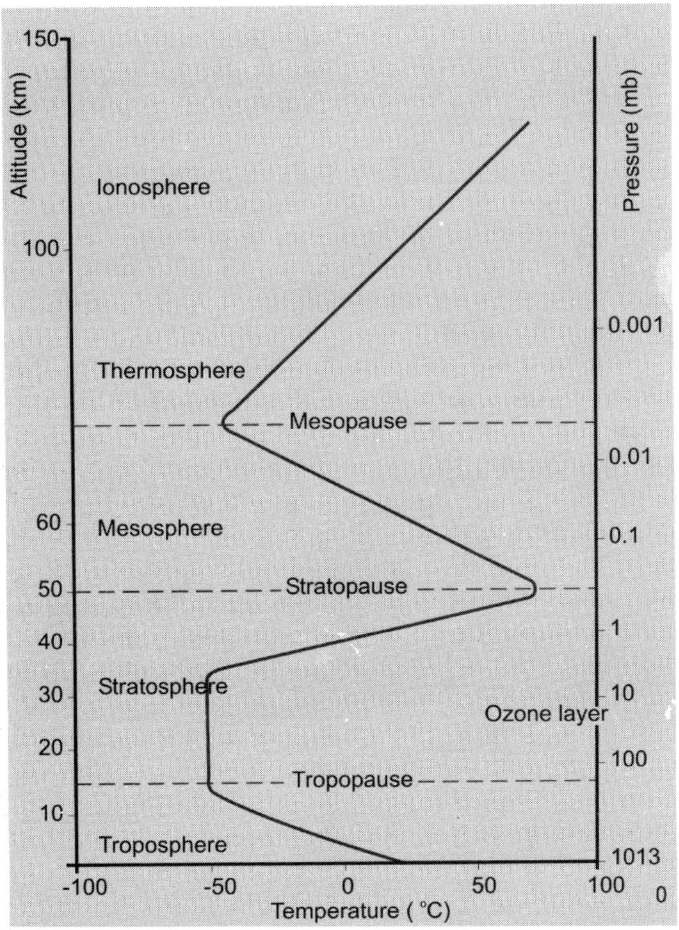

Atmosphere

stratosphere, above which is the mesosphere and the outermost layer, the thermosphere.

atmospheric circulation The movement of air in the form of planetary winds and pressure cells around the global surface and at higher levels in the atmosphere. The circulation is 'powered' by the global imbalance between incoming and outgoing radiation in which low latitudes received more than high latitudes. Thus, heat is transferred polewards by air masses and ocean currents.

atmospheric energitics It concerns with the energy content of the atmosphere and how it is changed from one form to another. Almost all energy is heat, latent heat and potential energy. Thermodynamic diagrams relate these. A small fraction of the total of sensible, latent and potential energy (known as available potential energy) can be converted into the kinetic energy of the wind.

atmospheric heat engine A term which has been applied to the functioning of earth's atmosphere system. The latter is 'driven' by inputs of heat for solar radiation (at a maximum in equatorial and tropical latitudes and equivalent output at a maximum in polar regions). Accordingly, there must be a compensatory transfer of heat from low to high latitudes, via the development of pressure cells and associated wind systems which comprise the earth's general circulation.

atmospheric instability It is a general system which is unstable if an introduced disturbance increases in magnitude through time. Conversely, the system is stable if the introduced disturbance is damped out through time. In the atmosphere, this

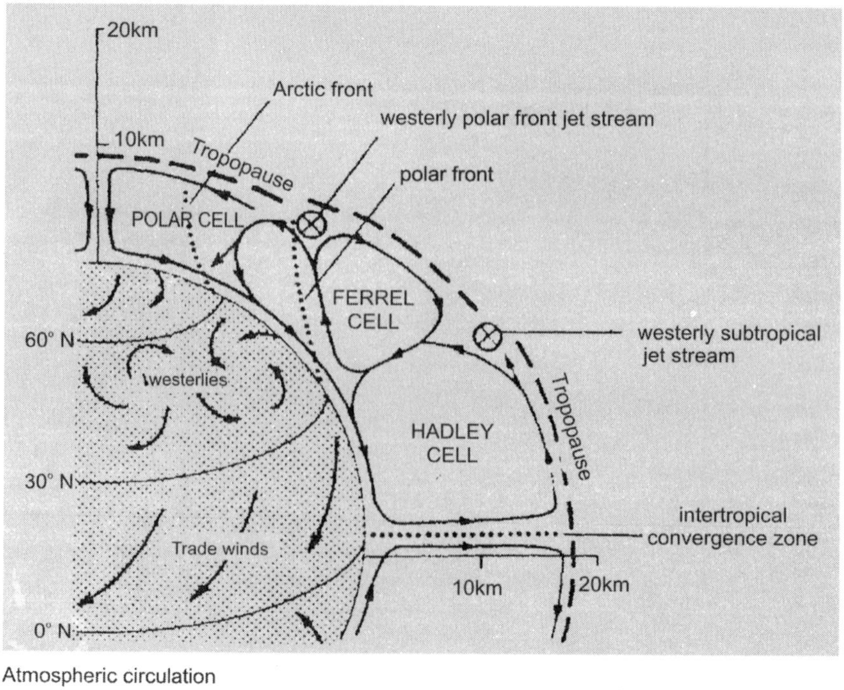

Atmospheric circulation

idea is applied at two main scales: cyclone scale and cloud scale. Cyclone scale instability is exemplified by the growth of cyclones within the extratropical westerly. They are a manifestation of baroclinic instability. Cloud scale instability results in vertical displacement of air parcels in an atmosphere initially in hydrostatic equilibrium. Cumulous clouds frequently result from this type of instability.

atmospheric layers The atmospheric layers are troposphere: 0–10 km above the earth's surface, the stratosphere: 1–25 km, the ozonosphere: 25–6 km, the mesosphere: 60–100 km. And between 1000 to 500 km, the thermosphere and the donosphere.

atmospheric pollution Solid and gaseous contamination in the atmosphere which occur as dust, smoke or sulphur dioxide and other gases particularly from the combustion of fossil fuels and certain industrial processes. Air pollution is most marked in urban areas.

atmospheric predictability A property of the type of information predicted as well as certain length of time ahead. Whether systems of middle latitudes take only a few days to grow from unobservable beginnings; clearly this means that explicit weather prediction must in turn be limited to the few days. Over longer time-scales (weeks, months, seasons) interaction with the sea and land surface begin to be important and current atmospheric models do not predict conditions well. Some economically important weather is incredibly difficult to forecast (e.g., thunderstorms).

atmospheric pressure The pressure exerted by the weight of the atmosphere on the earth's surface. The average pressure at the sea level is 1,013.25 mb (1,000 mb = 1b = 10 pascals). The highest atmospheric pressure recorded in the world = 1079 mb.; the lowest recorded in the world = 877 mb.

atmospheric shimmer The effect observed when light passes through moving air masses, which have differing refractive indices. Result is effects such as star twinkling and it ultimately limits the resolution of any remote sensing system.

atmospheric window Those parts of the long wave, infrared radiation spectrum, with wavelengths of 8.5–14 microns, in which radiation is not absorbed by water vapour and CO_2. Thus, when there is little cloud cover, terrestrial radiation of this wavelength is lost into space, allowing a cooling of the earth's surface. Remote sensing systems view the earth's surface through these windows.

atoll A coral reef, surrounding a central lagoon, found commonly among the islands of S. Pacific (for example, in the Gilbert and Ellice Islands). It is believed that atolls have been formed above former islands (sometimes volcanoes). As these have

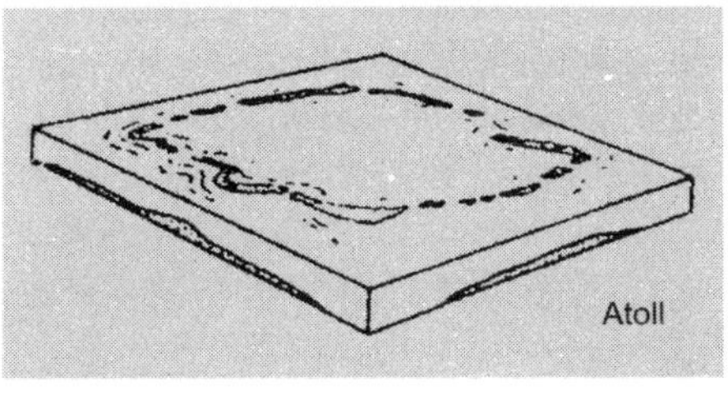

become submerged either as a result of subsidence of the sea floor or rises of sea level (including that at the close of the Pleistocene period). They provided a base or platform for upward coral growth. The corals have been most active on the outer walls of the reef which has thus grown not only upwards but outwards; the inner and inactive sites of the reef mark the boundaries of the enclosed or partially enclosed Lagoon. Such a hypothesis would explain why the coral in atolls extends to depths where light and water temperature are now unsuitable for coral growth.

atollon A small atoll which lies on the plank of a larger one.

attached dune A sand dune which occurs in deserts as a result of accumulation around a rock or other obstacle.

attenutaion (1) In general, the removal of wave peaks. (2) A term which described the loss of electromagnetic energy (solar radiation or radio waves) as it passes through the atmosphere owing to absorption and scattering of atmospheric particles and molecules.

attitude (1) Of a bed or other planar feature, the deposition with respect to the horizontal and compass bearings; that are obtained by measuring the dip and strike of the bed respectively. (2) Of a fold, the overall deposition; this is defined by measuring the dip and strike of the axial plane, and the trend and plunge of the hinge line.

attrition The mechanism by which the particle size of any material is reduced by friction during transport by ocean waves, wind or running water. It differs from abrasion and corrosion.

attuberg limits Series of thresholds, which are observed when the water content soil is steadily changed. The contraction limit occurs when sufficient water is added to a dry soil for contraction tracks to close. The addition of further water leads to plastic deformation at the 'plastic limit'. The liquid limit occurs when just enough water is then added for the soil to behave like a liquid. Knowledge of these limits is important for understanding and predicting hill slope failure. The different in percentage water content between the liquid limit and plastic limit is called the 'plastic index'.

aufeois A German term which has been adopted to describe a surface accumulation of ice under periglacial conditions, such as when ground water freezes seasonally where it emerges from springs and when rivers flood surrounding country side during periods of freezing. In general, aufeis is a phenomenon of winter, disappearing during the spring thaw.

auge A hot, dry wind which blows from the south of France to the Bay of Biscay.

aulacogen A long sediment filled graben oriented at the high angle to either a neighbouring modern ocean or a neighbouring orogenic belt.

aureole The luminous white or bluish disc, surrounded by a brown ring, sometimes observed directly surrounding the sun or moon. The term is also used to describe the bright area with no definite boundary commonly seen surrounding the sun in a clear sky.

aureole, metamorphic The zone of metamorphosed rock adjacent to an intrusion of igneous rock.

aurora Illumination of the sky, sometimes in brilliant colours, as a result of high speed solar particles entering the ionosphere (at a height of 100–130 km) and releasing electrons from air molecules by excitation. The re-establishment of molecules leads to the emission of light, especially red and green coloured light, e.g., in arcs or bands over large areas. The effect is called 'aurora borealis' or northern lights in the northern hemisphere and 'aurora australis' or southern lights in the southern hemisphere. Such atmospheric disturbances occur in relation to disturbances on the sun in the course of the sun spot cycle.

authigenic A geological term to describe a phenomenon which has developed *in situ*. It is particularly used to describe sedimentary material which forms after the formation of the rock in which it occurs, e.g., the growth of calcite on shell fragments and the secondary overgrowth of quartz on sand grains (contrast allogenic).

auto consequence A theory accepted by some geomorphologists that during a fall in base level (e.g., sea level) rivers may be superimposed onto underlying solid rocks from their own deposits of alluvium.

autobrecciated lava A viscous, commonly silica-rich lava flow with a congealed crust which has been broken up and fragmented by the continued movement of molten lava within the flow interior. Stressed and deformed by the movement, the crust may fracture in a brittle manner, producing angular, smooth-faced blocks able to weld together if they are hot enough but which otherwise become incorporated into the moving interior of the flow.

autochthonous, autochthone Denoting any feature which is non transported, i.e., *in situ*. An autochthonous sediment is one in which the main constituents have been formed *in situ* (e.g., coal evaporate). In studies of alphine structures, the term autochthone has been introduced to describe rock formations that have not been displaced by major thrusting although they have been folded and faulted.

autoclast A chart formed by friction from flowing lava or autobrecciation.

autocyclic mechanisms Events responsible for the accumulation of sediments that are part of the sedimentary system itself (e.g., size and configuration of a river channel).

autoecology The study of an individual plant or animal species in its reaction to environment and more specially to habitat.

autogenic succession The process of community change (succession), caused by the reaction of organisms, particularly plants on their own environment. By the reaction mechanisms, organisms may so change their environment that other species are given a competitive advantage. The original organisms are eventually replaced, having brought about their own destruction.

automatic picture transmission (APT) A term used in remote sensing studies referring to the continuous transmission of picture information from an orbiting satellite to any suitably equipped receiving station within the line of sight. It is of great importance in weather forecasting.

autometamorphism The type of metamorphic change that takes place during the cooling of an igneous rock.

autotroph A self-feeding organism (e.g., a plant) capable of combining solar energy with salts derived from the soil to form sugars via photosynthesis. Autotrophs form the first tropic level in the food chain.

autum Strictly, the transitional period between the autumnal equinox and the winter solistice (21 September to 22 December in the N. Hemisphere; 21 March to 21 June in the S. Hemisphere). In colloquial use, it refers to the two months of September and October in the UK while in the USA (termed the fall) includes September, October and November.

available nutrients Any element or compound in the soil solution that can be absorbed readily into the plant roots and that function as nutrients of growing plants. The available amount is usually much less than the total amount of the plant nutrient in the soil.

available relief The vertical distance between the highest points of a dissected land surface and the valley floors of the local rivers. Sometimes referred to as the amplitude of relief. The information is used in morphometry.

available water In soil, the water that can be absorbed readily by plant roots. It is usually taken to be water held in the soil under a pressure of 0.3 to about 15 bars.

avalanche The rapid descent of a large mass of rock, ice and snow (sometimes all three) down a steep mountain slope, avalanches occur most commonly in winter and spring. Snow avalanches in particular form either from large masses of recent uncompacted snow occupying the mountain side, or from partially thawed layers of older snow during the warm spells of weather (e.g., with the onset of Fohn winds in the Alps). Ice avalanches may continue through the summer, particularly on steep ice falls as deeply crevassed and broken box of ice are disturbed by glacial flow. Many avalanches developed along well-used avalanche tracks can be mapped and precautions taken to minimise damage. However, avalanches sometimes follow previously unused paths and thus can constitute a serious hazard to life and property.

avalanche wind The blast of air which proceeds an avalanche; can be very destructive causing building to explode before the falling masses of snow and/or ice reach them.

aven A French term which has been universally adopted to describe a deep shaft liable in limestone terrain, leading down into extensive cave systems (karst). In Britain, the term is confined to a vertical joint in a cave roof which narrows upwards and need not necessarily give access from the surface.

avulsion Lateral displacement of a stream from its main channel into a new course across its flood plain. Normally, it is result of the instability caused by channel aggradation. The avulation of a stream into an adjacent valley may explain some cases of apparent river capture.

axial plane The plane that bisects the angle between the two limbs of a fold.

axial plane cleavage The cleavage within a fold which is systematically oriented with respect to the axial plane, and which lies parallel or sub-parallel to the axial plane, particularly in the hinge region. Away from this region, the cleavage commonly deviates from its parallelism and forms a symmetric, convergent or divergent fan about the axial surface.

axial surface The surface which joins the linger lines on adjacent folded surfaces. Where it is planer, it is called the axial plane.

axial trace A line that marks the intersection of the axial surface of a fold with any other plane or surface. In practical terms, it is the intersection of the axial surface with the surface of the earth, or with a vertical profile through a fold sequence.

axis (1) The earth's axis is the diameter around which the planet rotates, tilted at an angle of 66.5° to the plain of the ecliptic or 23.5° from a line drawn perpendicular to that plain. (2) One of the two lines at right angles on a graph—the axis of the graph. (3) One of the lines which determines the dimension of a geometrical figure. (4) The central line of a structural fold (either the crest or the trough) from which the strata dip (downwards in an anticline, upwards in syncline) in opposing direction.

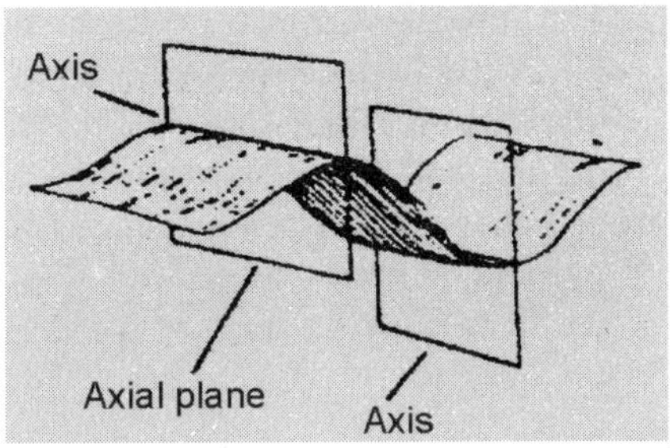

azimuth (1) In British, surveying the term relates to the horizontal angle measured in a clockwise direction from true north (true azimuth) or from magnetic north (magnetic azimuth) to another point, e.g., the azimuth of a point due east = 90° and of a point due south = 180°. It should be noted however, that the US coast and Geodetic survey measures the azimuth clockwise from a zero, which in their case is taken as due south not due north; thus a US azimuth of 90° = due west and an azimuth of 180° = true north. (2) In astronomy, the term relates to the angle between the plane of the meridian of the observer and the vertical plane passing through a heavenly body.

azimuth resolution In radar terminology, the width of ground area illuminated by each pulse of electro-magnetic radiation. The azimuth increases with the distances from the radar. The azimuth resolution and the slant range resolution govern the resolution of a radar.

azoic (1) Without life pertaining to the period of earth history before organic life evolved. (2) Portions of the seas and oceans where organisms cannot exist.

azonal soils A soil which has undergone limited pedogenic development (in other words, has been little affected by the process of reaching alluviation, illuviation, etc.). There are no well-developed soil horizons, and the soils are not associated with particular climatic vegetational zones (contrast zonal soils). A zonal soils include lithosols (formed on scree and glacial moraires, usually on steep slopes) regosals (on dry bands and gravels in deserts) and alluvial soils (on low lands prone to flooding).

azores high Semi-permanent anticyclones region with subsiding air over the Atlantic ocean at around 30°N latitude. Movement of the system poleward in summer has a major impact upon the climate of Europe. The aridity to the subsidence of air in the high pressure system.

azotobacter The principal nitrogen fixing bacteria. An aerobic bacteria which obtains energy from carbohydrates in the soil zone.

B

back reef Area behind, or to the landward of a reef. This zone usually includes a lagoon between the reef and the land.

back shore The section of a beach, which extends back from the level of normal high spring tides, as far as the cliff line or sand dune if one exists. The back shore is only affected by waves during severe storms or exceptionally high tides.

back slope The gentler slope of a cuesta in contrast to the steeper escarpment not to be confused with dip slope, a term with which it is often used synonymously but wrongly for the two may not be the same.

back swamp Area of low, in-drained ground on a flood plain away from the main channel. It stands slightly lower than adjacent alluvial fans extending from the valley sides.

back thrust A thrust in which displacement in an opposite direction to that of the main thrust propagation. Back thrusts are thought to form as a result of layer parallel shortening in a late stage of thrust sequences.

back wall The steep, precipitous rock wall at the back of a glacially-eroded hollow.

back wash The return flow of water down a beach, after a breaking wave has send swash up the beach. The back wash is more powerful with plunging breakers, formed when steep waves collapse with a strong over-turning motion.

back water That part of a river in which current velocity is low and the water virtually unmoving or stagnant. It is commonly found in braided rivers.

back wearing The retreat of slopes without loss of steepness. Back wearing is associated with particular climatic and geological conditions. For example, it is operative in many semi-arid regions, where rapid removal of fine-weathered material from slopes maintains renewal of exposer and constant weathering over the whole slope. It is added here by laterally eroding streams which undercut slopes and help to maintain steepness. Back wearing also occurs in tropical areas where dissected sheets of latuite form plateaus and mesas; the underline rotted rock exposed on the slopes, is readily removed by wash and creep, again maintaining the slope angle.

back-arc basin The zone of thickened sedimentation and extential tectonics which lies behind an island arc. Sedimentation in back-arc basins can be very varied, ranging from pelagic through turbidities to alluvial fans, although volcaniclastics are common. Various convective systems have been postulated to explain extension which is oriented parallel to the compression in a destructive margin. Back-arc basins has been classified into retro arc basins and inter-arc basins. Examples include the Japan Sea (between Japan and Korea) and the basins west of Mariana trench.

back-arc spreading The formation of the oceanic crust of the marginal basins, which is thought to occur by a process similar to that of normal oceanic crust, but involving a convective system developed over a subducting lithospheric plate. In most known instances, the injection of new oceanic crust appears to be diffuse, rather than concentrated into spreading ridges.

backing (of wind) A term used to describe a change of wind direction when it moves progressively in an anticlockwise direction, e.g., from N to NW to SW.

back-set bed A term denoting a sand deposit which accumulates to the windward of a dune, i.e., on the gentler slope.

bad lands A landscape of steep-sided gullies, which are difficult to cross and too steep for cultivation. Such areas usually underline by clays and shales, have been intensely dissected by surface streams and rivulets. Bad lands develop owing to a combination of rock impermeability, a sparse vegetation cover and rapid surface run-off from brief but heavy rainstorm. They are particularly characteristic of semi-arid region, but can occur under other climatic conditions.

baguio The name given to a tropical cyclone that forms in the vicinity of Indonesia and the Philippines (Baugio is the name of a town in Luzon, Philippines).

bai-u-season A season of rainfall maximum (June-July) in Japan during the period of the SE monsoon. It is the first of the two rainfall maxima which characterise the annual rainfall regime in Japan.

bajada (bahada) An alluvial formation in a semi-arid or arid region. In the SW USA, bajadas comprise coalescent fans of boulders, gravels and sand formed at the base of the mountain front. In this zone, ephemeral streams debouch from narrow canyons onto the gentler surface of the pediment, lose velocity and transporting power, and deposit bad load in large quantities. The individual fans of the bajada may, when comprising very coarse debris, and a slightly concave profile are more typical. In some areas, the bajada forms an intermediate zone of deposition between a mountain front and the rock pediment; but where deposition has been very extensive, it may extend as far as the peripediment, and thus obscure the rock pediment, giving a concealed pediment.

balanced sections A method for reconstructing graphically the original, deformational geometry of folded or thrust terrains from their present appearance. Balanced sections allow the amount of regional shortening to be measured and the sequence of thrusting to be deducted. Sections are usually made parallel to the main axis of compression and originate from a reference line within under form strata. Three principle methods are used to balance different properties line-length balancing; areal balancing and strain balancing.

ball and pillow structure A sedimentary structure occurring on the base of some sandstone which are interbedded with mud stones, and characterised by globular protrusions and isolated pillows of sandstones found in the 'underlying mud stone'. These structures form the differential settling of the unconsolidated sand into less dense mud below.

ball clay A sedimentary usually lacustrine, caolinitice clay, derived from the intense weathering of granites and other rocks to give a unique clay sediment with both plasticity and strength.

baltica (baltoscanida) The continental mass of North Western Europe (including most of what are now the UK, Scandinavia, European Russia, and Central Europe) that formed the southeastern margin of the Iapetus ocean. During the Caledonian orogeny event, this continental mass was brought into juxtaposition against N. America and Greenland by subduction of Iapetus ocean during the Silurian and early Devonian.

bank (1) A colloquial term for a slope or hillside. (2) A shoal-like feature covered intermittently by shallow sea water, composed of muddy, sandy or shelly (but not rocky) deposits, e.g., Grand banks of Newfound land. (3) The margin of a river channel beyond which lies the flood plain, which in turn occupies a river valley and where fluvial erosion is active, the river bank may be vertical if it is being attacked by the current. This is especially true on the outside of a curve when undercutting will cause slumping of the alluvial material into the channel where it is carried away, such collapse is termed bank caving.

bank erosion Removal of material from the side of the river channel. This may be accomplished by several processes: particle by particle removal following surface wash, frost heave, groundwater sapping and the dislodgement and fall of material and the subsequent entertainment of particles by flowing water, abrasion by transported ice; and mass failure of the bank.

bank full stage The state of a river's flow, or discharge at which the channel is completely filled from the top of one bank to the other. Beyond this point the channel cannot cope and over bank flow occurs.

bank storage Water retained in permeable channel side deposits as soil and groundwater. Drainage of such waterly affluent seepage may contribute to the maintenance of river flows at low water, while under dry conditions there may be a downstream decrease in river discharge because of percolation into bank storage. This is particularly important in semi-arid environments, in allogenic streams and on regulated rivers where low flow discharges are augmented by reservoir releases. A loss in the volume of water transmitted downstream has to be allowed for because of the bank storage factor.

banket An African term which relates to the gold-bearing pebble conglomerates of the African continent, especially the Transvaal in which the individual reefs are found.

banner cloud These are motionless, flag-like clouds commonly of lenticular (lens like) shape, forming to the lee (eddy zone) side of a hill or mountain peak. The cloud extends downwind in a strong current of humid air. Many distinctive mountain peaks (e.g., the Matter horn and Table mountain) are associated with a characteristic banner cloud.

bar (1) A deposit of sand or mud in a river channel (riffle). (2) An elongated deposit of sand, shingle or mud, occurring in the sea, more or less parallel to the shoreline and sometimes linked to it. It can be intermittently covered by tides (tidal bar) or permanently submerged (submerged bar), especially across the mouth of a river or harbour (Harbour Bar). If the bar is permanently exposed above water level it is strictly a barrier-beach, although a bar which links two coastal headlands is referred to as a Bay-bar. (3) A unit of atmospheric pressure (e.g., millibar).

bar-and-suale A term used to describe that part of a river flood plain in which the micro-relief of the alluvial surface is characterised by bars and trough (suales) formed in earlier depositional phases in an area of meander growth.

barat Local term for a fierce north westerly wind common from December to February on the northern coast of Celebeas.

barbed drainage A pattern of drainage in which the confluence of the tributary stream with the main river is characterised by a discordant junction, as if the tributaries intend to flow upstream and not downstream. The pattern is thought to result from capture of the main river which completely reverses its direction of flow, while the tributaries continue to point in the direction of former flow.

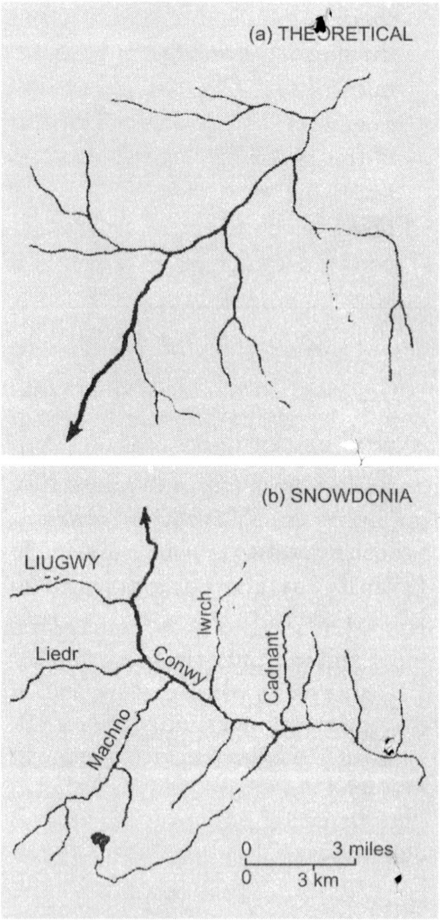

Barbed drainage

barchans (barkhan) One of the most common type of sand dune found in deserts. The barchan is a crescentic dune formed at right angles or transversely to the wind (hence transverse dune), particularly where this blows consistently from one direction. The dune begins as a small mound of sand formed on the lee side of an obstruction to air flow, such as a rock or bush once in existence, the dune will continue to trap sand blowing by the wind. This fresh sand will be transported up the windward slope, over the dune crest, and on to the lee face of the dune; this will cause downwind migration of the dune. The rate of migration will be slowest at the centre of the dune (where it is highest); as a result the dune will develop 'horns' pointing downwind. In profile, barchans are asymmetrical. The windward slope is gentle, but the lee slope is continually steepened by inputs of sand. However, the lee slope angle does not usually exceed 34° (the angle of repose of dry sand); and tendency for the dune face to steepen beyond this is countered by steepage of sand (hence slip face).

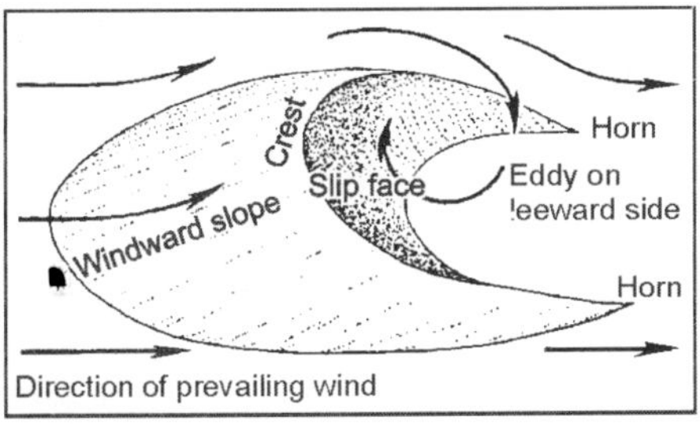

Barchan

baroclinic (1) Applied to an atmospheric condition in which isobaric and constant density surfaces are not parallel, e.g., in a frontal zone. (2) Applied to a state in the ocean in which the surfaces of constant pressure inversed surfaces of constant density. In this situation, the water density gradient depends on water properties (temperature and salinity) as well as pressure (depth). This can be contrasted with the barotropic situation.

baroduric Capable of withstanding high pressures.

barograph A self-recording aneroid barometer.

barometer An instrument for measuring atmospheric pressure, invented in 1643 by E. Torricelli who discovered that the weight of the atmosphere would balance a column of mercury supported in a tube.

barometric gradient Expresses the degree of spatial change in atmospheric pressure from one place to another, i.e., between areas of high and low pressure as revealed by the spacing of isobars on a synoptic chart. Where the isobars are closely spaced, so that the change in pressure from one point to another is considerable, the barometric gradient is steep where they are widely spaced, so that the change in pressure from one place to another is slight, the gradient is gentle. A steep barometric gradient (commonly associated with mid-latitude frontal depressions or tropical cyclones) produces gale-force or hurricane force winds; a gentle gradient (as is usually found under anti-cyclonic conditions) is associated with gentle breezes.

barometric tending The degree or amount of increasing or decreasing atmospheric pressure over a particular time span generally three hours.

barothermograph Device for the continuous measurement of both pressure and temperature, on a revolving chart.

barotropic Applied to a state in a water mass in which the surfaces of constant pressure are parallel to the surfaces of constant density. In this situation the density gradient depends on depth only, as in an isothermal fresh water lake.

barranca A Spanish term used particularly in USA and Latin America, to describe a deep ravine cut by stream action on the slope of a volcano.

barred basin A partially restricted sedimentary basin, where free movement of waters is impeded by the presence of a rock sill or sediment barrier. This restriction often results in anoxic or oxygen-poor waters, or, in arid areas, in evaporate deposition.

barrel A measure of oil and petroleum. One barrel = 32 Imperial gallons = 42 American gallons = 159 litres.

barren lands An early description of the Tundra regions of N. Canada, characterised by permafrost, sparse vegetation, long winters and inhospitable climate now rarely used.

barrier beach, barrier island An elongated sand or shingle bank which lies parallel to the coastline and is not submerged by the tide (barrier bar). If it is high enough to permit dune growth it is termed a barrier island. Both are separated from the main land by a lagoon or in which coastal marshes or mangrove swamps may occur. Examples of barrier beaches include those on the Castein coast of the USA from Florida to New Jersey; barrier islands include the Lido of Venice and the Frisian islands of Denmark, N. Germany and the Netherlands.

Barrier-beach ridges

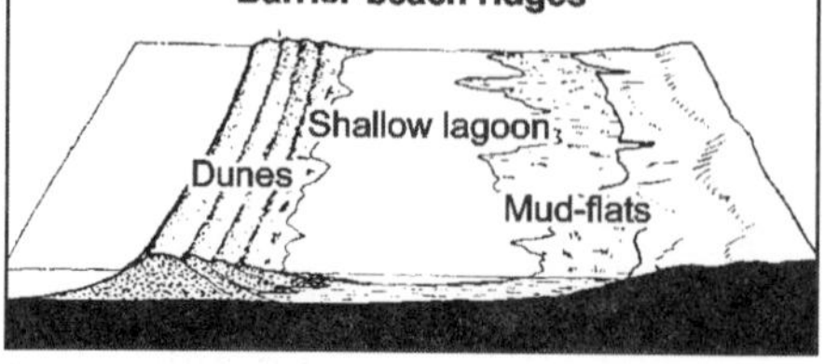

barrier island An elongated island of sand or gravel formed parallel to a coast.

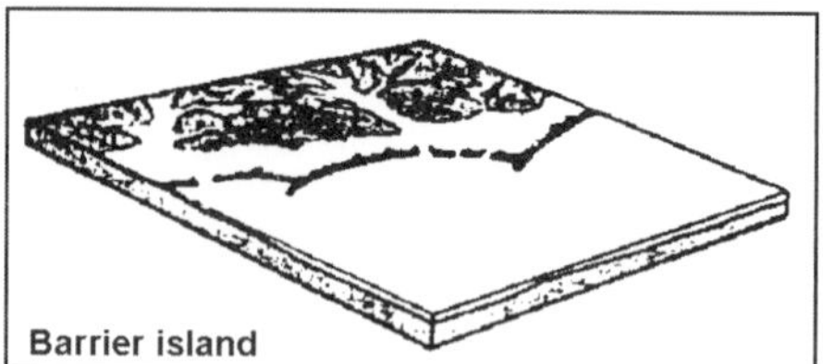

Barrier island

barrier lake Any lake created by the impounding of a river by a natural dam. It may be created by (i) a volcanic lava flow, (ii) a landslide, (iii) an ice-dam, (iv) a moronic-dam, (v) deltic deposits, (vi) a vegetation dam, (vii) a calcium carbonate deposit in a Karstic region, (viii) an avalanche.

barrier reef A large coral reef running parallel to the port line, from which it is separated by an extensive lagoon. The most famous example is the Great Barrier Reef of E. Australia, which extends for over 2000 km southwards from the Torrer strait to the islands of the Capricorn and Bunker groups. This is a massive and complex structure, comprising an outer line of coral reefs, in numerable cays (accumulations of coral sand just breaking sea-surface) and larger islands lying some 50–200 km. Offshore together with many inner reefs, as at Cook-town, Queensland.

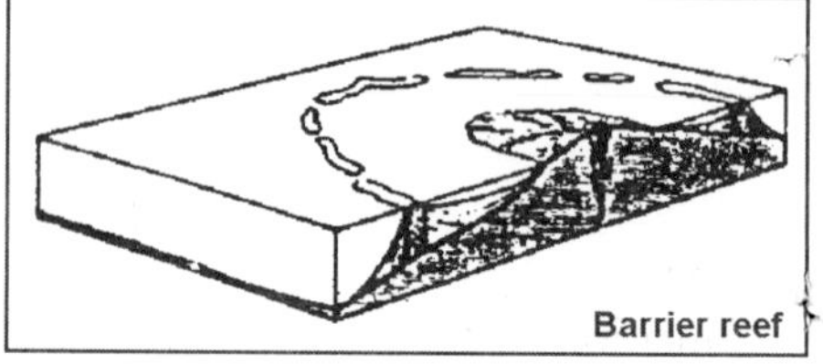

Barrier reef

The whole formation rests on the continental platform (maximum depth 200 m), and seems to have accumulated during a long period of sea bed subsidence, which allowed the reef building corals to grow upwards and outwards, keeping pace with the relatively rising sea-level and extending even further from the Browned coastline.

barysphere In strictest usage, the term refers only to all of the earth's interior which lies beneath the lithosphere. It therefore includes the core, the mantle and the Asthenosphere. Some writers have used the term loosely to describe only the core or only the mantle.

basal complex A general term for the rocks which make up the ancient 'shield' areas of Precambrian age.

basal conglomerate (1) A conglomerate deposited at the base of a sedimentary sequence, e.g., at the base of a channel fill deposit. (2) A conglomerate deposited above an unconformity surface.

basal ice A relatively thin layer (with a vertical extent of up to tens of metres) of debris—rich ice present at the base of glaciers, which is produced at and interacts with glacier bed. As a result of its different mode and environment of formation it may differ from normal glacier ice in terms of overall extent and properties (including its rheology), debris, solutes and gases.

basal melting The process during which the bottom surface (hence bottom melting) of a glacier exceeds the melting point, causing melted water to be generated and incorporated debris (till) to be deposited. It is associated with stagnant and decaying ice.

basal platform A term used to describe the unweathered granite surface above which the Dartmoor tors rise of some 10–20 m. The platform has been interpreted as representing the former surface of the water table, at a time when the upper layers of granite were being chemically rotted by vadose water to give cruss with corestones. However, this view is now in doubt; a clearly defined water table is not likely to develop in a variably jointed rock such as granite, nor this chemical weathering operate only above the water table.

basal sapping Erosion concentrated along the base of a slope, causing undermining and recession of that slope. Basal sapping may be particularly active in some tropical environments. For example, (1) where laterite-capped slopes are undermined by springs and seepages in underlying weathered sands and clays, or (2) at the foot of scarps and inselbergs, where concentrated moisture (from run-off on the slopes, or ground-water held by talus) results is rapid chemical decomposition. The recession of the back walls of cirques may also involve basal sapping whereby disintegrated rocks are incorporated within the ice and removed.

basal sliding The process by which a temperate glacier moves over its bed. It involves three mechanisms: relatively rapid creep in the basal layers; pressure melting whereby ice under pressure melts on the up-glacier side of a small obstacle and the released water freezes on the down-glacier side; and slippage over a layer of water at the bed.

basal surface of weathering The lower limit of a deeply weathered regolith, developed mainly in humid tropical environments but also found elsewhere. There is usually a very rapid change as the basal surface from rotted rock above to solid rock below. In the tropics, the basal surface may lie at depths of 30–60 m beneath the ground surface. Here the prevailing high temperatures and abundant groundwater have promoted intense chemical weathering, and removal of the resultant withered material has been hindered by the dense vegetation cover (in the tropical forests) and low relief (as on the African pediplains). However, in some areas removal of stream action has resulted from climatic change and uplift. The basal surface has then been exposed, giving rise to bare rock platforms and low rounded hills.

The basal surface of weathering

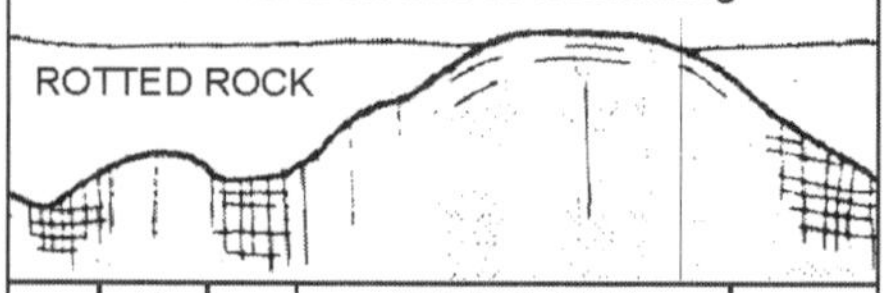

basal till Till carried at or deposited from the under surface of a glacier.

basalt A fine-grained, dark-coloured igneous rock. It is formed from extruded basic lava, containing a high proportion of plagioclase feldspars and ferromagnesium minerals but relatively little quartz (contrast acid lava). Basaltic lavas have a low melting point and low viscosity and flow freely to cover wide areas.

base exchange (or cation exchange) A vital soil reaction whereby bases or cations such as calcium, magnesium, sodium and potassium are made available as plant nutrients. These cations are usually loosely bonded on the surface of soil complex, and cation exchange takes place when hydrogen cations, derived from organic decomposition and atmospheric sources, replace the metal cations, releasing the latter into the soil water around the root hairs, where the plant may absorb them.

base flow The part of the discharge of a stream, which is provided by springs and seepages, i.e., not direct run-off; in other words, the streams are draining groundwater accumulated during a preceding period of replenishment.

base lap The discordant relationship marking the lower boundary of a depositional sequence, where upper beds cap out over the underlying surface. There are two types of base lap: onlap, where the overlying beds then out updip; and downlap, where younger, initially inclined, overlying beds then out downdip.

base level A theoretical plane surface underlying a land mass denoting the depth below which erosion would be unable to occur. Sea level provides a base level on a regional scale. Local base levels may be provided by the base of a hill slope, lakes or by the junction between a tributary and the main river.

base saturation The condition arising when the cation exchange capacity of a soil is saturated with exchangeable bases—calcium, magnesium, sodium and potassium—when expressed as a percentage of the total cation exchange capacity.

base surge A turbulent, dilute flow of ash and either water or stream, which expands readily as a collar-like cloud from the base of a vertically venting eruption column generated by the explosive interaction of magma and water. The surges are commonly cold and wet, consisting of ash mixed with water at temperatures below 100°C with a high magma: water mass ratio the surges can become dry and hot, consisting of ash mixed with steam. Base surges can be hazardous, as was the one that occurred during the eruption of Tall, Philippines, 1965. Radially expanding basal clouds observed in nuclear explosions are a type of base surge.

basement Any mass of ancient igneous and metamorphic rocks occurring beneath an unconformable 'cover' of stratified, unmetamorphosed sedimentary rocks. Most commonly the term is used to describe a basement complex of Pre-cambrian age but should not be restricted simply by age.

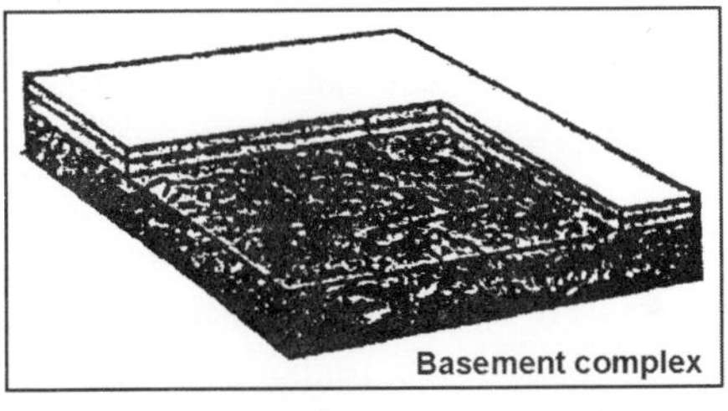

Basement complex

basement complex A broad term for the older rocks, usually Archean igneous and metamorphic rocks, which underlie more recent sedimentary rocks in any region. Such rocks are a feature of the ancient shield areas of the earth's surface.

basic grassland A name given to the grasslands of limestone terrain, where the dominant species are the fescues (e.g., Festuca ovina and Festuca rufra).

basic lava A flow of molten igneous material from a fissure eruption or a central vent of a volcano, in which the silica content is low and the ferromagnesian elements are high. Since it is very fluid, owing to its low melting point, it flows great distances to form lava plains or low angle volcanic cones. Basic lava constitutes some 90 per cent of all lavas, contrast acid lava.

basic rock Rock with a relatively high concentration of iron, magnesium and calcium and with 45–53 of silica by weight. Examples include gabbro which is a coarse-grained basic intrusive rock, and basalt, which is a fine-grained basic volcanic rock.

basin Several uses of the term are: (i) a structural downfold in the earth's crust in which the younger rocks occupy the centre and older rocks may be exposed by erosion on the planks, (ii) a small subsidence depression in the land surface due to solution of underlying deposits such as salt or gypsum, artificially or naturally created, (iii) a very large depression occupied by sea water, i.e., an ocean basin; (iv) a large sediment filled depression enclosed by higher land, with or without an outlet, (v) the catchment areas of a river system.

basin and range terrain A landscape characterised by a number of faulted block mountains interspersed with basins. Typically, the fault blocks are tilted to form a symmet-

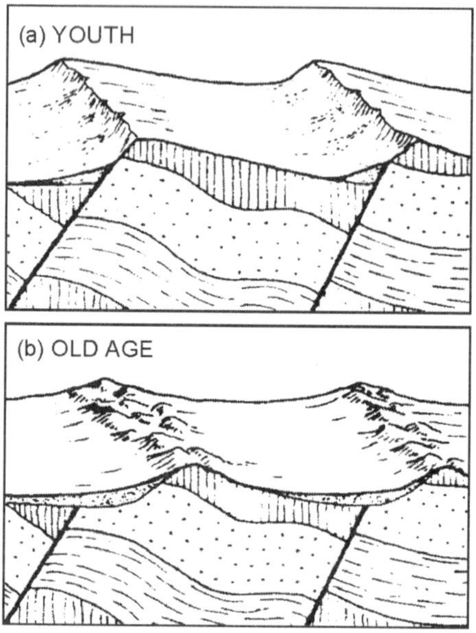

rical range with steeper and gentler slopes. The states of Nevada and Utah, USA have a basin-and-range topography.

basin and swell sedimentation A form of sedimentation in a region of differential subsidence, where thin, condensed, sedimentary sequences are deposited on slowly subsiding highs or 'swells', and thicker, usually muddier sediments accumulate in more rapidly subsiding basins between the swells.

basin discharge The total flow of water through a river cross-section as it represents drainage from the catchment area or drainage basin above that point. Total basin discharge depends upon the water balance of the drainage basin and this is climatically controlled but the temporal distribution of the discharge is a result of the ways in which water is routed through the basin hydrological cycle and this reflects both the climate and the catchment characteristics.

basin order A term used by geomorphologist to describe the hierarchy of stream drainage networks in a region. Thus, a first-order basin contains all of the drainage area of a first-order stream; a second-order basin contains all of the drainage area of a second-order stream.

basisol A type of tropical soil, characterised by its blackness, its low organic content and its accumulation of calcium carbonate in the form of a concretion horizon. It forms from the weathering of basalt under humid tropical conditions and is common in the trap landscape of the Indian Deccan.

basket-of-eggs terrain A colloquial term for a drumlin landscape.

basophilous The response exhibited by organisms adapted to life in an alkaline medium; e.g., many legumes such as alfalfa.

batholith (bathylith) A large body of intrusive igneous rock (usually granite) in which there is no observable bottom to the structure. Its surface manifestation may be in the form of a single pluton but often in the form of several distinct plutons linked at depth, e.g., the granite domes of SW England (Dartmoor, Bodmin Moor, St. Austell, etc.) are surface exposures of the same batholith which has been partly 'unroofed' by erosion. The batholiths

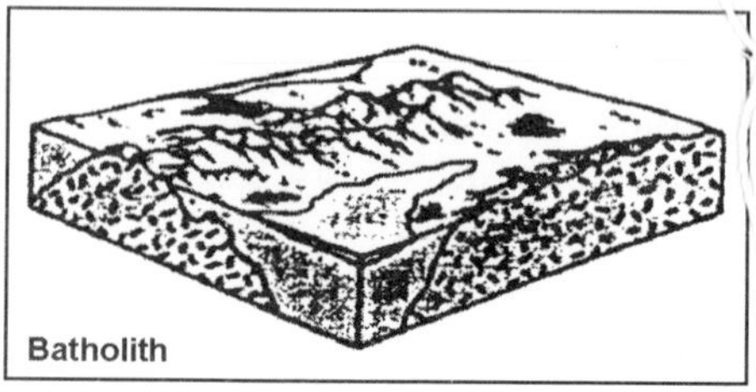
Batholith

are formed either by gradual igneous replacement of the country-rock at great depth or by stopping, during a mountain-building period. The junction of the igneous body and the surrounding country-rock is sometimes a sharp one but more frequently is marked by a transitional zone of altered rock in which varying degrees of mineralisation usually take place.

bathyl zone Oceanic zone at depths of 200–2000 m, lying to the seaward of the shallower heretic zone and landward of the deeper abyssal zone. The upper limit of the bathyl zone is marked by the edge of the continental shelf. In marine ecology, it is the region of the continental slope and rise. It may be geologically active and include trenches and submarine canyons, with underwater erosion producing avalanches.

bathymetry The measurement of the depth of the ocean floor from the water surface, the oceanic equivalent of topography.

bauxite A clay containing the mineral ore aluminium hydroxide ($Al_2O_2 2H_2O$) from which aluminium is obtained. It is formed from the tropical weathering of feldspathic rocks by the breakdown of feldspar into clay minerals and the removal of silica by leaching, leaving residual bauxite.

Baventian The lower Pleistocene, cold stage, marine silts and marine clays, that form part of the triparite division of deposits revealed in a bore hold at Ludham, in eastern England.

bay An indentation or inlet in the shore of a sea or lake. It is generally formed as a result of varying rates of erosion, the softer rocks having worn away faster than neighbouring harder rocks. Bay may provide useful sheltered anchorage for boats and are often the site of settlements.

bay bar A ridge of sediment deposited across the mouth of a bay and attached to the land at both ends. It consists of material which was deposited by a river or carried along the coast by long shore drifts. If it completely blocks the bay from the sea, deposits will pile up on the landward side of the bar. This will gradually form a delta of reclaimed land. A good example of a bay bar can be seen at Loe in Cornwell, England.

baydzharakh A type of thermokarst formed when the active layer is becoming deeper so that ice wedges begin to thaw and high-central polygons with conical surface expression form. They may be 3–4 m high and 3–20 m wide. Collapse of decaying mounds leads to the development of more continuous steep-sided depressions called duyods.

bay-head barrier A barrier beach protecting the head of a bay, but separated from it by a lagoon.

bay-head beach Sand or single beach in the low energy environment at the head of a bay. It is typical of irregular coastlines in which bays and promontories alternate.

bay-head delta A fan-shaped area of deposition formed when a stream abandons much of its load on reaching sea level at the head of a bay.

bay-ice An accumulation of floating sea ice formed in a bay during an exceptionally cold period.

bay-match barrier A barrier that partially encloses a bay at its entrance.

beach An accumulation of sand and gravel found at the landward margin of the sea or a lake. The upper and lower limits approximate to the position of the highest and lowest tidal-water levels. The angle of slope and the sedimentary structures of a beach are related to the grain size of the beach materials and to the nature of wave activity and other sedimentary processes active in the area.

beach cusp One of a series of regularly spaced crescent-shaped, structures forming local relief along a beach. The horns or 'headlands' of the cusp are composed of coarse sand or gravel, and point seaward down the beach. The intervening troughs or bays are made up of finer sand. Beach cusps are usually several centimetres high, although larger examples have been described. The size and spacing of cusps appears to be related to the nature of the wave breaking on the beach.

beach ridge Accumulation of sediment forming a prominent feature on many beaches. Beach ridges can take a wide variety of forms. Ridge may be formed near the top of shingle or sandy beaches (in sand these are called berms). Ridge features may also be formed in association with runnels near low-tide level on shallow gradient beaches. Multiple ridge forms often occur when sediment has been left by successive storm events.

beach rock Cemented beach sand deposit that develops within the intertidal zone by the precipitation of needle-like crystals of aragonite in the pore space between the grains. The concentration process is relatively rapid, taking as little as 10 years for a lithified rock to develop. The precipitation of the cement is favoured by a warm climate, and may be aided by algal or bacterial action.

beaded drainage A term given to a type of drainage pattern developed in a permafrost environment owing to differential surface melting. It is characterised by interconnected pools and short linking streams, having the overall appearance of a necklace of beads. It has been suggested that the pools may form by melting at the junction of ice-wedge polygons and that the streams may follow aligned ice-wedge patterns.

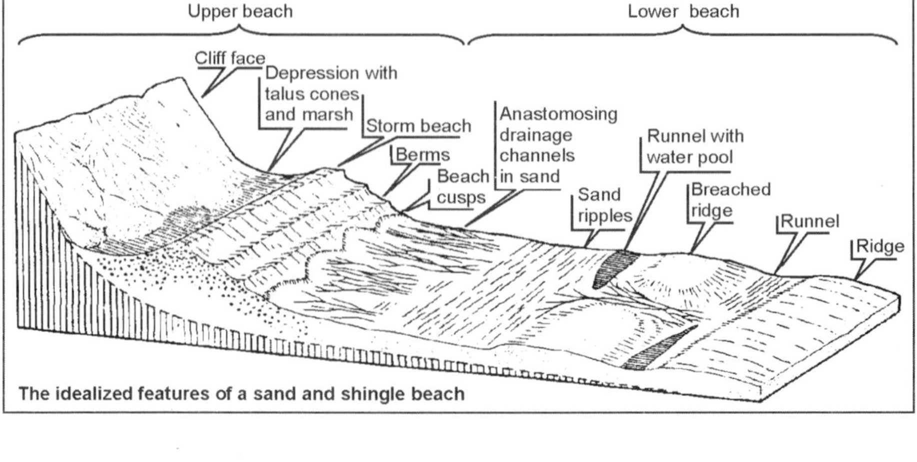

The idealized features of a sand and shingle beach

beaded esker A sinuous upstanding ridge of sand and gravel in which occasional wider hillocks on the ridge are linked by a narrower segment, creating the overall impression of a necklace of beads. Since the esker is known to be of glacio-fluvial origin it is suggested that the beaded form represents differential input rates of sediments from the melt water stream. The narrow sections may have been formed during colder periods when melting slowed down, and the 'leads' when melt water activity increased during warmer phases.

beaded valley A term to describe a valley in which wider parts alternate with narrower linking sections in the form of a string of beads. It is a rarely used term.

bearing capacity Maximum load per unit area a surface can support in safety without shear failure.

Beaufort scale Scale of values, from 0 to 12, describing wind strength, as defined by Admiral Beaufort in the nineteenth century. Each wind force is recognised by its common effects on objects in the landscape (dust, flags, trees, etc.) and on people in the open or on the state of the sea surface.

bed (1) The smallest layer (or stratum) of a stratified sedimentary rock which can be divided from its adjacent layers, both above and below, by a planar surface. It is the smallest of the lithostratigraphic units. (2) A layer of pyroclastic or volcanic material (e.g., an ash-bed or a bed of lava). (3) The floor of a water body, e.g., ocean, lake or river.

bedding plane The dividing line between two different strata of sedimentary rock. A bedding plane generally indicates a slight change of rock type, each stratum would have been deposited over a period of years, and then consolidated to form a layer of rock.

bedload A term describing the amount of solid material carried along river bed by rolling, pushing or bouncing. The bed load generally comprises the coarsest fraction of the carried load, in contrast to the finest fraction which is usually carried in suspension.

bedrock The rock which remains solid and relatively unweathered beneath the superficial layer of soil and weathered rock.

bedrock meander A term to describe the sinuous, curving course of a stream.

beestonian (1) A cold period during the middle Pleistocene. (2) Arctic, freshwater bed deposits of sand and silts represented at Beeston, Norfolt, England. Some gravels in Norfolk, Suffolk, and Essex, England have also been ascribed to this period.

belt (1) A spatial term applied to a distinctive zone of terrain, vegetation or climate. (2) A term applied to a narrow strip of sea water.

belt of no erosion A term devised by an American, R.C. Horton, in 1945 to describe the uppermost hilltop zone of a slope in which the gradient is insufficient to generate erosion in the form of hill wash. The zone, extending from the crest to a point at which running water attains erosion

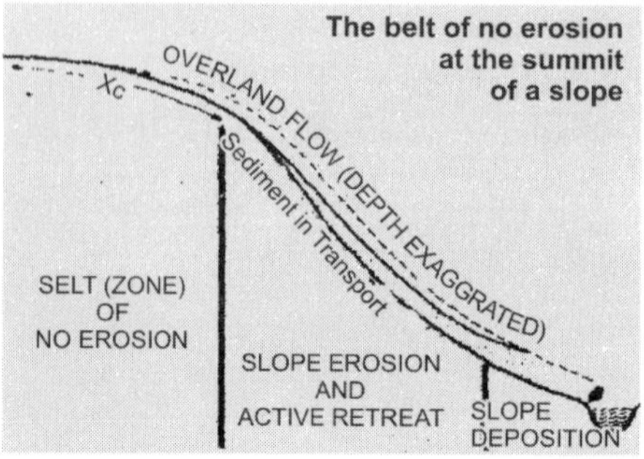

velocity, is termed the critical distance, which varies according to climate, lithology and vegetation cover.

belted-outcrop plain A planar topographic surface of low relief, which may be a peneplain or a wave-cut platform which truncates a parallel series of rock outcrops (e.g., the coastal plain of SE USA and the coastal plain of SE Ireland).

ben A Scottish term for mountain or peak.

bench A narrow flat or gently sloping ledge or step bounded above and below by steeper slopes. It may be formed by structural movement or by agencies of erosion.

bench-mark (BM) A virtually permanent point of reference used by a surveyor during a topographical survey. A specific point or level on the bench-mark (which may be artificially erected, cut into an existing building or in to the solid rocks) has a known elevation related to a datum at a fixed level.

Benguela current Oceanic water current that flows northward along the west coast of southern Africa between latitudes 15°S to 35°S. It is distinguished by an area of cold upwelling water, and is a relatively weak current, flowing at less than 0.25 m/s.

Benioff zone One of the zones of deep earthquake hypocentres whose existence was first demonstrated in 1927 by the Japanese seismologist Kiyoo Wadati. The zones were mapped in the 1940s and 50s by Hugo Benioff. They dip from near-surface to a maximum depth of approximately 700 km and are associated with oceanic trenches, island arcs, volcanic chains, and young fold mountains and are thought to indicate active subduction.

benthic Pertaining to plants, animals and other organisms that inhibit the floors of lakes, seas and oceans.

bentonite It is a montmosillonite-rich clay formed by the breakdown and alleviation of volcanic ash and volcanic tuffs.

berg (1) A term of German and Dutch derivation used locally to describe a single hill or range of mountains, respectively: in Germany, a single hill, e.g., Vogelsberg, in South Africa a range of mountains, e.g., Drakenberg. (2) A colloquial term for an iceberg.

berg wind Local wind, generically of the Fohn type, which blows offshore in S. Africa.

berg's index of continentality An empirical attempt to define the spatial limits of a continental climate. Berg relates the frequency of continental air masses (N), in the index of continentality; K = C/N (%).

Bergeron-Findeisen theory A theory, proposed around 1930 by T. Bergeron, and subsequently developed by W. Findeisen, that provides a mechanism for the growth of raindrops in ice water cloud. It is based on the differential values for saturation vapour pressure over ice and super cooled water surfaces. At cloud temperatures of –12 to 30°C, air can be saturated over ice but not over water particles, so evaporation can take place from the water droplets, and ice particles can grow by sublimation at the expense of water droplets. When they are large enough, the ice particles can fall from the cloud, melting as they pass through lower, warmer air. The process depends on these being a mixture of ice and water and so may operate in mid-.and high-latitude cloud but not in all clouds, e.g. not in tropical clouds which are at temperatures above freezing.

Bergmann's rule The idea, proposed by C. Bergmann (1847) which state that the size of homoeothermic animals in a single closely-related evolutionary line increases along a gradient from warm to cold temperatures, i.e., that races of species from cold climates tend to be composed of individuals physically larger than those of races from warm climates. This is because the surface area: body-weight ratio decreases as body weight increases. Thus, a larger body loses proportionately less heat than a small one. This is advantageous in cold climate but disadvantageous in a warm one.

bergschrund A wide and deep crevasse found between a cirque glacier and in head wall. It forms when the glacier has developed to the stage at which it pulls away from the rock slope on its upper side. A series of small bergschrunds (bergschrund crevasses) may form instead of the single feature.

bergsturz A German word which has been used to denote large scale falls (> 10 m^3), usually of rock from mountain sides. There is usually no implied mechanism and the released rock debris may or may not travel very long distances on an air cushion.

bergy lits Small pieces of floating ice less than 1m in diameter still visible above the sea; remnants of large icebergs, eroded by wave action and melted by increasing temperatures.

berm A ridge of sand parallel to the coastline commonly found on the landward side of steeply sloping beaches. It is a nearly horizontal feature formed by deposition at the upper limit of the Swash zone. When steep beaches are transformed into more shallow gradient beaches, due to a change in wave regime, the berm is removed and a long-shore bar deposited just below low tide level.

bermuda high Part of the subtropical high pressure cell of the W. part of the North Atlantic ocean, frequently linking up with the Azores anticyclone. Together they move a few degrees towards the equator in winter but move poleward again in summer.

bevel A flat, surface that has been cut across the upper termination of a cuesta on a hilltop. This is its definition in geomorphology, although strictly speaking a bevel denotes any change of slope-angle which departs from the vertical plane.

bevellel cliff A type of sea cliff in which the junction of the vertical (or near vertical) face with the horizontal plane at the cliff top is interrupted by an obliquely angled linear slope, frequently mantled by soil and regolith.

B-horizon The layer beneath the A-horizon in a fully developed or mature soil profile, characterised by a less advanced degree of weathering of constituent minerals, reduced humus content, and some degree of enrichment by compounds washed down from above. It is sometimes referred to as the illuvial horizon, and is often more yellow brown or red than the A-horizon. In soils where leaching and eluviations of the A-horizon are intense alleviation of the lower part of the B-horizon may be considerable, leading to the formation of hardpan.

biochore (1) A general term denoting any geographic area supporting a distinctive plant or animal life. (2) More precisely, it has been used to describe each of four major vegetation types: forest, grassland, savanna and desert.

bifurcation ratio It is a dimensionless number denoting the ratio between the number of streams of one order and those of the next-higher order in a drainage network. It may be a useful measures of proneness to flooding, the higher the bifurcation ratio, the greater the probability of flooding bight.

big bang theory The current explanation for the origin of the universe, in which it expands and evolves from an initial very high temperature condition about 15–20 billion years ago. The expansion time is given from the reciprocal of the Hubble constant (the rate at which galaxies are receding). All appervasive background radiation of 3K is considered to be residual from the big bang and is the strongest supporting evidence for the theory.

bight A large scale indentation of a coastline, generally of continental proportions; e.g., Great Australian bight, Bight of Benin. The term is also used to denote an indentation in the edge of the ice-sheets of polar latitudes.

bill A beak-like coastal promontory, e.g., Portland Bill.

billow clouds The parallel rolls of cloud with distinct clear areas between the cloud bands, often associated with the cloud variety undulates.

bimetallic thermograph An instrument for measuring air temperature in which the temperature sensor is a curved strip formed by welding together two metals with contrasting coefficients of expansion. The changes in strip curvature associated with temperature variations are used to actuate a recording pen through a lever mechanism.

biogeo-chemical cycles Mechanisms by which chemicals (e.g., carbon, oxygen, phosphorous, nitrogen and water) are moved through the ecosphere to be renewed over and over again. The three major cycles are gaseous, sedimentary and hydrologic.

biochemical oxygen demand (BOD) A useful measure of organic pollution (e.g., by sewage) whereby the amount of dissolved oxygen (in milligrams per litre of water) used up by microorganisms feeding on the organic matter is measured over five days at a temperature of 20°C. Water cleanliness may therefore be expressed in BOD_5 terms: clean river water will have a BOD_5 of less than 4, effluent a BOD_5 of 20 or 80.

bichorn A geological time unit equivalent to a biostratigraphic range zone. It was once regarded as synonymous with the term biozone, but this usage is now obsolete.

biocoenosis The mixed community of plants and animals (a biotic community) in a particular habitat. It may be artificially partitioned into three components: a plant community (phytocoenosis), an animal community (zoocoenosis) and a community of microorganisms (micro-biocoenosis). First introduced by K. Mobius in 1877, the word is normally used in the sense of an actual community in the landscape rather than in the sense of an abstract concept or community type.

biodiversity A term describing the variety of species, both flora and/or fauna, contained within an ecosystem. Some ecosystems are characterised by very high biodiversity (such as tropical rain forest), whilst others (such as Boreal forest) have much lower biodiversity.

biogeography The study of plant distributions (phytogeography) and animal distributions (zoogeography) together with their environments (ecology), studied over time.

bioherm A large, unstratified, dome-like deposit of reef limestone in which the organisms are in the position of growth.

biological productivity The rate at which growth processes occur either in an ecosystem or in an organism. It is normally expressed as; the weight of dry matter/unit area/unit time (kg/ha/year). Occasionally, it is expressed as grams of carbon/unit area/unit time (gc/m^2/day).

biomass The total organic matter contained by plants and/or animals, usually expressed in terms of oven-dry weight per unit area. Plants, both living and decaying, constitute by far the greatest proportion of the earth's biomass. Animal biomass (zoo biomass) is very small by comparison, and much consists of microorganisms in the soil, plant biomass varies considerably with climate and vegetation type. In the prairie grasslands of N. America, maximum biomass (at the height of the growing season) is approximately 10,000 kg ha^{-1}, in the tropical savanna it is 60,000 kg ha^{-1}, and in deciduous oak forest it is 250,000 kg ha^{-1}.

biome An organic community, or biotic complex, of plants and animals viewed within its physical environment or habitat, is the context of the ecosystem approach to biogeography. The savana woodlands and grassy plains of the E. Africa (e.g., the Serengeti plains), with their vast herds of browsing and grazing animals (elephant, giraffe, wildebeests, zebra and gazelles) and associated predators (lion, leopard, cheetah, hyena and jackals) constitute a biome.

biosphere The part of the earth's environment in which living organisms are found, and with which they interact to produce a steady-state system, effecting a whole plant ecosystem. Sometimes it is termed

Biomes

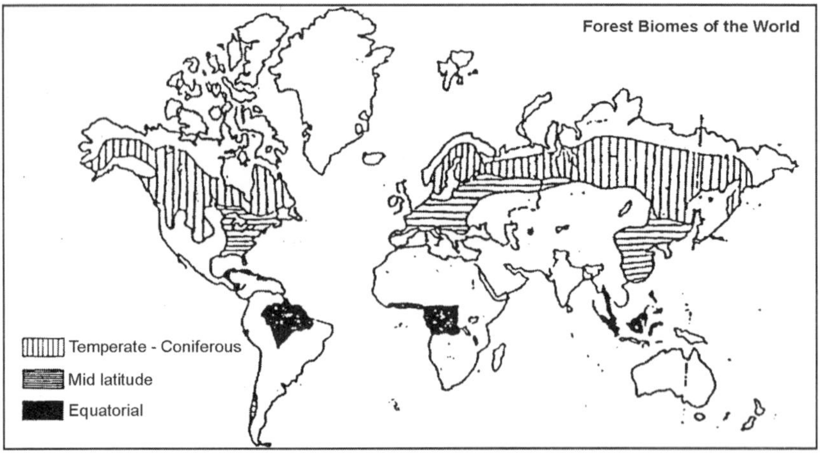
Forest Biomes of the World
Temperate - Coniferous
Mid latitude
Equatorial

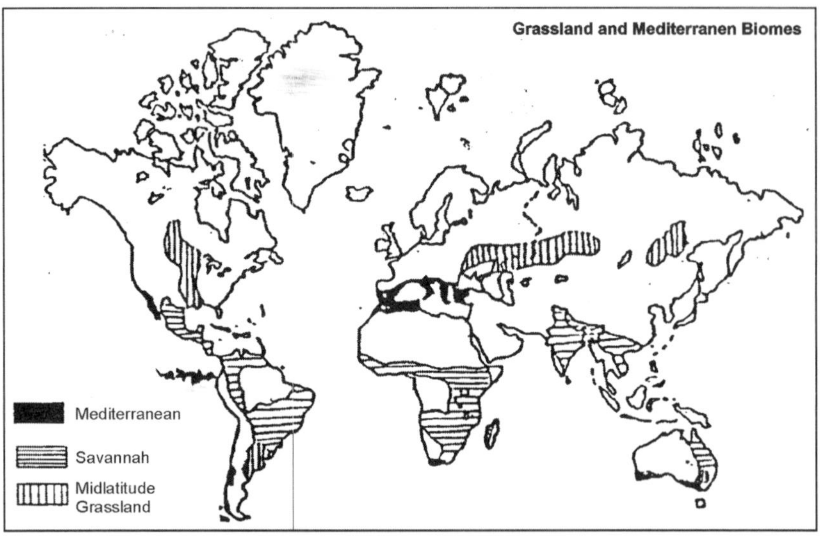
Grassland and Mediterranen Biomes
Mediterranean
Savannah
Midlatitude Grassland

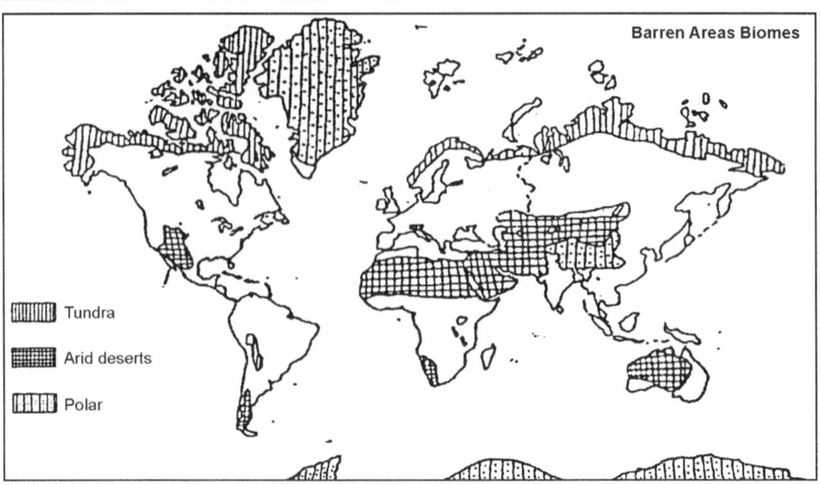
Barren Areas Biomes
Tundra
Arid deserts
Polar

'ecosphere' to emphasise the interconnection of the biotic and abiotic components.

biostasis A state of biological equilibrium between soil, vegetation and climate.

biota The living organisms occupying a place together, e.g., marine biota, terrestrial biota.

biotic complex The varied cover extending over much of the earth's surface, comprising an interacting complex of living organisms which exist in relation to the various environmental conditions.

biotic index Scheme for grading river quality according to the diversity and abundance of the river fauna present, with particular reference to key species with known pollution tolerances. The scale ranges from 10 (clean water with diverse fauna) to a grossly polluted water (with no fauna or with only a few anaerobic organisms).

biotite A common rock-forming mineral belonging to the mica group.

biotope An ecological term to designate an area in which all the faunal and floral elements are uniformly adapted to the environment in which they occur. The ultimate subdivision of the habitat.

bioturbation The breakdown and reworking of sediment by the action of its contained organisms, e.g., earthworms.

biozone A biostratigraphic unit including all strata deposited during the existence of a particular kind of fossil. The term was originally proposed as a geological time unit but this usage is now obsolete.

birch discontinuity A seismic discontinuity within the earth's mantle, some 900 km below the surface.

bird's foot delta A type of delta formed by the outgrowth of natural river levees into a body of water to form a finger-like pattern, reflecting the number of distributory streams.

biscuit-bound relief Topography which is characterised by a rolling upland or plateau into which glacial cirques have taken large 'bites', but where the cirques have not yet coalesced.

bise, bize A cold, dry northerly to north-north-easterly wind occurring in the mountains of Central Europe during the winter months.

bishop's ring A dull reddish-brown ring observed around the sun in a clear sky, attributed to diffraction associated with fine dust in the high atmosphere. That first observed after the volcanic eruption of Krakutoa (1883) remained visible until spring 1886.

bitter lake A saline lake which is rich in sulphates and carbonates, dominated by high concentrations of sodium sulphate.

bitumen A term applied originally to any of the flammable, viscous, liquid hydrocarbon mixtures. Now applied to all hydrocarbons, including gases, mobile liquids, viscous liquids and solids.

bituminous coal A type of coal between lignite and arthracite in rank, with 1835 per cent volatiles, which softens and swells on combustion, either caking or non-caking and burns with a smoky flame. It is used in town gas and coke manufacture.

black band ironstone A sedimentary rock comprising a mixture of carbonaceous (coal like) material and iron carbonate. This combination makes the iron ore virtually self-smelting. The best known is the black band ironstone of the coal measures in Staffordshire, England.

black body A body which absorbs electromagnetic energy perfectly. If the body remains at constant temperature, then it also radiates electromy nitric radiation.

black ice Type of glazed frost (e.g., on roads or the super structures of ships). Caused when water falls on a surface that is below freezing temperature. The thin sheet of ice is rather dark in appearance and, unlike white hoar frost or rime, may be hard to see.

blaky lava A surface flow of hot molten lava covered in a carapace of crystalline angular blocks, which tend to be smoothly faceted and may have dimensions upto several metres.

blanket bog An area of bog (waterlogged spongy ground, occupied by sphagnum, cotton, grass, etc. which decays to form highly acid peat), developed under conditions of high rainfall and swathing extensive areas of low relief landscape, e.g., Rannah Moor in Scotland, and much of the central plain of Ireland. Many blanket peat deposits in Britain are relict features dating from the Atlantic stage (approximately 7500–4500 BC) of the Post-Glacial period, when the climate was milder, cloudier and much wetter than at the present day.

B-layer The uppermost part of the earth's mantle, mostly above the first major phase change at about 370 km depth and below the Mohorovicic discontinuity.

blind valley A type of valley in a karstic, limestone terrain. It may be (i) occupied by a stream which disappears underground at the valley's lower end as it approaches an enclosing rock-wall, (ii) a dry valley, formerly occupied by a surface stream, which is closed at the lower end. Blind valleys may have been formed by roof collapse which exposes an underground stream-course.

blizzard A snow storm, either of falling snow or deflated snow, usually accompanied by low temperature and high winds.

block glide The sliding movements of a large block of rock over a surface that has been lubricated. Alternatively, the block may be carried down slope by the plastic deformation of underlying material.

block mountain An area of high relief which is bounded on most sides by faults and which has been either (i) uplifted by earth movements or (ii) left elevated by the sinking of surrounding areas; e.g., Vosges Mts. Black Forest.

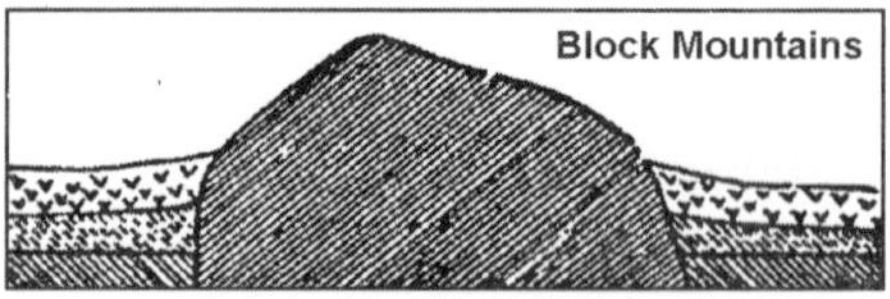

block-faulting The action of faulting on a section of the earth's crust in which parallel or semi-parallel faults produce a series of individual aligned blocks. Some maybe down-faulted, others raised while others may simply be tilted.

blockfield Spread of coarse, angular, frost shattered rock debris resting on a level or gently sloping upland surface, and found in present or former periglacial environments.

blocking anticyclone A system of high atmospheric pressure which remains stationary thereby preventing the passage of other pressure systems, especially in the westerly zone of mid-latitudes.

blood rain Phenomenon of reddish coloured rainfall, caused by dust particles that have been lifted up from arid areas and carried long distances by the wind before they are washed out in precipitation. Sahara's red dust sometimes occur in rainfall over parts of Europe even as far north as Finland.

blow-hole (1) A vertical or near vertical cleft in a coastal cliff linking a sea-cave with the cliff top and through which columns of spray are violently ejected below hold is formed by wave erosion along a fault or a joint at the back of the sea cave; when a wave surges into the cave the compressed air forces spray violently up the narrow orifice. In Scotland, the term gloup is synonymous with blow-hole. (2) A minute crater formed on the surface of thick lava flows.

blow-out Wind eroded section of a sanddune that has been largely stabilised by vegetation. Erosion results from a break in the vegetation cover due typically to over-grazing or to recreational pressure. A parabolic dune may result.

blow-well, blowing-well A colloquial term in E. England for an artesian well of either natural or artificial derivation.

blue band A layer of dense bubble-free ice in a glacial probably due to the freezing of melt water is a crevasse. When this layer is exposed at the surface it forms a darker ribbon-like pattern.

blue hole Like a sapphire set in turquoise, a circular, steep-sided hole which occurs in coral reefs. The classic examples come from the Bahamas (Dill, 1977) but other examples are known from Belize and the Great Barrier reef of Australia (Backshell et al., 1979). Although vulcanicity and meteorite impact have both been proposed as mechanisms of formation, the most favoured views that they are the product of karstic processes (e.g., collapse dolines) which acted at times of low sea level when the reefs were exposed to subaerial processes.

blue mud A submarine deposit occurring on the continental slope at depth of 228–5,183 m (750–17,000 ft). It contains upto 75 per cent of terrigenous material and derives its colour from organic matter and from iron sulphides. Its individual particles are smaller than 0.03 m in diameter.

bluff In general usage, a high, bold headland or steep, prominent cliff, but used specifically to denote a river-cut cliff or steep slope on the outside of a meander. A line of bluffs often marks the edge of a former flood plain.

bodden A German term for an irregularly shaped coastal inlet along the Baltic coast of E. Germany, formed by a rise of sea level in an undulating terrain. It is especially well-illustrated on the island of Rugen.

body wave A seismic wave which propagates through the body of a medium; P*-waves and S* waves are body wave. These contrast with surface waves.

bog An area of waterlogged ground characterised by thick accumulation of dead but not markedly decomposed plants, particularly mosses, which form acid peat.

bog burst The sudden disruption of a bog so that there is a release of water and peat which may then flow over a considerable distance.

bogaz A Serbo-croat term for a deep and narrow chasm in limestone country, especially in Karst. It is formed by carbonate solution following the penetration of water along a joint-plane or a fault line.

boghead coal A type of bituminous coal, similar to cannel coal in appearance, and characterised by a high percentage of algal remains.

bole A fossil indicative of a Paleeosl, now interbedded between two basaltic lava flows. The presence of bole indicates that tropical climates prevailed during its formation.

bolson Intermontane basin extending from the divide of one block-faulted mountain to the divide of the adjacent mountain. The surface form is made up of mountain front bajada, pediment and playa. It is classically described for the basin and range province of the western USA.

bomb, volcanic A globular mass of lava ejected from a volcano in a liquid or plastic from which solidifies in the air prior to hitting the ground. Ranging in length from 4 mm to over 1 m.

bone bed A richly forsiliferous but thin rock stratum containing many fragments of fossil bones, corproletes scales and teeth of vertebrates. The rapid accumulation and concentration of these fossil remains is thought to result from a catastrophic event, possibly an earthquake.

bora From Latin word boreas, means the north wind. A dry, cold, gusty north-east wind which affects the northern part of the Adriatic Sea and the Dalmatian Coast. Peak frequency occurs during the months from October to February and a maximum speed of 47.5 m s^{-1} has been recorded at Trieste. Two types of bora are recognised: anticyclonic and cyclonic, according to the atmospheric pressure field. The term bora is now applied to similar winds in other parts of the world, e.g., Oroshi in Central Japan.

bore Very rapid rise of the tide in which the advancing flood waters form a wave with an abrupt front. Bores occur in certain shallow estuaries and river mouths where there is a large tidal range, and suitably funnel-shaped regions, e.g., the Amazon, the Bay of Fundy, the Tsing Kiang River in China and the Rivers Seven, Trent and Oure in England.

boreal climate In Koppen's classification scheme, a climate which is characterised by a snowy winter and warm summer with a large annual range of temperature, such as obtains between 60° and 40° north.

boreal forest The northern coniferous zone of the Holarctic. The most northerly section, transitional to the tundra, is synonymous with taiga. The southerly sections are frequently made up of dense forest with closed canopies, permitting little light to reach the floor, which possesses a variable cover of lichens, mosses and herbaceous plants. Typical hardy and undemanding trees include spruce, fir and hemlock, with pine providing more open formations and having a denser ground cover.

borehole A hole drilled into the crust of the earth for exploratory purposes, especially during prospecting operations for oil, coal and natural gas.

bornhardt A German term, named from a geologist W. Bornhardt, for an isolated residual hill which rises abruptly from a plain. It is often taken as a synonym for inselberg.

boss A small batholith or any dome-shaped intrusion of igneous rock, especially one exposed at the surface through erosion of the less resistant host rocks.

bottom (set beds) The stratified layers of finer material carried out and deposited on the bottom of the sea or a lake in advance of a delta. Subsequently, as the delta grows seawards, they are buried by the foreset beds.

bottom current An underflow, or current denser than any part of the surrounding fluid and which flows along the bottom of the water body.

bottom water The water mass that lies at the deepest part of the water column in the ocean. It is relatively dense and cold, e.g., the North Atlantic bottom water has a temperature of 1–2°C. Bottom water circulation is slow moving, is greatly influenced by sea-bed topography and is driven by differences in water density.

Bouger gravity map A map of the gravitational field over an area after correction for the theoretical gravity, free-air, solid rock and topography; i.e., a map showing Bouger gravity anomalies.

Bouguer anomaly A gravity anomaly first observed by P. Bouguer in 1735, in which the deviation of the plumb line in gravitational observations is less than calculated. This deviation (or anomaly) is because of isostatic compensation resulting from a deficiency of mass in the earth's crust where it underlies mountain ranges.

boulder A large, rounded or subangular piece of rock lying on the surface of the ground, embedded in the soil or within a layer of boulder clay. Its diameter is larger than 200 mm by official British Standards but in the USA it should exceed 256 mm.

boulder clay Glacial deposit consisting of bounders of varying size in a clay-dominated matrix, and laid down beneath a valley glacier or ice sheet. Typically, it is unstratified and unsorted and characterised by rock types derived from the country crossed by the depositing glacier.

boulder train A term for the 'stream' of boulders generated from an identifiable source of bedrock and carried laterally in a more or less straight line by a former glacier or ice-sheet. Such trains can be used to identify former directions of ice movement, but since they rarely lie in exactly farewell lines they have sometimes been referred to as boulder fans, with their apexes pointing to the original rock-knobs, or exposure from which the erratic rocks were glacially quarried.

boulder-controlled slope A concept defined by a US geomorphologist K. Bryan, based on the belief that in arid environments the slope angle will be controlled by the angle of rest (repose) of the boulders which have been broken up by weathering but remain on the slope. Although the size of the boulders will slowly diminish by a granular disintegration, the slope angle will remain constant since the original size of the boulders resting upon it will remain the same when they are released by weathering, providing that the joint spacing remains unaltered.

A boulder-controlled slope undergoing parallel retreat

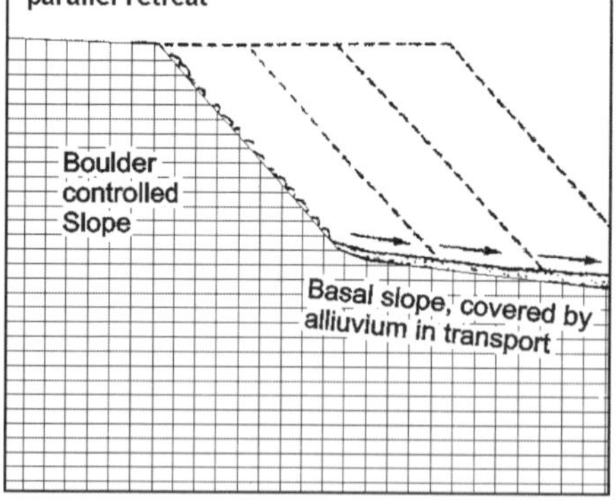

boulder-pavement (1) A surface of boulder rich till that has been abraded to flatness by the passage of a glacier. (2) The horizontal alignment of boulders when seen in section in a layer of till.

boundary current Northward or southward directed ocean-water current which flows parallel and close to a continental margin. Such currents are caused by the deflection of eastward and westward flowing currents by the continental landmasses. Boundary currents on the western margins of ocean-basins, such as the Gulf stream and the Kuroshio current, are deep, narrow, fast-moving currents; while currents along the eastern boundaries, such as the Canaries current and the California current, tend to be relatively shallow, broad, diffuse and slow-moving.

boundary layer (1) At the interface between a solid surface and a fluid, a thin fluid layer that is static because of friction between molecules of the fluid and the solid. (2) General term used to describe the atmospheric layer upto about 100 m above the ground surface. Mean velocities in the boundary layer are typically less than the free stream values.

boundary wave Seismic wave which travels along a boundary between two media of contrasting properties, rather than propagating through the bodies of the media where the boundary is a free surface (e.g., ground air, sea bed, sea water). These waves are called surface waves and give rise to ground roll and mud roll respectively. Boundary waves are of two types: Love waves and Rayleigh waves.

boundinage Minor structural feature in which competent strata resemble a series of sausages ('boudins') that form by the stretching of the competent units which unlike incompetent ones can not deform plastically. Initially, local thickening and thinning occurs ('pinch and swell' structures), as deformation proceeds, complete separation between boudins may take place.

bourne A local term for a stream in chalk country, exhibiting a sporadic or intermittent flow. After periods of heavy rainfall, usually during winter, the surface water percolates through the chalk until it reaches the saturated layers at the water table. After a long period of time the latter rises to a sufficiently high elevation to enable the bourne to flow on the surface of an intermittently.

Bowen's ratio The ratio of sensible heat to latent heat transport from the ground to the atmosphere which is generally calculated from the ratio of the vertical gradients of vapour pressure and temperature. It is used in the assessment of evaporation.

Bowen's reaction principle A concept first propounded in 1928 by Norman Bowen, which explains how minerals can respond to the changing equilibrium conditions when a magma is cooled, either by a continuous diffusion-controlled exchange of elements with the magma or discontinuous melting of the mineral.

box fold A rectangular, conjugate fold in which the inter-limb angles approximate to right angles. The fold has two angular hinges and three limbs.

braided stream A stream characterised by a network of interconnected converging and diverging channels resembling the strands of a braid. The whole river system is generally shallow so that the intervening shingle bars and alluvial material are exposed at low water. It is due mainly to the non-cohensence of the load, which leads to selective deposition and the subsequent diversion and reunification of the numerous channels. Braiding is common place where streams debouch from glaciers and flow across glaciofluvial deposits, and also where they cross alluvial fans.

brash A mass of fractural rock that has been weathered *in situ*, also applied to a mixture of shattered rock or ice.

brave west winds A nautical term used to describe the westerlies of the southern oceans, between latitudes 65°S and 40°S associated with strominess and rough seas.

Brazil current Warm water forming the oceanic boundary current that flows southward along the Brazilian coastal margin. It is marked by its high salinity (36.0–37.0 parts per thousand), its low velocity and its shallowness (100–200 m) in comparison to its northern hemisphere equivalent the Gulf stream.

breached anticline An anticline that has been deeply eroded along its central axis by river incision, thereby exposing older rocks in the central anticlinal valley, bounded by infacing scarps of younger rocks.

breached crater, breached cone Any volcanic crater in which the lave breaks through one wall of the crater to create a breach. Similarly, when lava breaks through the wall of a cinder cone it is termed a breached cone.

breackish A term applied to any water which exhibits salinity intermediate between sea water and fresh water. It is variously defined as containing between 15 and 30 parts of salt per 1000. Degrees of breackishness in decreasing order of salinity are termed: polyhaline, pilolhaline, marsholing, miohaline, oligopaline.

bread rust bomb A volcanic bomb with a cracked and crusty exterior, created when the skin shrank rapidly on cooling.

break of slope A clearly visible, sharp change of steepness in a slope profile or river long profile. The change of slope in the long-profile of a river-bed is termed a knick point.

break point bar A marine bar which form offshore in the zone where waver first break.

breaker Wave that is collapsing or breaking as a result of the wave approaching the shore and reaching shallower water. The decreasing water depth causes the wave length and speed to decrease and the wave height to increase. Consequently, wave steepness increases and the wave becomes unstable: it breaks when the wave height is about 0.8 times the water depth.

breccia A general term for coarse, clastic, sedimentary rock, the constituent clasts of which are angular. 'Breccia' literally means 'rubble' and implies a rock deposited very close to the source area. The term may also be applied as a volcanic deposit made up of pyroclastic material ejected during an eruption.

brick earth Fine-grained silty deposit occurring in South-eastern England. Of complex origin, it probably resulted from the reworking of loss, either by hill slope washing or by redeposit ion in standing water.

brick fielder A hot, dry northerly wind which blows from the Australian interior out to the SE coastlands during the summer, generally accompanied by quantities of dust and very high temperatures (>33°C/100°F). Most common in Victoria, normally associated with the advancing edge of a low-pressure trough.

brine Any water deposit which is strongly impregnated with salt.

brocken spectre The shadowy effect and surrounding rings of coloured light projected onto a cloud or bank of mist as a diffraction effect when the sun throws the shadow of an observer onto a bank of fog, called after the summit of the Brocken Mt. in Germany.

brodel A highly contorted and irregular structure in soils which has been subjected to churning by frost processes.

brow (1) The top of a hill or mountain slope: the point where a steep slope cases into a gentle slope. (2) The frontal portion of an overfold in a nappe structure.

brown alluvial soils One of the subdivisions of the brown earths which are themselves part of the soil classification of England and Wales. They are typically non-calcareous soils developed on recent alluvium.

brown earth Freely draining, and only slightly horizonated, soil profile type. It has a mull humus in the surface horizon and very little differentiation of horizons below. Brown earths are well-weathered and slightly leached soils, with a cambic horizon in the middle part of the profile (also known as braunerde and now included in the inceptions).

brown podzolic soil Freely draining, leached soil profile that has developed acid surface horizons, a mor surface humus, and a clearly visible enrichment of translocated iron oxide in the middle or B-horizon. The profile has been leached to the early stage of podzotisation, identified by the movement down-profile of iron and aluminium compounds.

bruckner cycle Tendency for the cyclical recurrence of runs of wet years at about 35 year intervals, with warmer and drier years in between reported in 1890 by A. Bruckner. Much attention was paid to it for a time, but its operation is obscured by the greater magnitude of the irregular year-to-year variations, and by cyclical recurrences of other periods.

buckle folding It occurred due to sideways deflection from a median line of competent layers in a less competent matrix due to mechanical.

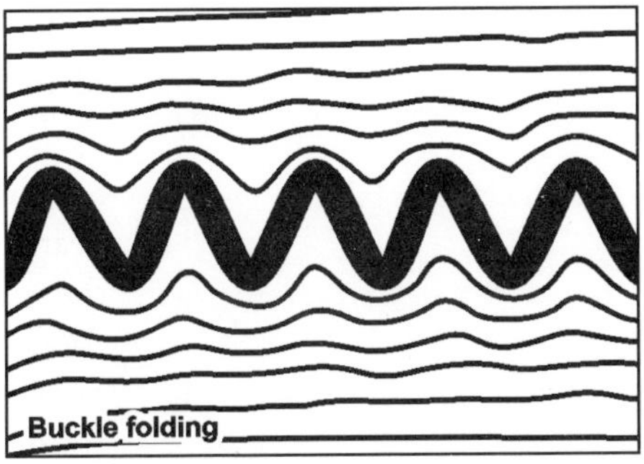

buran A strong north-easterly wind in Russia and Central Asia, occurring in any season but most frequently in winter, when it is very cold and accompanied by blizzards. The winter blizzards are also called 'Purga.'.

buried topography An overstepped erosion surface (unconformity) which shows topographic relief of hills and valleys. Such relief has a strong effect on the thickness and attitudes of the overstepping strata.

bushveld A type of savanna vegetation in Africa in which the ratio of trees to grassland varies from near-forest conditions to open, grassy parkland.

butte Small, isolated, flat-topped hill resulting from the erosion of near-horizontal strata. The diameter of the caprock is less than the height of the land from above the surrounding country. Buttes are commonly found in semi-arid regions dominated by duricust horizons. A butte is an erosional remnant of formerly more extensive slope.

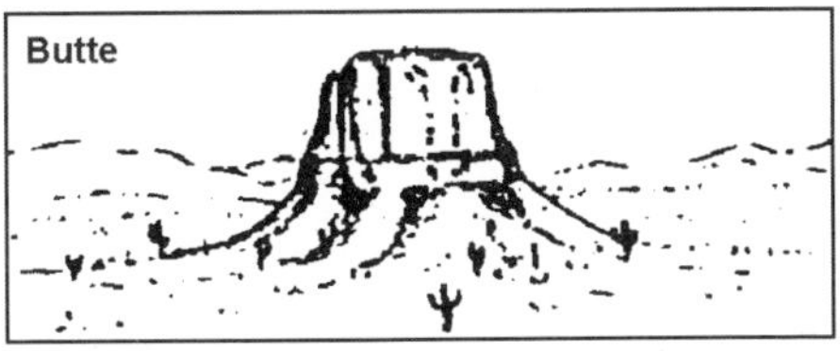

Buys Ballot's law A law, enunciated by Buys Ballot of Utrecht in 1857, that in the northern hemisphere the winds blow anticlockwise around centres of low pressure and clockwise around centres of high pressure. In the southern hemisphere, both these tendencies are reversed.

bysmalith A platonic plug or mass of igneous rock which has been forced up into the overlying rocks causing them to dome up and fracture.

C

caatinga A form of thorny woodland found in areas such as north-east Brazil, and characterised by many xerophytic species.

caballing The mixing of two water masses of identical *in situ* densities but different *in situ* temperatures and salinities, such that the resulting mixture is denser than its components.

cainozoic (Cenozoic) A Greek term meaning 'recent life'. A geological era spanning the Palaeocene, the Eocene, the Oligocene, the Miocene, the Pliocene and the Pleistocene. It was a time of climatic decline, possibly associated with the breaking of the super-continent of Pangea into the individual continents.

cairn A Gaelic word used: (i) to denote an artificial pile of stones erected on a mountain summit or along a mountain path as a guide, (ii) as a place-name or part of a place-name in mountainous districts; e.g., cairngorm.

calcareous Containing calcium carbonate ($CaCO_3$).

calcareous ooze This is a deep sea, fine-grained pelagic deposit containing more than 30 per cent calcium carbonate. This ooze is the most extensive deposit on the ocean floor but is restricted to water depths less than about 3500 m.

calcareous soil The soil that contains enough free calcium carbonate to effervesce visibly, releasing carbon dioxide gas, when treated with cold OIN hydrochloric acid, and which could also be regarded as a basic or alkaline soil.

calcification A soil-forming process in which calcium carbonate accumulates in the B-horizon, particularly characteristic of arid and semi-arid climates.

calcite The crystalline form of calcium carbonate and the principal constituent of limestone. It forms on stream beds and in caves as stalactites, stalagmites, and tufa.

caldera A large, circular basin shaped volcanic depression created in one of three ways: (i) by destruction of the upper part of the volcanic cone by an eruption of great force, e.g., Katmai in Alaka; (ii) by collapse of the volcanic cone inwards; (iii) by gradual reduction of an extinct or dormant volcano bay erosion.

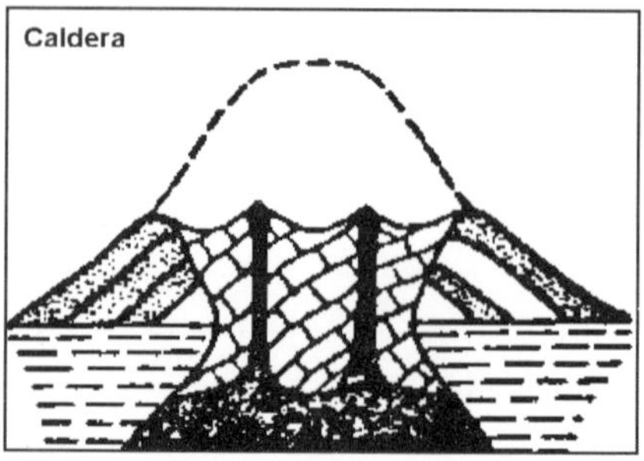

calms Winds with a velocity of less than one Knot and which are represented by a force of zero on the Beaufort scale.

calving The breaking away of a mass of ice from a floating glacier or ice shelf to form an iceberg or brash ice (small fragments). Large tabular icebergs calve from ice shelves while smaller icebergs and brash ice are commonly produced by valley or outlet glaciers.

cambering Consistent dip of strata towards local valley floors, in conflict with the general regional dip. It is well-displayed in the English midlands where ironstone beds above clays are cambered down as much as 30m below their original level. It is probably due to large-scale structural disturbance when permafrost thawed and when the plastic clays allowed overlying massive beds to flow towards valleys.

Camber

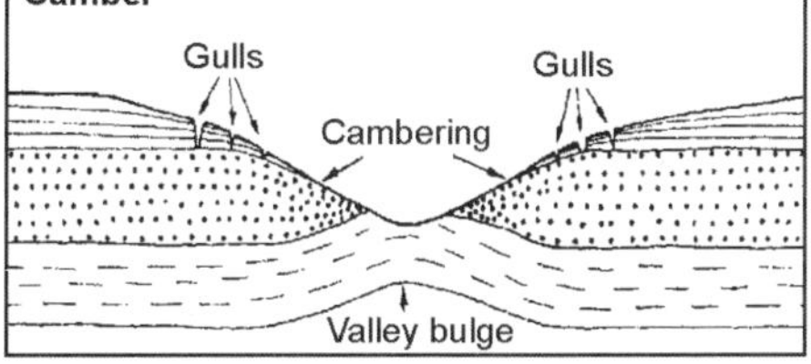

cambrian The first of six periods of the Palaeozoic era which began about 570 million years ago and ended about 510 million years ago. Sediments deposited during the period include the first organisms with mineralised skeletons. Common fossils include brachiopods, trilobites, ostracods, and late in the period, graptolites. Trilobites are important in the stratigraphic subdivision of the period.

Campbell's Law A law relating to tectonic upwarping and its effect on river drainage systems. It states that, where two streams that head in opposite directions are affected by an even lengthwise tilting movement, the one whose declivity is increased cuts down more vigourously and grows in length headwards at the expense of the other. If the tilting that affects them is part of a general warping, the divide migrates towards the axis of upwarping. This general law that axes of upwarping become drainage divided is inapplicable in the case of an antecedent river.

campo A savanna vegetation in Central Brazil, when the ratio of trees to grass allows a division into campo cerrado, which has scattered trees, and campo lempo, which has tall grass and virtually no trees.

canaries current Cool, oceanic water current which flows south along the continental margin of Spain, Portugal, and W. Africa. This slow-moving eastern boundary current is the cause of frequent sea jogs off north-west Spain and Portugal. The water is further cooled by upwelling of cold water off W. Africa.

cancer, Tropic of The tropic of the hemisphere, at 23°32′ N latitudes, at which the sun's rays are vertical at midday about 21 June.

cannon shot gravel—A coarse gravel deposit characterised by large, almost perfectly rounded stones. Thought to have been laid down by Pleistocene melt water streams at ice-sheet margins.

canopy The layer of foliage in a woodland, formed by the crowns of fully grown trees, and often effectively shading the ground beneath. Indeed it is calculated that the light intensity at the floor of a tropical rainforest may be only one per cent of that outside the forest. A well-developed canopy will therefore restrict the growth of small trees and shrubs so that forest regeneration may be possible only at points where the canopy has been breached, owing to the death and fall of individual trees.

canyon A steep-walled gorge, ravine or chasm cut by river action, in which the depth considerably exceeds the width.

cape A headland or promontory of significant size jutting into the sea.

capillatus A cloud species characterised in the upper portions by the presence of fibrous or striated cirrus in the form of an anvil or plume or mars of hair. Cumulonimbus capillatus is usually accompanied by a thunderstorm with hail and violent squalls.

cap-rack A hard stratum which protests underlying weaker racks and gives rise to plateaus and associated landforms (mesas and buttes) and cuestas. In tropical savanna regions, hardened sheets of lateritic ironstone frequently act as cap-rocks.

capricorn, tropic of The tropic of the S. hemisphere, at 23°32′ S latitudes, at which the sun's rays are vertical at midday about 21 December.

capture, river The diversion of the head waters of a river system into a neighbouring system, leaving a beheaded stream in the original system. The point at which capture takes place may be characterised by an anomalous bend in the beheaded stream, termed an elbow of capture.

carbon Non-metallic element, chemical symbol C. Carbon is extracted from gaseous carbon dioxide by plants during photosynthesis, is incorporated in living matter, and when organic matter decomposes its carbon is oxidized and so returned to the atmosphere as carbon dioxide. Pure carbon occurs as diamond, graphite, fullerene, and as the amorphous carbon black.

carbon cycle The cyclic pattern of the movement of carbon through the

A Simplified Carbon Cycle

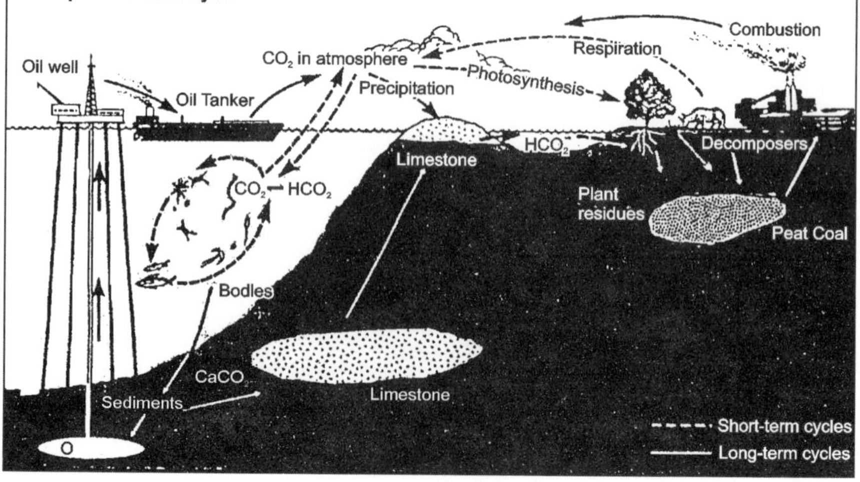

surface, interior and the atmosphere of the earth. The major movement of carbon results from photosynthesis and respiration with exclaims between biosphere, atmosphere and hydrosphere.

carbonation A form of chemical weathering involving a reaction between dilute carbonic acid, derived from the solution in water of free atmospheric and soil-air carbon dioxide and a mineral. The best example is the reaction between limestone (calcium carbonate) and carbonated water (carbonic acid).

carboniferous A period of the Palaeozoic era, following that of the Devonian and preceding that of the Permian. It ranged from about 345 million years to about 230 million years BP.

carbonisation The process of converting to carbon by the removal of other ingredients, and by slow decay or reduction following the death of a plant or animal.

cardinal points The four main compass directions: north, east, south and west.

caribbean current Warm, ocean water current which flows westward through the caribbean sea, passes into the Florida current, and thus contributes to the Gulf stream.

carlsberg ridge The slow-spreading oceanic ridge which separates the African plate and the Indo-Australian plate.

carr (1) A former bog and marsh land that has been reclaimed by drainage. (2) An isolated islet or rock, especially off the north-east coast of the British isles.

carr A woodland in a waterlogged area where the pH is not too acid and where the soils are not too deficient in mineral elements. The main trees of sown are alder, willow and sallow.

carse A Scottish term for the low-lying alluvial lands along the river estuaries of eastern Scotland. The carse clays are thought to represent deposits in part laid down by a marine transgression in post-glacial times.

cascade (1) Either a small waterfall or a stepped series of small falls or rapid. (2) A gravity-collapse structure.

cataclasis The process of rock deformation accomplished by the fracture and rotation of mineral grains, as in the production of a crush breccia.

castellanus (castellatus) clouds Clouds which tower into castle-like profiles, with cumuliform protuberances in the form of turrets.

castle Kopje A small-scale inselberg in which the joint pattern of the deeply weathered and subsequently exhumed rock has influenced the shape of the summit. Thus, the African term Kopje, which means isolated hill, is described as a castle Kopje when its summit blocks and joints create a castellated appearance.

cataclasis The process of rock deformation accomplished by the fracture and rotation of mineral grains, as in the production of a crush breccia.

cataract A series of stepped waterfalls, generally large than a cascade, created by a river as it crosses a hard rock band.

catchment The area drained by a river defined by a surrounding watershed or divide. Within the catchment all surface run-off, through flow and groundwater will eventually find its way into the river.

catchment The area from which a surface watercourse or a groundwater system derives its water. Catchments are separated by divides. A surface catchment area may overlie an aquifer system, but may be unconnected with aquifer rock itself if there are intervening impermeable aquicludes. In US usage, a catchment is often termed a watershed.

catena Topographic sequence of soils, of the same age and usually on the same parent material, that is repeated across larger landscape transects. Individual soil profile types are related to site conditions and to position on a slope. The term was introduced in E. Africa in the 1930s, and is mainly applicable in certain non-glaciated landscapes, particularly those with small hilly relief, e.g., loess areas.

catenary curve The curve formed by a chain of uniform density hanging freely from two fixed points not in the same vertical line.

cauldron subsidence Collapse of a volcanic crater due to the evacuation of a large magma chamber and marked by a ring fracture, or riny-dyke. There are many examples of cauldron-subsidence (e.g., Glen Coe, Scotland) although the various mechanisms of their formation are not necessarily identical.

cay (key) Small, flat, marine island formed from coral-reef material or sand. The term is applied, for example, to the low-lying sparsely vegetated islands off the coast of Southern Florida.

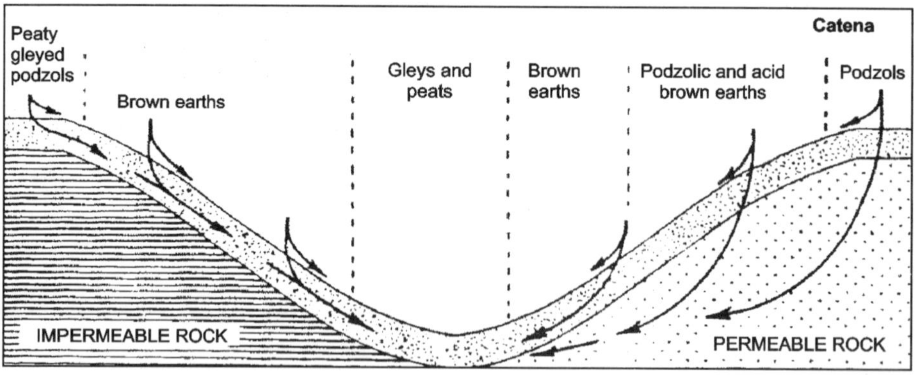

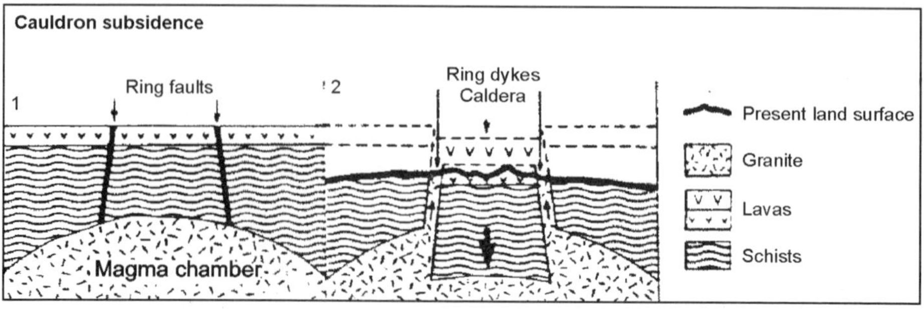

Celsius scale A temperature scale devised by the Swedish astronomer, A. Celsius, in 1742. It divides the interval between the freezing point and the boiling point of water into 100 parts, the lower fixed point being marked 100. The present system in which freezing point is marked 0 and boiling point 100, was introduced by Christin in 1743.

centigrade scale A temperature scale in which 0°C represents the melting point of ice and 100°C the boiling point of water, officially replaced by Celsius since 1948. °C = 5/9 (°F–32°); 100°C = 180°F (212°–32°).

central eruption A volcanic outburst or eruption from a single source or a closely linked group of vents, as opposed to the volcanic activity of a fissure eruption in which material is ejected not from a point but from a linear fracture.

central meridian The line of longitude upon which a map projection is constructed. In general, it is the central axis of the projection.

central-vent volcano A volcano which has one single orifice or vent around which volcanic debris has accumulated to build up a more or less symmetrical cone.

centrifugal force A body moving on a highly curved path is subject to centrifugal force acting radially outwards from the centre.

centripetal acceleration Acceleration of a body moving on a curved path, equal and opposite to the centrifugal force, directed to the centre of curvature.

centroclinal A geological term relating to a structure in which the strata dip inwards to a central low point, sometimes referred to as a centroclinal structure of centrocline.

cerrado A type of savanna found in Brazil, in which trees (9–15 m in height) are intermixed with tall grasslands.

challenger expedition (1872) The first expedition to explore the deep oceans was led by John Murray, in the British naval ship HMS challenger.

chalybeate Spring-water which contains a high concentration of iron compounds and is reputedly of therapeutic value. Thus, it is commonly utilised in spa towns.

chandler wobble The small wobble of the earth during rotation around its axis, discovered in 1891 and found to have an amplitude of 0°5′ and a periodicity of fourteen months. Although it has not been explained it is thought to be linked with earthquake activity.

channel flow That part of the surface run-off of water which is confined in a river channel in contrast to that which spreads laterally over wider and less confined areas.

chaparral A type of evergreen scrub vegetation, common in SW USA and New Mexico. It comprises poor grasslands interspersed with evergreen oaks and its name is derived from a Spanish term marquis.

charrian A stage of the upper Proterozoic of charnwood forest, England. The charnwood sequence contains impressions of very early Metaphyta or Metazoa.

chart (1) A specialised map of a coastline and its adjoining seas to facilitate marine navigation. It depicts depths of water light houses. (2) A specialised map to assist aviation—termed an aeronautic chart. (3) A weather map of aid weather forecasting.

chart datum Datum, or plane, to which depth measurements on a chart are referred.

chasma Originally, a very large canyon on Mars generally of structural origin. This term is now used for any similar large valley on a planetary or satellite surface.

chattermark Small (less than 5 mm) crescentic scar typically found on the surface of rocks of rock particles, and of rounded beach pebbles.

chelation The process by which rocks and soils decompose or disintegrate through the action of organisms or organic substances.

cheluviation A term derived from the combination of chelation and eluviation, whereby water containing organic extracts combines with metallic cations in the soil to form a chelate.

chemical weathering The process which lead to the decomposition or breakdown of solid rocks by means of chemical reactions. These comprise—carbonation, hydrolysis, oxidation, reduction and solution.

chenier A beach-ridge which has been formed above swamp deposits in a delta as a stationary phase or 'still-stand' of a prograding shoreline.

chernozem This is a type of zonal soil developed in mid-latitude continental grassland, such as, the Russian steppes or N. American Prairies

chernozem (tschernosem) A term used in soil science, meaning 'black earth' in Russian and so derived because of its very widespread existence on the steppes of the USSR, chernozems are associated with grassland vegetation in temperature climates.

chert (1) Chalcedonic variety of cryptocrystalline silica, SiO_2 that occurs as nodules or irregular masses in a sedimentary environment often in association with black shales and spilites. (2) A fine grained rock consisting of beds of cryptocrystaline silica.

chestnut soil This is also a type of zonal soil associated with drier climatic conditions. Chestnut soils are found in areas of less than 250 mm annual rainfall and contain less humus.

chevron crevasses A crevasse or fissure in a glacier which has taken on the form of a chevron owing to differential movement of the glacier ice.

chevron fold A very sharp V-shaped fold in layered rocks.

chine Precipitous ravine found along eroding coastlines.

chinook Warm dry, westerly wind of the Fohn type which blows on the eastern side of the rocky mountains. The quick onset of the wind and the sudden large rise in temperature is associated in spring with rapid melting of the snow.

chott A seasonal lake, often very saline, flooded only during the winter months. Applied especially to the tectonically formed lake basins of North Africa.

chron The smallest interval of geological time, equivalent to a chronozone, which is the body of rock formed during the time period of one chron. When chrons are grouped together they form an age.

chronomere Any interval of geological time, of no standard uniform duration.

chronosequence A sequence of related soils that differ from each other in certain properties, primarily as a result of time as a factor in soil formation.

chronostratigraphy One of the branches of stratigraphy. It is concerned with time rather than spatial distribution and lithology. It includes a set of terms which are applied to intervals of geological time as shown in table.

chute A narrow channel with a swift current, applied both to rivers and to the straits between the mainland and islands.

cinder cone A conical-shaped hill surrounding a volcanic view, formed by an accumulation of fragmented lava which has solidified as it flew through the air. The cinders are really

Era	*Period*	*Epoch*	*Absolute time-scale*
CAINOZOIC		Holocene	12,000 B P
			Million B P
	QUATERNARY	Pleistocene	c.1.5
	TERTIARY	Pliocene	c.7
		Miocene	26
		Oligocene	38
		Eocene	54
		Palaeocene	65
MESOZOIC	CRETACEOUS	Chalk	100
		Upper Greensand and Gault	105
		Lower Greensand	112
		Wealden	136
	UPPER JURASSIC	Purbeck and Portland	
		Kimmeridge, Corallian, Oxford and Kellaways Beds	162
		Cornbrash, Great Estuarine series and Oolites	172
	MIDDLE JURASSIC	Lias	195
	LOWER JURASSIC	Rhaetic	
	TRIASSIC	Keuper, Bunter (New Red Sandstone)	225
PALAEOZOIC	PERMAIN	Magnesian Limestone and Sandstones (New Red Sandstone)	280
	CARBONIFEROUS	Upper: Coal Measures, Millstone Grit	325
		Lower: Carboniferous Limestone Series, Calciferous Sandstone Series	345
	OLD RED SANDSTONE (Devonian)	Upper	359
		Middle	370
		Lower	395
	SILURIAN		
	ORDOVICIAN	Generally referred to as Lower Palaeozoic	440
	CAMBRIAN		500
			570
PROTEROZOIC & ARCHAEOZOIC	PRE-CAMBRIAN AND ARCHAEAN	Including Dalradian (Upper Dalradian is of Lower Palaeozoic age)	Pre-600

glassy and vesicular ejecta, varying in size but mainly between 3 and 4 mm in diameter.

circadian rhythm The approximately 24-hour rhythm of activity exhibited by most living organisms—man, higher animals, insects and plants. Organisms isolated from external stimuli will continue to display circadian rhythms of temperature, respiration, hormone level etc. for sometime, but may get out of phase and need a diurnal cycle to rest their internal clocks.

circalittoral zone The area of the continental shelf sea bed that lies below the zone of periodic tidal exposure. It is approximately equivalent to the sub-littoral zone.

circumference of earth Because the earth is an oblate spheroid and its circumference at the equator differs from that between the poles. The equatorial circumference is 40,076 km (24,902 miles) while that at the poles is 40,008 km (24,360 miles).

circumferential wave A seismic wave that travels parallel to the earth's surface.

circum-Pacific belt A seismic belt, which girdles the Pacific ocean, in which some 75 per cent of the world's earthquake occur, hence its popular name is the Ring of Fire.

cirque (corrie, cwm) Half-open, steep-sided hollow in a mountain region that has been or is being glaciated. Its form is due to a combination of glacial scouring (which deepens the floor and often produces a reversed long-profile gradient), and glacial erosion by basal sapping and gelifraction (which acts on the cirque's head and side walls).

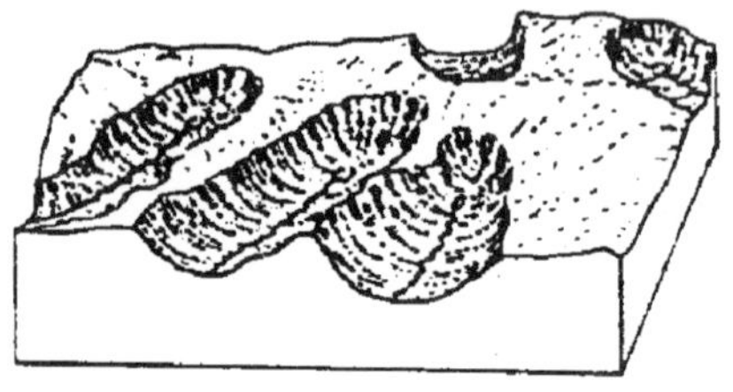

Cirque

cirque glacier A relatively small body of ice and snow, occupying an armchair-shaped hollow in bedrock. It is generally wide in relation to its length. It is actively supplied by drifting snow and therefore shows vigorous behaviour, involving rotational sliding.

cirque lake A small waterbody within a glacially eroded arm-chair shaped rock-basin. It may be retained entirely by the lip of the over-deepened rock hollow. It may be dammed by glacial moraines which mark the limits of the former cirque glacier.

cirrocumulus Cloud type comprising shallow, high-level cloud in stable air. The form is sheeted or layered with small-scale billows, or ripples. As freezing of the super-cooled droplets of young cloud occurs so the fine-patterned features of the cloud become more diffuse.

cirrostratus Cloud type comprising a semi-transparent veil of fibrous or smooth appearance, often covering the sky and identified by halo phenomena when associated with cold front, the cloud sheets have distinct edges.

cirrus A type of cloud [symbol (i)], white in colour, occurring as silky, fibourous bands or wisps, detached from each other. Its bare can occur between 5,000 and 13,700 m ice at high altitudes. Long-drawn out wisps known as mare's tails are produced by strong winds in the upper atmosphere.

clapotis A type of stationary wave (standing wave) in a water body created when an ocean wave meets a reflected wave head-on as it returns seaward after impinging upon a vertical cliff, barrier or very steep beach.

clarke A term referring to the average percentage of an element in the earth's crust. The clarke of concentration is a measure of the amount of an element within a particular deposit.

clast Particle of broken-down rock. These fragments may vary in size from boulders to silt-sized grains, and are invariably the products of erosion followed by deposition in a new setting.

clastic (1) Pertaining to fragments such as mud, sand and gravel, produced by the mechanical breakdown of rocks. (2) A sedimentary rock composed chiefly of consolidated clastic material.

clastic texture The texture of sedimentary rocks consisting of fragments of minerals, rocks and organic skeletal remains.

Clastic texture

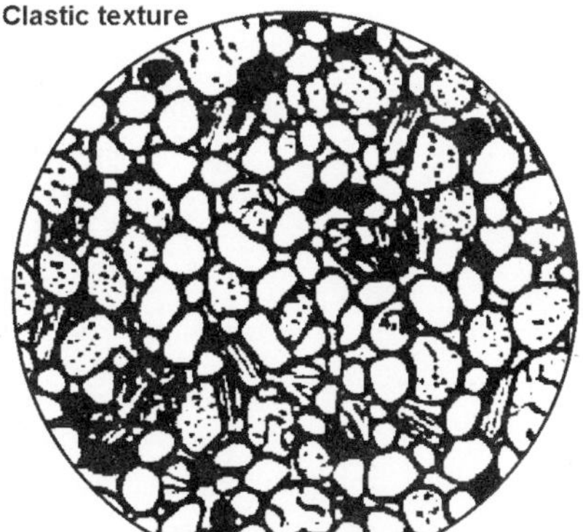

clear-air turbulence (CAT) The variable pattern of up-and downdraughts, or turbulence, sometimes occurring in the absence of any cloud. It is caused by strong wind shear, especially associated with jet streams in the upper troposphere and lower stratosphere. The phenomenon is significant for aircraft.

cleavage Of minerals, the plane along which a crystal can be split owing to its internal molecular arrangement. Also those planes of weakness in fissile rocks which are not related to jointing or bedding.

cliff A steep coastal declivity which may or may not be precipitous. The slope angle being dependent partly on the jointing, bedding and hardness of the materials from which the cliff has been formed, and partly on the erosional processes at work.

climate The average weather conditions at a specific place over a lengthy period of time, including absolute extremes, means and frequencies of given departures from these means.

climatic region A specific area in which various combinations of climatic elements can be recognised. (1) Based on latitudinal temperature zone. (2) Based on climatic effects (i.e., vegetation responses). (3) Based on climatic causes.

climatograph A circular graph depicting seasonal mean-temperature changes plotted on a monthly basis.

climatology The scientific study of world climates.

climograph A graphic diagram to illustrate the general climatic character of a particular place. This term was first used by T. Griffith Taylor with reference to human comfort and humidity.

clinometer An instrument for measuring vertical angles, either of surface slope or of geological dip.

clint Joint-bounded surface unit of a hard, almost horizontally bedded limestone. The diameter varies typically from 1 to 2 m. It may be shallowly dissected by smooth gutters, or runnels, called grikes.

clitter A local term given to the blockfields of granite boulders which surround the tors on Dartmoor, England. Clitter is thought to have been formed by frost action during the Pleistocene ice age.

cloud A visible mass of tiny particles of water and ice held in suspension by vertical motion of air, formed by condensation around nuclei, such as dust and smoke particles, pollen or salt. Different forms of vertical motion give rise to different cloud types: Vigorous local convection causes convective clouds; forced a scent of stable air produces layer clouds and forced lifting over topographic feature gives rise to orographic clouds. The international classification scheme (WMO, 1956) groups clouds into four basic categories: **cumulus**—a heap or pile; **stratus**—a layer; **nimbus**—rain; and **cirrus**—a filament of hair. In terms of height the clouds are identified as: **strato**—low level; **alto**—middle level; and **cirro**—high.

cloud burst A popular term for a violent, torrential shower of rain or hail, generally short-lived but capable of considerable damage because of the large volume of water precipitated. It is associated with violent upward and downward air currents.

cluse A French term for a steep-sided valley that cuts transversely across a limestone ridge. It may have been formed by an antecedent river. It is

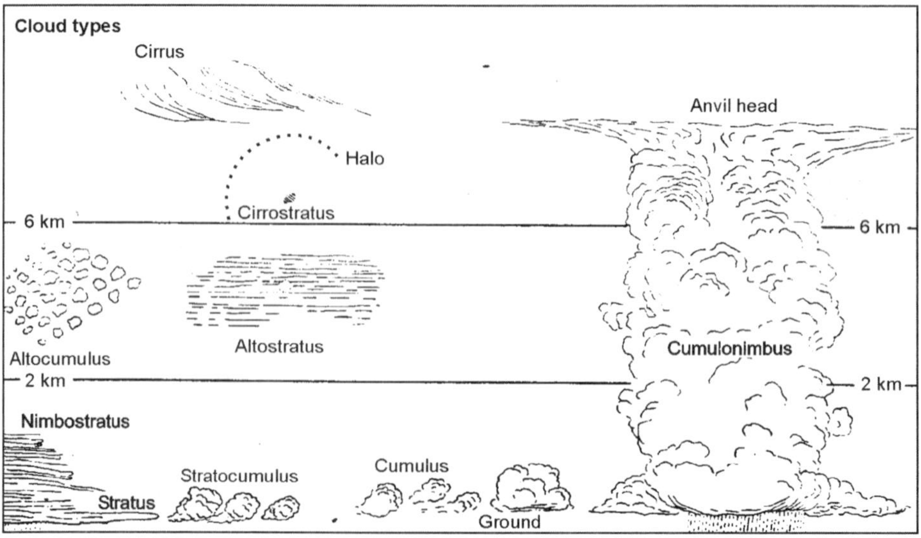

commonly found in the jura and the limestone ridges of the Savoy Alps.

coastal erosion A term which describes the actions of marine waves, through their four-fold processes of hydraulic action, corrosion, attrition and solution, when they combine to attack the coastline, thereby causing it to retreat.

coastal plain Any gentle sloping plain or lowland which borders the landward side of a coastline and which often continues offshore as the submarine continental shelf.

coastal wetlands Shallow coastal shelves and margins that are normally flooded or marshy, they include tidal flats.

coastline (1) In general usage, the term refers to the line forming the boundary between the land and the water. (2) More specifically the term refers to the highest limit reached by the swash of storm waves during high water spring tides.

cobble A rounded to sub-rounded rock fragment between 64 and 256 mm in diameter. It is smaller than a boulder but larger than a pebble.

cockpit karst A type of tropical karst landform characterised by deeply incised hollows (cockpits), which have developed from alluvial-florred dolines due to the fluctuating water-table in the limestone. Between the cockpits are smoothly crested steep-sided hills, often conical in shape.

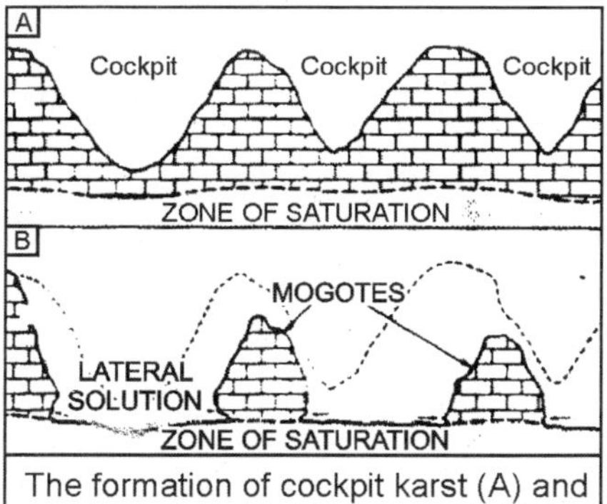

The formation of cockpit karst (A) and tower karst (with mogotes) (B)

cocos plate One of the present-day minor lithospheric plates that lies beneath the Pacific ocean. A remnant of the Farallon plate. It is subducting under the North American plate and Caribbean plate and has constructive boundaries with the Pacific plate, and the Nazca plate.

col (1) High pass or saddle in a ridge. It may mark the line of a former stream valley or of a former glacier, and so provides evidence of an early stage in the development of the landscape. (2) The saddle region of the atmospheric pressure field between two high and two low-pressure centres.

cold desert A term generally used as a synonym for tundra and polar regions where plant life is inhibited by low temperatures and physiological drought.

cold front The front boundary of a mass of advancing cold air where it undercuts and replaces the slower-moving warm air mass, thus forcing it to rise. In this way, a wedge of cold air replaces the warm air at ground level, and is accompanied by a marked fall of temperature. Rainfall associated with the passage of a cold front is often heavy and sometimes thundrey.

cold glacier A glacier in which the ice temperatures may be as low as –30°C throughout the year, with no appreciable surface melting. Sometimes referred to as a polar glacier.

cold low An atmospheric depression in which the central low pressure cell is composed of cold air because of the column of cold air at the core of the depression, the low pressure cell at sea-level with intensify with increasing elevation into the troposphere.

cold occlusion A climatological term referring to an occlusion in which the advancing mass of cold air has lower temperatures than the air mass in front. A secondary depression can form at the point where the cold and warm fronts unite to form a cold occlusion.

cold pole The area or point in each hemisphere that has the lowest mean temperatures. Vekhoyansk, in north-eastern Siberia is in the region of the northern hemisphere cold Pole.

cold sector Zone of colder air surrounding the narrowing wedge of warm air, the warm sector, in a developing depression. At the occlusion stay the whole of the surface layer forms a cold sector, with the warm air lifted off the surface.

cold wave The condition associated with air of continental polar origin often dominated by an anticyclone behind a cold front, that moves south into central and eastern parts of the USA.

collision theory of rainfall One of the attempts to explain the generation of raindrop growth. Initially it was believed that the collision of cloud particles by turbulence would cause particles to coalesce, thereby due to atmospheric turbulence within a cloud caused not only coalescence but many instances of rain drop disintegration, so that alternative explanations had to be sought.

colluvium A heterogeneous mixture of weathered materials transported down slope by gravitational forces and deposited at the foot of the slope. Thus, they can be contrasted with alluvium.

columnar jointing A system of fractures that splits a rock body, into long prisms, or columns. It is characteristic of lava flows and shallow intrusive igneous flows.

combe rock A structureless mass of unstratified rubble, consisting of flints, broken chalk fragments, compacted chalk mud and some sand. It was produced by solifluction during cold phases of the Pleistocene period and has accumulated in great thickness in the valley bottoms of the chalk lands of S. England.

comber Deep-water wave that has a breaking crest blown forward by a strong wind. The term is also applied to a long period spilling breaker.

comet A small body composed of meteoric dust and frozenices (H_2O, CO_2, CO, HCHO) in a highly elliptical or parabolic orbit around the sun.

competence The maximum size of rock particle transported by a particular flow of ice, water, or air. Ice has a high competence, because of its viscosity; flowing water has a lower competence though this increases sharply as velocity increases. Wind has the least competence.

composite volcano A lava volcanic cone, built of alternating layers of lava and cyroclastic material, over a lengthy time period.

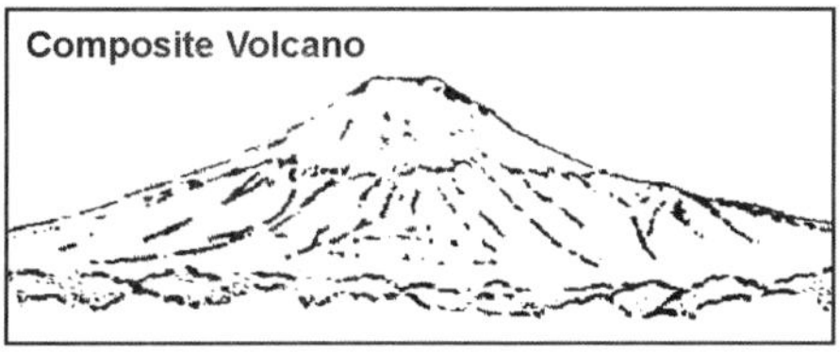

compressional wave A moving disturbance is an elastic medium characterised by volume and density changes caused by pressure. Any displacement of particles is in the direction of wave propagation.

conard discontinuity A seismic discontinuity in the earth's crust at which velocity increases from 6.1 to 6.4 or 6.7 km/sec. It is thought to mark the junction of an upper granitic and a lower basaltic layer and can occur at a variety of depths. It occurs only beneath the continents, since beneath the ocean floors the upper layer is missing.

concordant Pertaining to the harmonious correspondence between a morphological feature and its underlying geological structure.

condensation The process by which vapour changes into a liquid or solid form, either by cooling air below its dew point or by saturating air.

cone of depression A conical depression of the water table surrounding a well after heavy pumping.

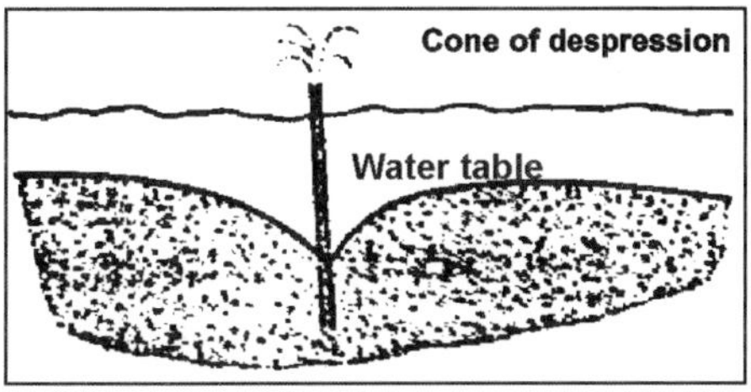

conglomerate Coarse-grained rock with rounded clasts that are greater than 2mm in size.

connate water From the latin conatus meaning 'born together'. Water that has remained trapped in a sedimentary rock since the original sediments were laid down in that water, prior to lithification. Connate water may be very old and saline.

consequent stream A stream in which the direction of flow is controlled by or dependent upon the original slope of the land.

constant slope The part of a slope profile which stands above the

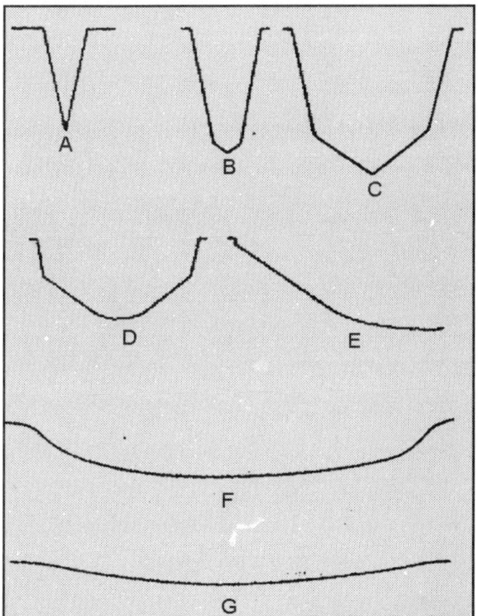

constant slope wood's diagram of the slope cycle. A = free face only; B and C = constant slope froms; D and E = waning slope develops; F = waxing slope forms, waning slope rises up side of constant slope, alluvial filling represented by dots; G = constant slope has been consumed, alluvial fill deepens, slopes gradually flatten and approach a peneplan.

waxing slope and below the free face and the waxing slope. The term constant slope was first used by Wood (1942) to define the straight part of a hillside surface, lying below free face and having an inclination determined by the angle of repose of the talus material forming it.

constructive wave A water wave breaking on a beach, with a powerful swash and a weaker backwash. Thus, it tends to move material up the slope of the beach.

continent A large land mass rising fairly abruptly from a deep ocean floor, although its peripheral areas may be submerged, as a continental shelf beneath shallow seas.

continental air mass An air mass, usually of low humidity, formed over a continental interior where high atmospheric pressure prevails. It may originate either from a low-latitude source-region (tropical, Tc) or form in a high-latitude source-region (Polar, Pc).

continental climate A type of climate characteristic of the interior of large land mass of middle latitudes. The major effects of continentality are to produce extreme seasonal variation of temperature, with dryness.

continental crust The part of the earth's crust which is composed of sial and constitutes the continents. Although its average thickness is 33 km it achieves thickness of over 50 km beneath high mountain regions.

continental drift Hypothesis proposed around 1910 to describe the relative movements of continental masses over the surface of the earth. A. Wegener (1915), suggested that continents can move around the earth's surface because of the weakness of the sub-oceanic crust.

continental ice sheet (continental glacier) An ice sheet of great extent, large enough to cover a continent. Today, only that of Antarctica survives, but during the Pleistocene large ice sheets covered parts of Europe and N. America.

continental island An island which is geographically near and geologically related to a continent having once been part of the mainland but now being detached in almost all cases by shallow sea water. Thus, the British isles and Sri Lanka are continental islands.

continental margin A zone which combines both the continental shelf and the continental slope and is distinct from the deep sea floor. It extends from the coastline down to depth of some 2,000 m.

continental sea A partially enclosed sea, linked with the open ocean but lying wholly within the area of a continental shelf; e.g., the Black Sea, Hudson Bay.

continental shelf Gently seaward sloping surface that extends between the shoreline and the top of the continental slope at about 150m depth. The average gradient of the shelf is between 1:500 and 1:1000 and, although it varies greatly, the average width is approximately 70 km.

continental slope The relatively steeply sloping surface that extends from the outer edge of a continental shelf down to the continental rise. The total relief is substantial, ranging from 1 km to 10 km but the slope is not precipitous and ranges from 1° to 15° of slope. Along many coasts of the world the slope is furrowed by deep submarine canyons, terminating as fan-shaped deposits at the base.

contorted drift A term used to describe the crumpled and folded appearance of glacial till at any location where it has been deformed by an overriding ice sheet or by the pressure of an ice front.

contour A line connecting points on a map or chart, all of which exhibit the same height above or below a given datum.

convection Transmission of heat energy by the rise of buoyant hot material and sinking of cold material.

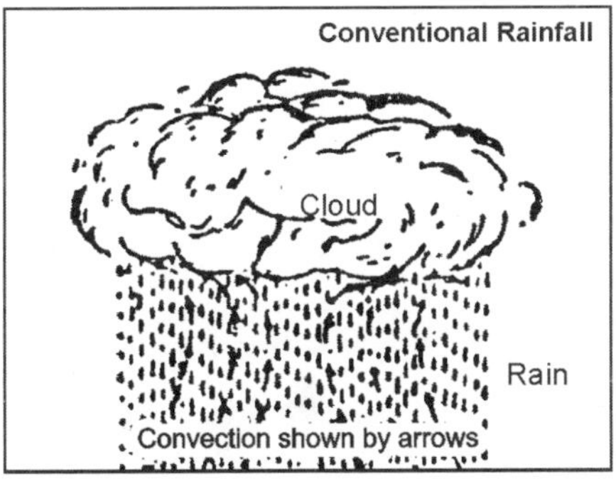

conventional rain Rain which is caused by the process of convection in the atmosphere. When moist air is warmed by conduction at a heated land surface it expands its density,

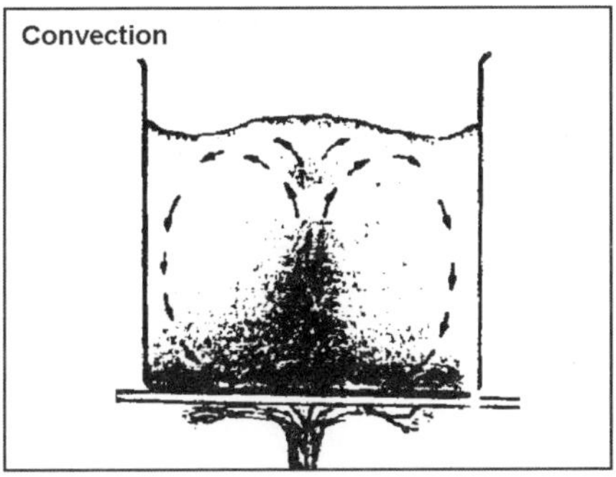

falls, and it rises conventionally. It is thereby cooled adiabatically and its temperature falls below the dew point, thereby forming clouds, often of the cumulonimbus form, from which heavy rainfall is generated.

coquina A limestone composed of an aggregate of shells and shell fragments.

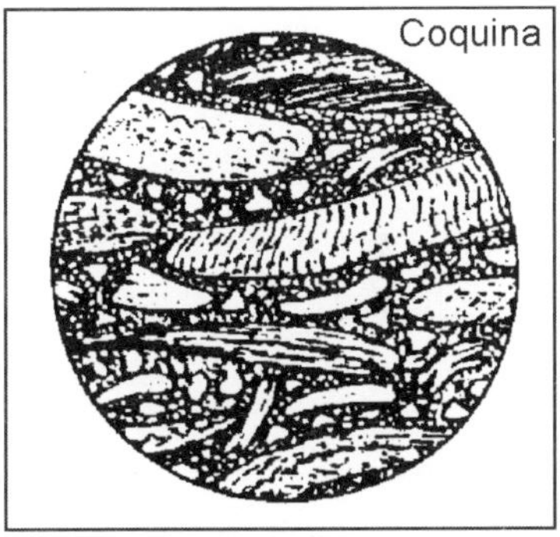

coral A marine polyp capable of secreting calcium carbonate to build an external skeleton. It flourishes in clear, warm, tropical oceans, usually between 30°N and 30°S of the equator.

coral reef A reef created by the coral polyp along the shallow shores of some tropical seas.

cordillera A Spanish term referring to a major system or group of mountains. It includes not only the parallel mountain ranges but also the intervening valleys, plateaux, intermontane basins, etc.

core The central part of planet earth, probably consisting of a dense nickel-iron alloy with a temperature estimated at about 2,700°K. The outer perimeter of the core commences at the Gutenberg discontinuity, 2900 km from the surface, where the outer core may be liquid.

Coriolis force A deflecting motion or force discussed by G.G. de Coridis in 1835 and developed by W. Ferrel in 1855. The rotation of the earth causes a body moving across its surface to be deflected to the right in the N. hemisphere and to the left in the S. hemisphere.

corona The coloured rings, which surround the sun or moon, comprising a range from blue (inner) through green and yellow to red (outer). Corona is produced by diffraction of light by water droplets. It must be contrasted with the halo, which is due to refraction in ice crystals and in which the colour sequence is opposite.

corrasion The process of mechanical erosion of a rock surface caused when materials are being transported across it by running water, by glaciers, by wind, by marine waves, or by gravity movement down slope.

co-seismal line A line which connects those points at the earth's surface where seismic waves arrive at the same time.

cotidal line Line joining points at which given tidal levels (such as mean high water or mean low water) occur simultaneously.

cove A small coastal bay, often with a narrow entrance.

covered karst A limestone terrain in which the characteristic karstic features are buried beneath a cover of superficial materials alluvium, blown sand, glacial till, etc.

crag and tail Land-form consisting of a small rocky hill (crag) from which extends a tapering ridge of unconsolidated debris (tail). The crag is a residual feature left by selective glacial erosion. While the tail is drift-deposited by ice on the lee side of the obstacle.

crater General term for a circular funnel-shaped depression, upto 1 km in diameter, produced by volcanic processes by which gases, tephra, and lava have been ejected.

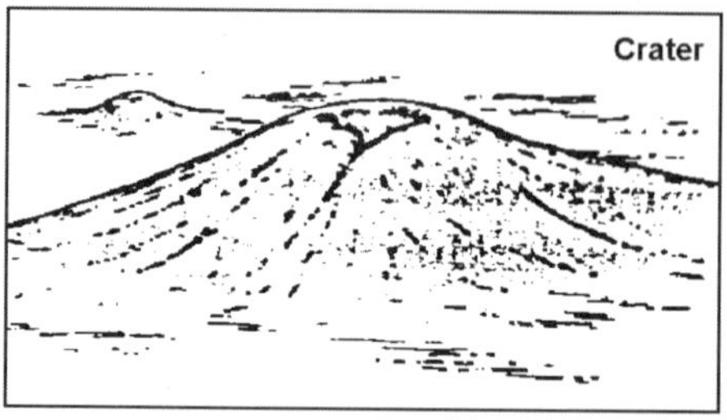

Crater

crater lake A fresh water lake occupying the crater of a dormant volcano. It is generally circular in shape and often of a brilliant colour.

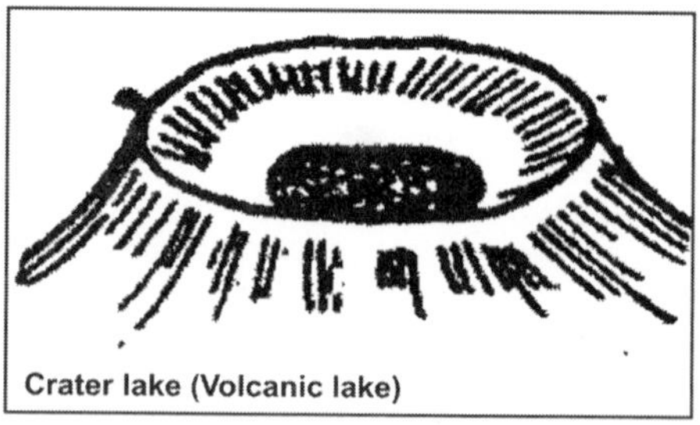

Crater lake (Volcanic lake)

craton, kraton A large section of the earth's crust that has remained relatively stable and unaffected by mountain building (orogeny) for a considerable period of geological time. It is made up of ancient rocks which were once parts of very old mountain ranges but have remained immobile since Precambrian times. Any feature relating to a craton is termed cratogenic.

creep The imperceptibly slow down slope movement of material as a result of gravity.

crevasse A deep fissure of variable width in the surface of a glacier, caused by differential movement within the ice resulting from tensional forces.

cross-bedding Stratification inclined to the original horizontal surfaces upon which the sediment accumulated. It is produced by deposition on the slope of a dune or sand wave.

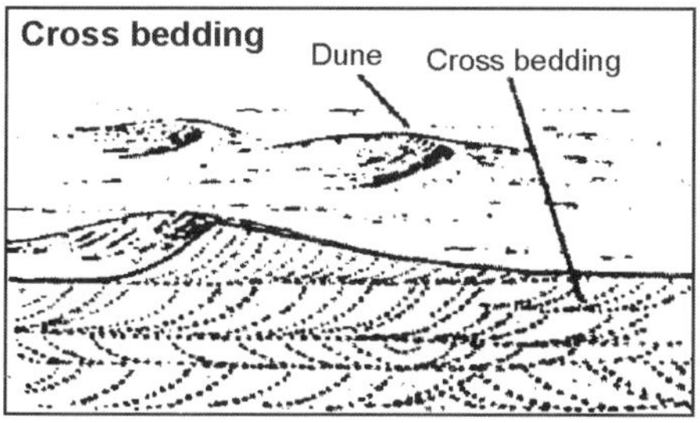

crust The thin outermost solid layer of the earth. It represents less than 1 per cent of the earth's volume and varies in thickness from approximately 5 km beneath the oceans to approximately 60 km beneath mountain chains.

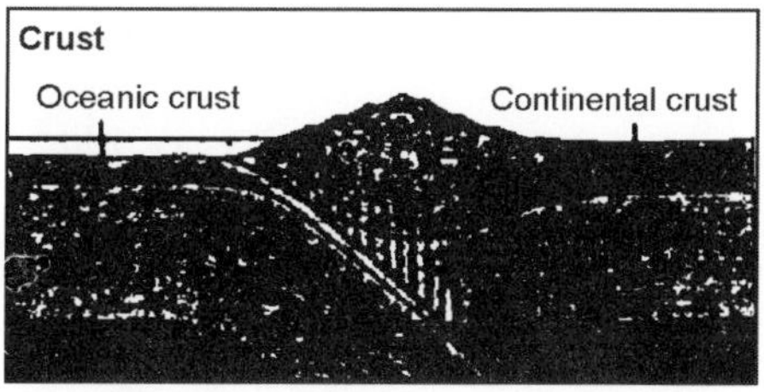

crystal A solid, polyhedral bounded by naturally formed plane surfaces resulting from growth of a crystal lattice.

crystal face A smooth plane formed by growth of the surface of a crystal.

cuesta A Spanish term that has been widely adopted to describe an asymmetrical ridge, produced by differential erosion of gently dipping strata, in which the long, gentle slope is generally accorded with the dip of the resistant strata which form the cuesta.

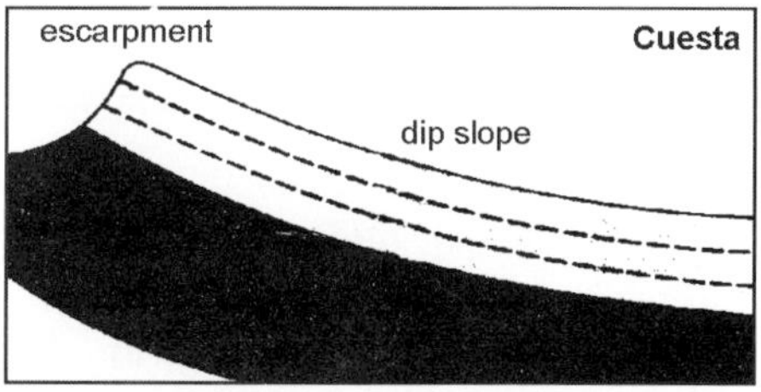

cumulonimbus A type of cloud (Cb), dark, grey when viewed from beneath but dazzling white when seen from the side, cha racterised by its considerable vertical extent, where it rises into towering forms to heights of over 10 km.

cuspate delta A symmetrical delta, usually formed where a river deboucher on a straight coastline, and

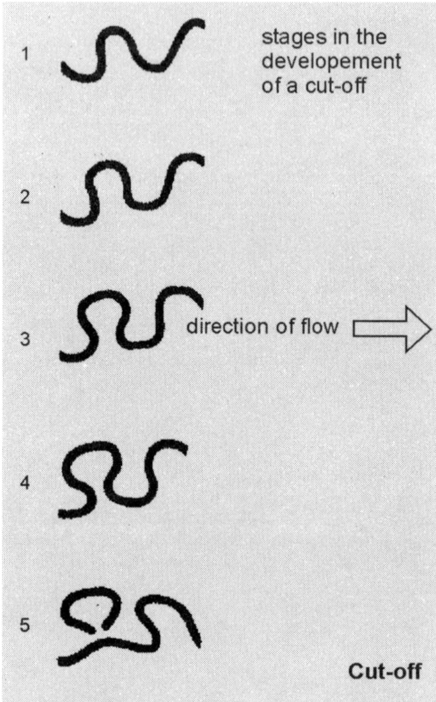

in which the sedimentary material is deposited evenly on either side of the river mouth.

cut-off (ox-bow) The section of a river channel that no longer carries the discharge. It has been developed during meander formation, when lateral channel migration across a flood plain, channel length is shortened by contact at the neck of a loop, which becomes a cut-off.

cycle of erosion The continuous process by which landscapes are formed and then gradually transformed by erosion. In other words, it is a concept relating to the evolution and modification of the physical landscape, in which the various stages of erosion are believed to be parts of a cycle process. The concept of cycle of erosion was first introduced by W.M. Davis in 1889, which suggests the three stages are: rejuvenation and incision of the drainage pattern, i.e., **youth** or **adolescence stage**; the second stage is the **maturity** and the third stage is the **old age** when the characteristic land form was low, featureless peneplain. This cycle of erosion was termed by Davis the **normal cycle of erosion** and is also known as Davisian Cycle, the geographical cycle and the **geomorphic cycle**.

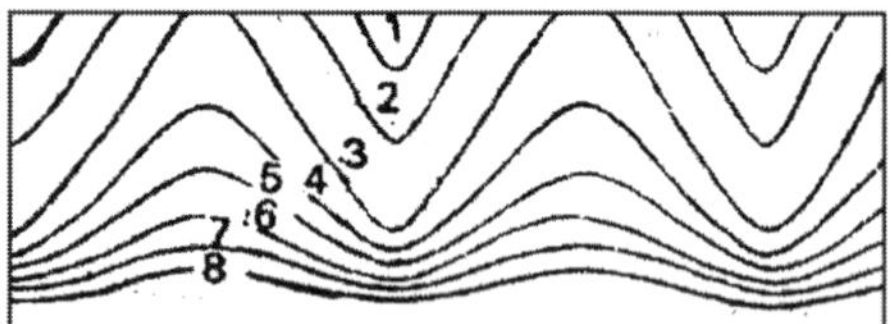

The uplifted land subjected to erosional cycle is represented in the three stages of erosion :- Youth.....1-2, Initial Maturity.....3, Middle Maturity...4-5, Final Maturity....6, Old Age....7-8.

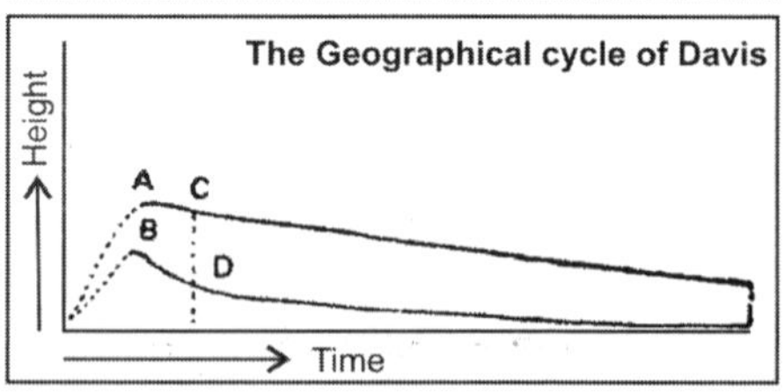

cyclone A system of low atmospheric pressure in which the barometric gradient is steep. Winds circulate blowing inwards in an anticlockwise direction in the N. hemisphere and in a clockwise direction in the S. hemisphere.

cyclostrophic wind A type of air movement which follows a strongly curvilinear path around a low pressure system in which the coriolis acceleration is negligible compared with the centripetal acceleration.

cymatogeny A term relating to large scale tectonic warping of the crust to produce a type; of basin-and-swell structure in which the crests of the swell (domes) collapse to form graben. The term was devised in 1961 by L.C. King.

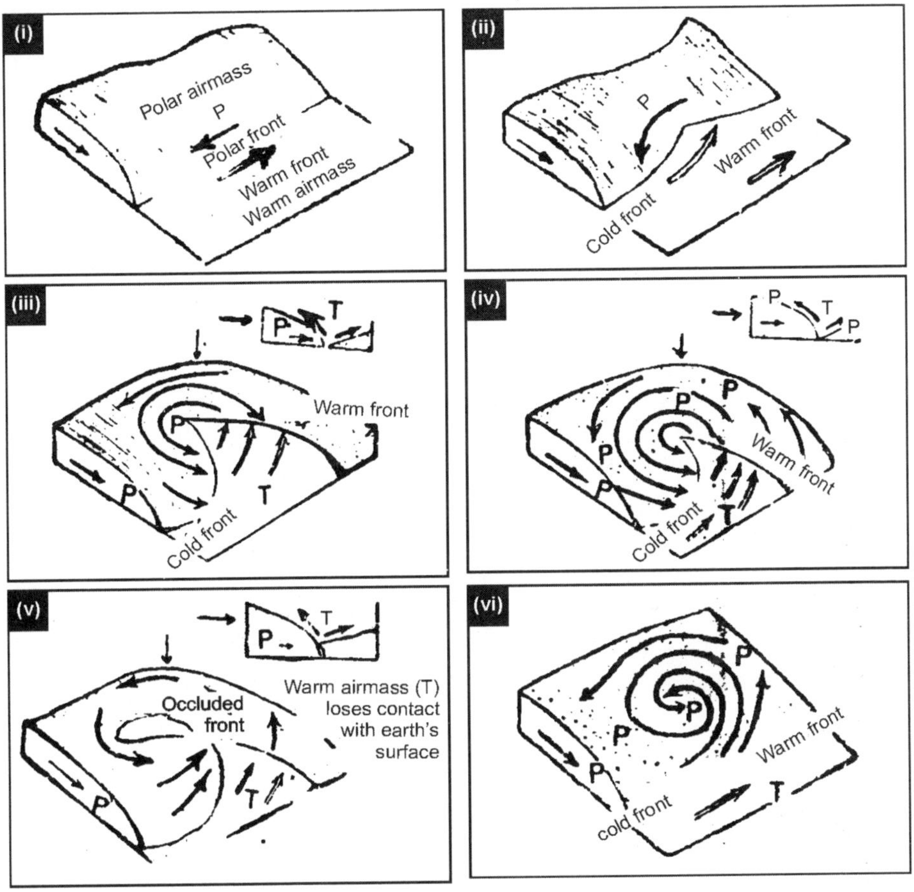

The birth and development of a Cyclone

D

dal A local term in S. Scotland and N. England for a valley in an area of uplands.

dalmatian coast A term derived from the Yugoslavian Adriatic Coastline in which the coast runs parallel with the lineament of the topography and probably with the underlying geological structure. A rise of sea level has drowned the coastal area, resulting in a coastline of narrow peninsulas, lengthy gulf and channels and a number of linear islands.

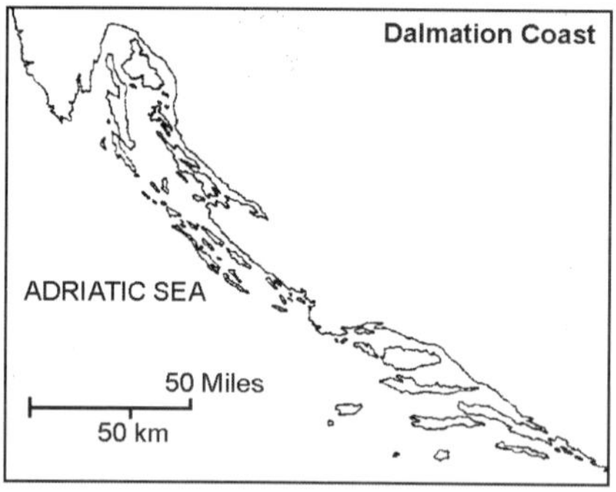

dambo A shallow streamless depression at the headwaters of a drainage system, in which no actual channel can be identified.

darcy A standard measurement unit of permeability. It is equivalent to the passage of 2 cm^3 of fluid with a viscosity of 1 centipoises flowing in 1 second under a pressure differential of 1 atmosphere through a porous medium, the areal cross-section of which is 1 cm^2 and whose length is 1 cm.

datum level A surface or level which is regarded as a base from which other levels can be counted. For example, sea level is often used as a datum level against which the height of land and depth of the sea bed are measured.

debris A French term for 'wreckage' which has been adopted for geological use to describe a superficial collection of broken rocks, earth and other inorganic material that has been moved from an original site by streams of water or ice and redeposited at other localities.

deciduous forest A type of forest comprising trees experiencing annual leaf fall, generally in the autumn or fall. These trees are mainly broad leaves such as oak, beech, ash, birch and maple. The important areas of these forests are located in W. Europe and N. America.

declination Solar declination is the angular measurement of the sun in relation to the celestial equator and the celestial poles. The declination of the sun varies from 23.5° N on about 21 June to 23.5° S on about 22 December.

decomposer An organism which feeds on plants and animals, breaking them down and in the process releasing energy and nutrients. Decomposers play a major role in soil fertility by aiding the decay of leaves and other plant material at the surface.

decomposition The weakening and breaking down of rocks by chemical weathering.

deep, ocean A trench of considerable depth on the deep-sea plain of the ocean floor, that is 3000 m or more deeper than the adjacent deep-sea floor.

deepening A term used in meteorology to indicate a decrease in atmospheric pressure at the centre of a depression, or system of low pressure. It generally denotes worsening of the weather.

deep-focus earthquake An earthquake that originates at a depth greater than 300 km.

deep-marine environment The sedimentary environment of the abyssal plains.

deep-sea fan A cone-shaped or fan-shaped deposit of land-derived sediment located seaward of large rivers or submarine canyons. Synonymous with abyssal cone, abyssal fan, submarine cone.

deep-sea plain A far-spreading virtually level area on the ocean floor in the abyssal zone from which the ocean deeps descend and the seamounts rise.

defile A narrow, gorge-like pass in a mountainous region.

deflation The removal of material from a land surface by aeolian processes. It is most effective where extensive unconsolidated materials are exposed, e.g., on beaches and on dry lake and river beds.

deflation basin A shallow depression formed by wind erosion where groundwater solution activity has left unconsolidated sediment exposed at surface.

deflation hollow A large scale basin or depression formed by the action of the wind in arid or semi-arid lands.

deflection (1) A term used in geophysics, referring to the angle between the geoid and the spheroid. (2) A sudden change in the trend of a lineament, especially that in a mountain range.

deglaciation A term used to denote the slow wastage of an ice-sheet and the uncovering of a formerly ice-covered terrain. It is used specifically to describe an event which took place during an earlier episode of geological time, e.g., the Pleistocene ice sheets of N. Europe.

degradation A term referring to the gradual lowering of a land surface by denudation in which erosive force (rivers, ice, etc.) erode material, transport it and deposit it elsewhere. Also used as environmental degradation.

delta A fan-shaped alluvial deposit at a river mouth formed by the deposition of successive layers of sediment. As the river current enters the sea (or lake) the bulk of the coarsest fraction of the load is deposited immediately, but the fine material is carried farther out by the divergent river channels.

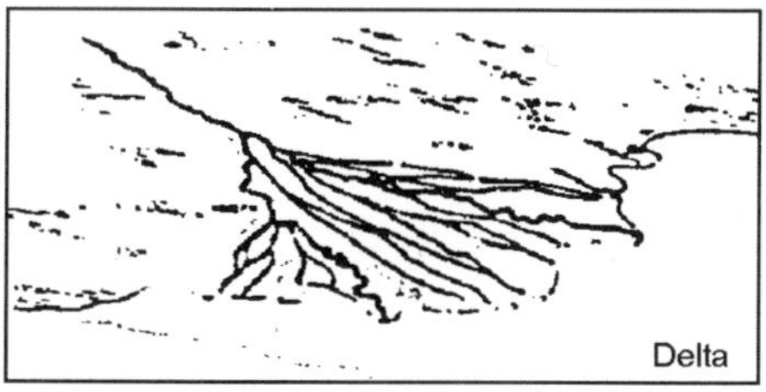

Delta

Fine clays, carried in suspension, are deposited by flocculation when they enter saline water.

demoiselle A French term derivation used to describe an earth pillar in which boulder has protected the underlying material from weathering and now survives as a cap-stone precariously perched on the pinnacle of the slender pillar.

dendritic drainage A type of drainage pattern which develops as an entirely random network because of the absence of structural controls. Thus, a dendritic pattern will be characteristic of terrain which is of uniform lithology and where faulting and jointing are insignificant, e.g., massive crystalline rocks or thick clay plains. Because of its branching nature the stream network has been termed dendritic from the Greek term dendron (tree).

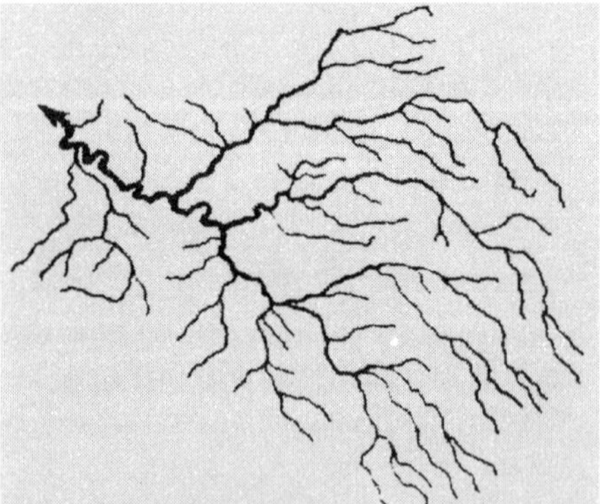

Dendritic drainage pattern

density current An ocean current caused by the different temperature and degrees of salinity exhibited by adjacent water masses, which therefore posses contrasting densities, e.g., cold currents flow south from the Arctic ocean where they descend beneath the warmer waters of the North Atlantic drift as a density current.

denudation From the Latin denudare, to strip bare, and involving the process of weathering, transportation, and erosion.

denudation chronology Branch of landform studies that deals with the historical development of landscapes by denudation, especially during pre-quaternary time. Evidence for developmental stages is provided by studies of erosion surfaces and their mantling deposits, drainage patterns, stream long-profiles, and geologic structures.

denudation slope A type of slope from which material is being lost by erosion and weathering leading to a net loss of ground. This form of the slope is controlled by the rate of denudation.

deposition The laying down of material that has accumulated, after having been eroded and transported by various physical processes (wind, ice, running water, marine waves and ocean currents).

depression An enclosed area of low pressure revealed by the pattern of pressure distribution. Depressions are also described as cyclones or cyclonic systems, and have a characteristic pattern of wind circulation.

depth hoar The upper surface zone of a snowfield, which loses heat naturally owing to radiation. Because of changes in the ice-crystal structure of this surface zone, a thin and fragile layer of ice forms above the underlying snow layers.

deranged drainage A distinctively disordered drainage pattern formed

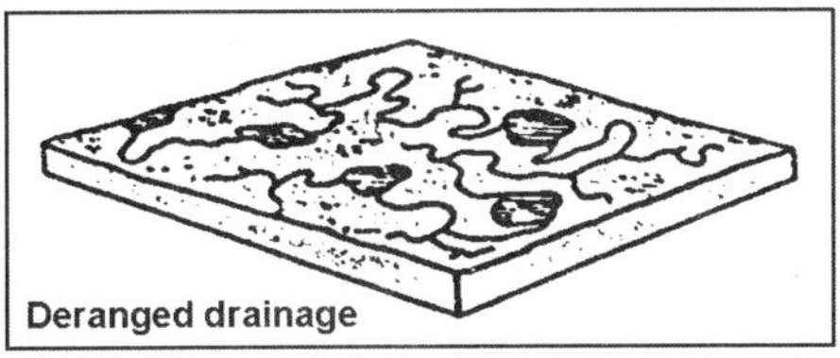

in a recently glaciated area. It is characterised by irregular direction of stream flow, few short tributaries, swampy areas, and many lakes.

desert An arid region characterised by little or no rainfall in which vegetation is scanty or absent, unless specially adapted or where groundwater conditions are favourable. This term is now often used to describe mid-latitude low-rainfall areas of the continental interiors and also regions of perennial ice and snow of high latitudes where vegetation cannot exist because of low temperatures and physiological drought rather than deficiency of precipitation.

desert pavement An extensive area of bare pebbles or larger stones that occur at the surface as a wind-polished tightly packed crust in an arid environment.

desertification The extension of typical desert landscapes landforms and processes to areas where they did not occur in the recent past. Such changes take place in arid zones bordering the deserts, in areas, which

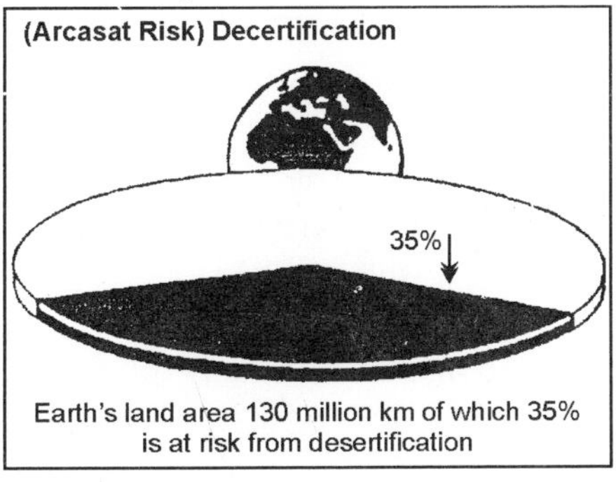

experience average annual rain falls of 100–300 mm. In fact, desertification is the process of transforming arid land into a barren desert. Often induced by human activities or climate change.

desiccation A progressive increase of aridity, often as a result of climatic change. It may be due to natural charges such as a decrease in precipitation or an interference with a river region or it may result from human interference like, overgrazing, deforestation, irrigation failure, etc.

desillication This is a pedogenic process, involving removal of silica from the upper to the lower soil horizons. This occurs in all humid regions.

destructive margin The contact between two lithospheric plates which are moving towards each other and where oceanic crust is being destroyed by subduction. Destructive margins, a type of convergent margin, are marked by shallow to deep focus earthquakes and typically andesitic volcanicity and most are also marked by an oceanic trench.

destructive wave A storm wave in the ocean which when breaking upon a beach, has an almost vertical plunge of water which combs beach material seawards. Thus, its backwash is stronger than its swash.

dew The moisture deposited in the form of water droplets on the surface of vegetation and other objects located near to ground level. It forms when nocturnal terrestrial radiation causes heat loss from the earth's surface thereby cooling the lowest layer of the atmosphere to below the dew point and leading to condensation.

dew-point The temperature, with constant pressure and water vapour content, to which air must be cooled for saturation to occur. Dew point may be determined from the dew point hygrometer which indicates initial condensation on a cooled surface.

D-horizon A term relating to a soil profile, although, in fact it refers to the unweathered and undisturbed solid rock occurring beneath the true soil.

diaclinal A geomorphological term introduced in 1875 by J.W. Powell to describe those river courses which cross the strike of the geological structure at right angles.

dialation (1) In general use, the act of making wider. (2) In geomorphology, the expansion of fissures in a rock by ice, after conversion of water to ice by freezing. (3) In geology, the expansion or widening of an initial aperture by the intrusion of magma.

diatom A microscopic single-celled marine or freshwater plant. A subdivision of the algae.

diatom ooze A soft, deep-sea, siliceous deposit, composed predominantly of the remains of diatoms.

differential weathering erosion The relatively greater rate of rock disintegration in certain parts of a rock mass than in others. The differential erosion occurs where (i) rocks of widely varying hardness, and (ii) clearly defined lines of weakness such as joints and faults.

diffluence (1) A term used in climatology to describe the condition during which the streamlines depicting a horizontal flow of air diverge from each other. Diffluence causes a decrease in the velocity of air particles. (2) A term used by gemorphologists in glacial geomorphology.

diffraction A term used in meteorology to describe the process of radiation spreading which results when light rays are bent by an obstacle. It is responsible for a wide range of atmospheric optics.

diffuse radiation Indirect solar radiation received at the earth's surface after selective (normal $< 0.4\ \mu m$) by atmospheric molecules and minute dust particles. Also termed skylight or sky radiation.

diffusion A term used in meteorology, to denote random mixing of air bodies either by turbulence or by slow molecular diffusion.

dike A tabular intrusive rock that cuts across strata or other structural features of the surrounding rock.

Dike

dip The angle of inclination of a planar feature measured from a horizontal datum. The true dip is always measured in a vertical plane perpendicular to the strike of the plane of either bedding or cleavage. The angle of dip measured in any plane not perpendicular to the strike is an

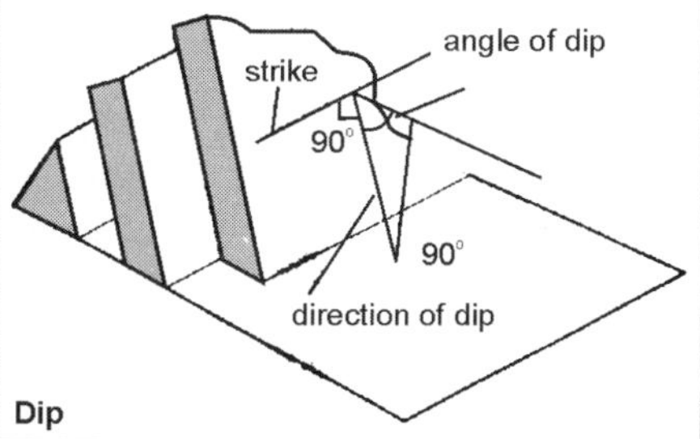

Dip

apparent dip and will always be less than the true dip.

dip fault A fault that has a strike approximately at right angles to the strike of the rock bedding planes and is therefore, parallel with the true dip of the strata.

dip slope A geomorphological topographic surface which dips in the same direction and often by the same amount, as the true dip or apparent dip of the underlying strata. Dip slopes are commonly found in cuesta and vale topography.

disappearing stream A stream which changes from a surface course to an underground course when it passes onto an area of limestone rocks.

disconformity Unconformity such that the beds above and below the surface are parallel. Some disconformity surfaces are highly irregular, whereas others have no obvious relief.

discontinuity In geophysics, a term denoting a change in the physical properties of the earth's interior with increasing depth. A seismic discontinuity is one in which the velocity of seismic waves changes as they cross the boundary of discontinuity.

discordance A geological term referring to a lack of parallelism between contiguous rock strata.

discordant coast A coastline, which cuts across the structural 'grain' of the region, e.g., S.W. Ireland. It is also known as an Atlantic coastline.

discordant drainage A drainage pattern which has not developed a systematic relationship with the underlying structure, in contrast to an accordant drainage pattern. The reason for its discordance may be antecedence.

disintegration The process by which rock breaks down into smaller particles by weathering and erosion.

distributary A separate river channel created when a river splits, but one which does not rejoin the main channel. Thus, distributaries are common in delta, but the term should not be used to describe a braided stream or an anastomosing stream.

diurnal range A measure of the difference between the maximum and minimum values of temperature, recorded within a 24 hour period.

diurnal tide A type of tidal pattern found in a few areas of the world where, because of the configuration of the coastline (e.g. the Philippine Is.), only one high tide and one low tide occurs every 24 hours.

divagation A geomorphological term relating to the lateral shifting of a river channel caused by the excessive accumulation of sediments on the stream bed. It is always found in the process of meander formation.

diversity The physical or biological complexity of a system.

divide The area of high ground which separates two different drainage systems.

Doab A local term for a low alluvial plain between two converging rivers in the Indian subcontinent.

dogger (1) A large spherical concretion occurring in a sedimentary rock. (2) An obsolete name for the upper Jurassic in Europe.

doldrums The zone of light, mainly westerly winds or calms in equatorial latitudes, applicable largely to the oceans. It moves a few degrees N and S of the equator seasonally, following the passage of the overhead sun accompanied by violent thunderstorms and squalls.

doline A term for a circular hollow or depression in the surface of Karstic terrain, in which the funnel-shape may or may not lead down into a vertical shaft descending into the limestone. It varies in size from 10 m in diameter and is initially caused by solution, thus known as **solution doline**. Others are **collapse** doline from where the roof of a sub-surface cavern collapsed.

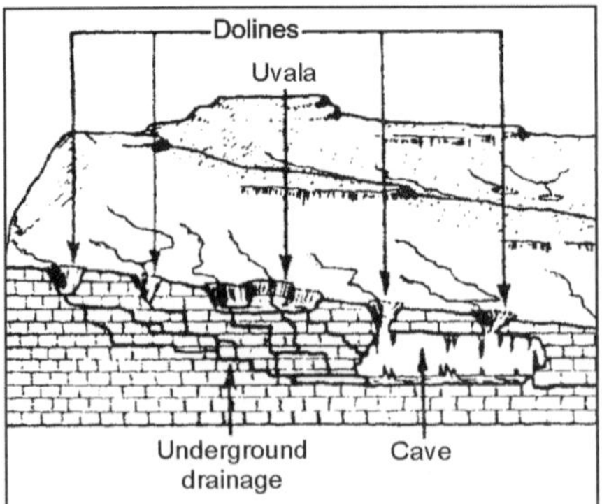

Dolines and uvala (Karstic landforms)

dolomite (1) A sedimentary rock composed primarily of the mineral dolomite. (2) A mineral composer of $CaMg(CO_3)_2$.

dome (1) In geology, a fairly symmetrical uphold in which all the beds dip away from a central point. Strictly, it is an anticlinal structure which plunges in all directions. (2) Topographically, the term refers to a smoothly rounded mountain summit or hill. (3) A rounded snow or ice accumulation on a mountain summit. (4) An arid volcano, e.g., Puy de Dome in the French Auvergfne. (5) A rounded intrusion of igneous rock.

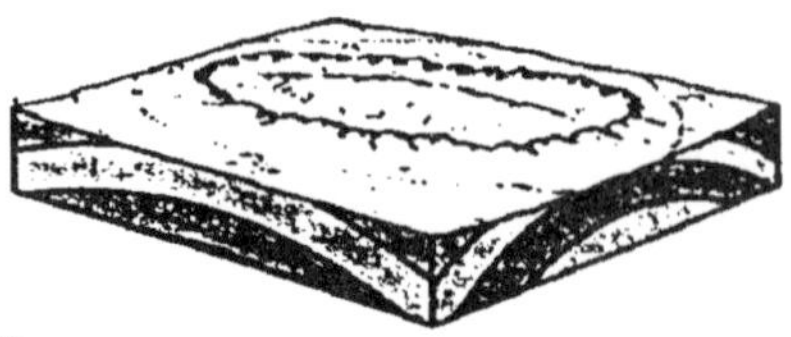

Dome

domed inselberg A term applied to a composite inselberg in which a smaller summit inselberg rises from a larger and broader inselberg to give a dome-on-dome effect.

dominant wave The wave which has the greatest influence on a particular shoreline.

dominant wind A term introduced to describe the wind which plays the most significant part in a local situation, in contrast to the prevailing wind.

dormant volcano A volcano which although not extinct, has not been known to erupt within historic time.

double tide A tidal phenomenon caused by the deformation of a progressive wave either by the shape of the coastline or by the shallowing of an estuary. Consequently, the semi-diurnal tide has a quarter diurnal tide imposed upon it such that the tide rises to a maximum and retreats slightly before rising to a maximum and retreats slightly before rising to a second maximum some two to three hours later. Alternatively, depending on the phasing of the tidal harmonic curves, a double low tide may result.

down warping A slight downward deformation of the earth's crust caused by sagging under the weight of an overlying burden, leading to a down warp and in some cases to the beginning of a geosyncline.

down wasting A term used by geomorphologists to describe the regional process of ablation as an ice-sheet diminishes in thickness over a period of time.

downs (1) A plain of temperate grassland in the South Island of New Zealand and in Australia. (2) A terrain of gently undulating hills, generally of chalk, especially in Southern England. (3) In the USA, an area of coastal sand dunes.

down-valley migration A term referring to the way in which a meander belt moves slowly down-valley owing to lateral sifting of the river channel.

draa (dzraa) The largest accumulation of sand in the Sahara, often with small dunes on its summit, and thought to be a coalescence of seif-dunes.

drainage The process by which water is discharged from an area by a river or by sheet flow.

drainage basin That part of the land surface which is drained by a unitary river system. Its perimeter is marked by a drainage divide or watershed.

drainage pattern The spatial relationships of all streams within a drainage system. The actual pattern will depend on a number of variables, including soils, geology, structure, present climate and human interference. The various types of drainage patterns are dendritic, parallel,

Drainage Pattern

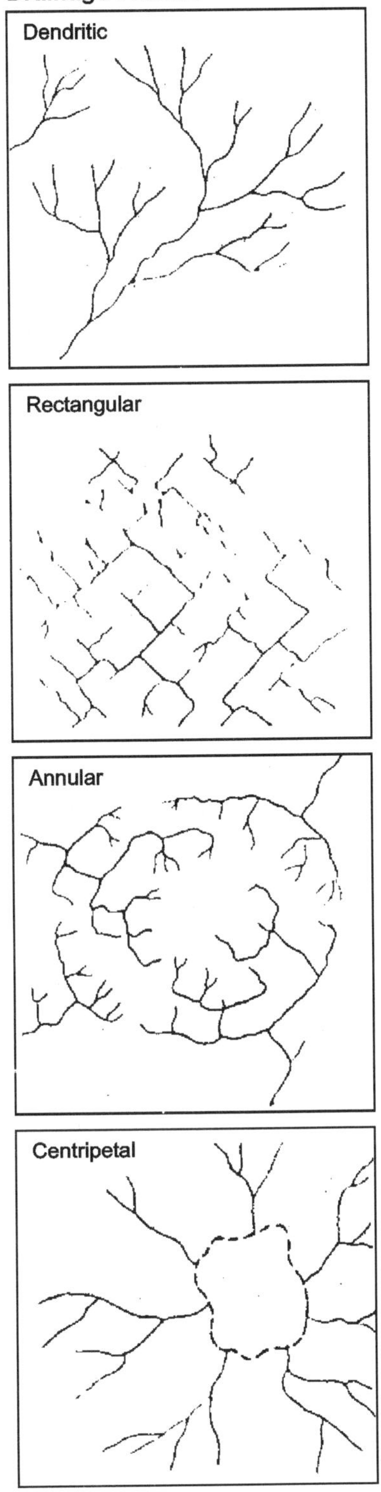

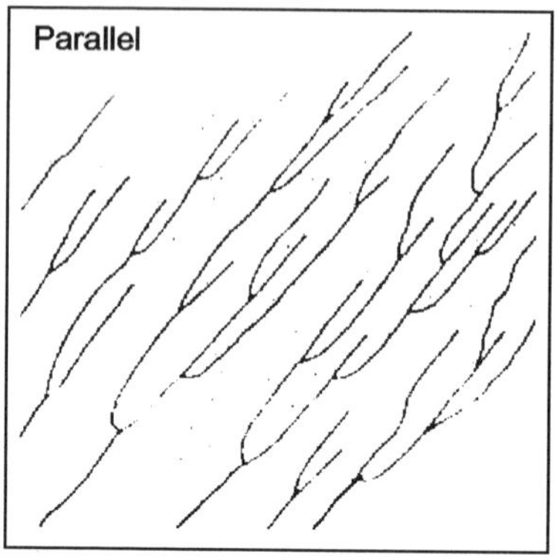

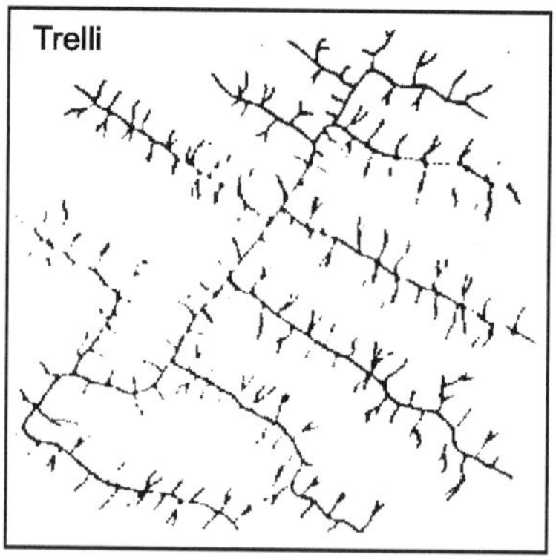

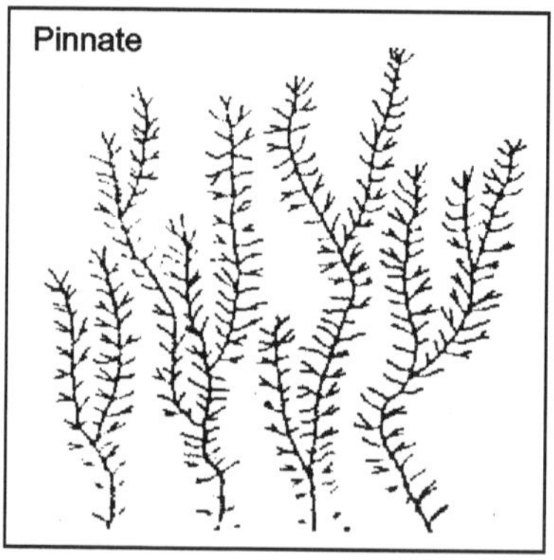

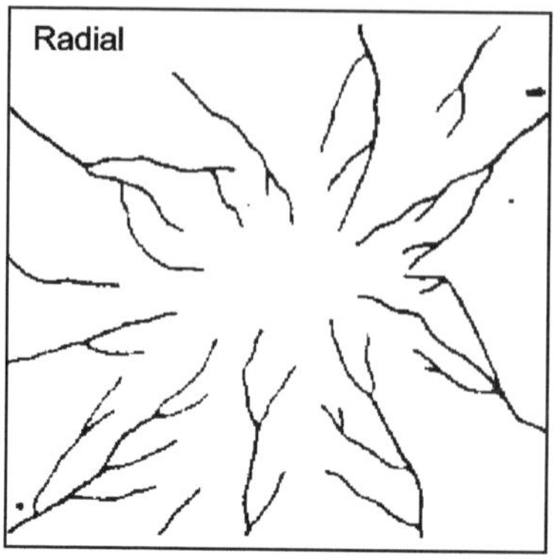

pinnate, radial, rectangular and trellis.

drainage system A river and all its tributaries within a single drainage basin.

dreikanter A German term widely adopted to describe any pebble that has been shaped by aeolian sand blasting into a faceted ventifact displaying plane faces with three sharp angles or edges bounding them.

dripstone A term given to calcite material (tufa) deposited by dripping water. It is synonymous with both stalactite and stalagmite.

drizzle Light continuous rainfall, the droplets, upto 0.2 mm, i.e., like a fine spray. Generally associated with a warm front and gentle orographic uplift. Because of the small size of the droplets they would be unable to fall to earth if there was strong vertical air motion.

drought A continuous and lengthy period during which no significant precipitation is recorded.

drowned valley A former valley that has been inundated by a rise of sea level.

drumlin Smooth, streamlined, oval-shaped land form, one end of which is blunt and the other tapered. Drumlins may occur singly but more commonly they are found within a large group, called 'drumlin field' or 'drumlin swarm'.

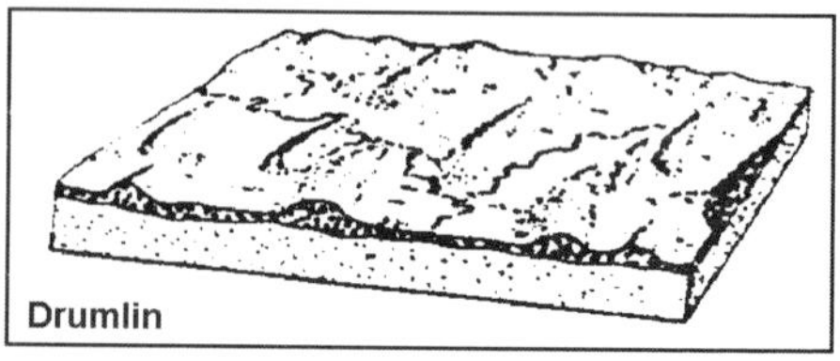

dry adiabatic lapse rate The rate at which dry (i.e., unsaturated) air cools when rising adiabatially through the atmosphere as a result of the utilisation of energy in expansion. It is 9.8°C/km.

dry valley Linear depression that lacks a permanent stream but that shows signs of past water erosion. It is a common land form in areas underlain by permeable rock, e.g., the chalk of Southern England. The dry valley was eroded during an episode of surface drainage, perhaps due to permafrost conditions to greater precipitation, or to a higher water table.

dry-bulb thermometer Thermometer that registers normal air temperature. It may be used in conjunction with a wet-bulb thermometer; the relative humidity can be found by measuring the depression of temperature registered by the wet bulb.

dujoda A periglacial term from Yakutia, USSR, referring to a linear, steep-sided depression with an uneven floor developed in a thermokarst environment. It forms by the coalescence of individual surface hollows after the baydjarakh stage.

dune A land-form produced by the action of wind on unconsolidated sediment, normally sand. Aeolian dune forms range from small ripples less than one cm in height, to the draa forms of the Sahara which rise to more than 300 m. Such dunes may be divided into three basic categories **barchans**; **longitudinal** or **seif dunes**, which parallel the wind direction; and **transverse dunes** which are aligned normally to the dominant wind. **Transverse dunes** are initial forms on sandy coastlines in temperate regions.

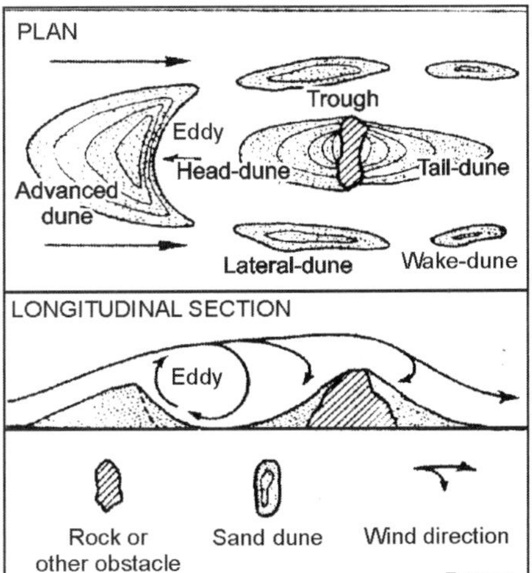

Dunes

dust Solid particles, the size of clay and silt particles, that can be raised and carried by the wind.

dust devil Very localised whirlwind in a desert area, where strong convection uplifts dust and sand often to a height of a few tens of metres.

dust storm A wind carrying sufficient dust for visibility to be reduced to less than one km.

dust well Small pit in the surface of a glacier caused by a patch of dust particles which have melted the ice because the dust has absorbed solar radiation to a much greater degree than the surrounding ice, thereby raising its temperature.

dust-bowl An area of the Great Plains region, USA, where a combination of drought and inappropriate farming particles, especially an expansion of wheat production, led to severe deflation and soil erosion during the middle of 1930s. More generally, any region where deflation of cultivable land occurs.

dyke Discordant or cross-cutting, tabular intrusion. Most dykes are vertical or near vertical, having pushed their way through the overlying country rock (See dike).

dyke set A suite of dykes whose alignment is parallel or sub- parallel, reflecting their emplacement from a common source and under a common stress regime.

dynamic climatology A study of the dynamic relationships which gives the general circulation of the atmosphere in contrast to synoptic climatology.

dynamic metamorphism A type of metamorphism in which only localised stress is involved, causing rocks to break up, occasionally into a rock flour or mylonite. It is sometimes termed dislocation metamorphism.

dynamic rejuvenation A term referring to the rejuvenation of a river due to the uplift of the land or a fall of sea level. Either of these changes of base-level will initiate a new cycle of erosion in which river incision increases, because of the river's greater energy following rejuvenation.

dysphotic (disphotic) zone A zone between 80m and 200m depth in a water body in which there is some light but of the wrong wavelength for photosynthesis to occur.

E

earth The third planet in the solar system outwards from the sun. The mean distance of the Earth from the sun is 149.6×10^6 km. The earth has a mean radius of 6371 km, density of 5517 kg/m^3 and a mass of 5.99×10^{27} g. The oceanic (5–7 km thick) and continental (40 km thick) crusts are separated by the Mohorovicic discontinuity from the silicate mantle, which extends to the Gutenberg discontinuity at 2900 km depth and overlies a molten, iron-rich core. The oldest rocks are about 3980 million years old, and the earth formed about 4600 million years ago.

earth flow Flow of unconsolidated material down a hill slope normally resulting from an increase in pore-water pressure, which reduces the friction between particles. Flow velocities vary from slow when behaviour is plastic to rapid, when behaviour is more liquid, and reflect variations in water content. Dry flows may occur when an earthquake shock breaks inter-granular bonds.

earth hummock A type of patterned ground due to frost heaving in the soil under periglacial conditions.

earth science A combined name given to studies related with the physical characteristics of the earth. It includes the following disciplines: **geology**, **geochemistry**, **geophysics**, **climatology**, **geomorphology**, **hydrology**, **meteorology**, **oceanography**, **geodesy**, **physical geography**, **soil science**, **remote sensing**, etc.

earth slide A downhill movement of a mass or superficial material due to slope failure often as a result of water reducing the friction along a shear plane in the soil mantle. With an increasing addition of water the slide will probably turn into an earth flow.

earth temperature The temperature of the ground surface, derived largely from long wave solar radiation and partly by conduction from the earth beneath. The surface temperature fluctuates both seasonally and diurnally, with the latter short term change leading to the formation of dew and ground frost.

earth tremor A slight earthquake.

earth-movement A differential movement of the earth's crust caused by diastrophic forces. It may be a slow movement, or a rapid movement. The term should be strictly confined to movements caused by internal forces within the crust (endogenetic) and should never be used to describe denudation process acting upon the surface of the earth (exogenetic).

earthquake A shock or series of shocks due to a sudden movement of crustal rocks, generated at a point (focus) within the crust or mantle. The point where the shock waves reach the surface is termed the epicentre around which lines of equal seismic intensity can be drawn. The strength of an earthquake is known as its magnitude which can be measured on the Richter scale.

east Australian current Oceanic water current that flows along the east cost of Australia. This narrow (100–200 km wide) current forms the westerly part of the anticyclonic circulation in the S. Pacific. The flow velocity varies in the range 03–5 m/s. It is an example of a western boundary current.

east Pacific rise The oceanic ridge which separates the Pacific plate from the Antarctic, Nazca and cocas plates. The east rise is a fast spreading rate calculated at 4.4 cm/yr and its topographic profile is relatively smooth compared with slow-spreading ridges, e.g., the Mid Atlantic ridge.

easterly wave A shallow linear trough disturbance in the trade wind flow of the tropics, more evident in upper-level winds than in surface pressure patterns. The wave moved westwards transporting bands of convergence with pronounced cumulus cloud and heavy showers. It develops from an intense steepening of the lapse-rate associated with the intertropical convergence zone. Some waves (about 10%) act as initiators of a tropical cyclone.

eastonian A stage of the upper Ordovician of Australia, underlain by the Gisbornian and overlain by the Bolindian.

ebb tide The retreating tide, or the outgoing-tidal stream, immediately following the period of high tide and preceding the flood tide. In coastal areas where the tidal range between high and low water is large, the ebb tide may create a very powerful tide current.

eccentricity The amount by which a point varies in distance from the centre of a circle or orbit. For example, the orbit of the earth around the sun is not circular, i.e., the sun does not lie in the exact centre of the earth's orbit, the divergence is termed the eccentricity of the orbit.

echo-sounding A method for determining water depth by using the elapsed time(t) between the transmission of a pulse of high frequency sound and the return of the echo from a reflector, e.g., the sea floor.

eclipse The partial or complete obscuration of one heavenly body by another, as perceived by an observer on one of the bodies. The proper description of an eclipse also refers to the period of time involved.

ecliptic The plan of the orbit of the earth around the sun. It forms an angle of 23°27′ with the earth's equator. The orbits of the planets all lie within 3.4′ of this plane, except for those of Pluto (17.2°) and Mercury (7°).

eclogite Very rare, coarse-grained, igneous rock with a chemical composition similar to that of basalt.

ecological niche The specific part of an environment to which a particular organism is best suited.

ecological succession The slow change in an ecosystem during which one community of organisms is gradually replaced by a different kind of community.

ecology The scientific study of the interrelationships of organisms with each other and with the environment. Some of the definitions of ecology are: "Ecology is the science of all the relations of all organisms to their environment" (Taylor). "Ecology is the science which treats the organisms in relations to their environment" (Moore). "Ecology is the study of plants and animals in relation to their environment" (Haggett). The actual rate or environmental lapse rate, where temperature decreases with increasing altitude at a given place at a specific time. It averages about 0.6°C/100 m (3.5°F/1000 ft) of ascent.

ecosystem Term first used by A.G. Tansley (1935) to describe the interdependence of species in the living world with one another and with their non-living (abiotic) environment. Fundamental concepts include the flow of energy via food chains and food-webs, and the cycling of nutrients biogechemically.

ecotone Narrow and fairly sharply defined transition zone between two or more different communities.

ecotope (1) The area occupied by an ecosystem. (2) Sometimes the term is used synonymously with ecosystem.

ecoulement The gravitational sliding of large masses of rock, as a result of tectonic deformation.

edaphic A term referring to the soil, particularly with respect to its influence on organisms.

eddy Motion of a fluid in directions differing from, and at some points contrary to, the direction of the larger scale current. In air, eddies vary in size from small scale turbulence to large scale movements (e.g., cyclone and anticyclone cells) within the general global circulation of the atmosphere.

eddy currents The alternating current induced in a conductor when it is subjected to a time-varying magnetic fields in accordance with Lenc's law. The eddy currents then generate their own secondary electromagnetic field which is of considerable importance in electromagnetic exploration techniques.

einkanter A German term referring to a type of ventifact which has only one facet cut by wind-blown sand.

ejecta blanket The blanket of debris surrounding an impact crater. It is composed of material ejected from the crater during its formation and is laid down with stratigraphy inverted from that of the bedrock.

Ekman effect The tendency for upwelling ocean currents to be deflected by atmospheric winds as they approach the ocean surface. The surface flow is complicated by the earth's rotation and the configuration of the continents.

Ekman spiral The Ekman effect in the atmosphere has been termed the

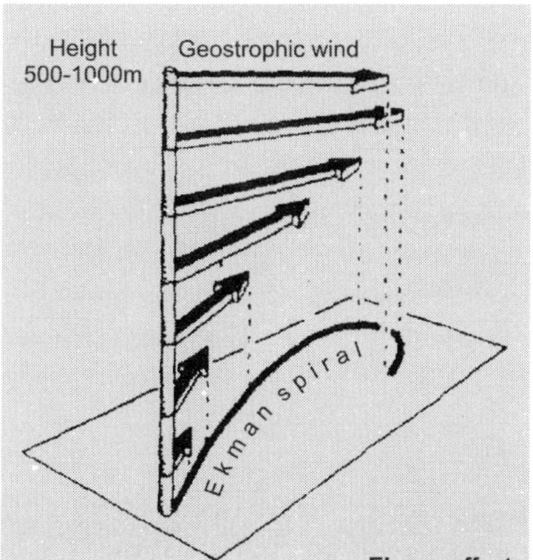

Ekman effect

Ekman spiral, which refers to the frictional drag of the earth's surface or circulating air currents. Its frictional effect decreases with height, thereby increasing the wind speed units, the geostrophic wind becomes dominant between 500m and 1000 m.

El Nino A name given to the occasional development of a warm ocean current along the coast of Peru as a temporary replacement of the cold Peru current (Humboldt current) which normally operates. El Nino is an extension of the equatorial current and leads to an increase in surface water temperatures of 10°C and a decrease in Plankton which thrive in the colder current. Due to El Nino, the tropical atmospheric circulation is seriously disturbed, resulting in major climatic anomalies such as floods, draughts, etc. An El Nino is only one element of dub phase oscillating ocean—atmospheric system. El Nino usually occurs at intervals of two to seven years.

elastic rebound The action by which rocks that have been stressed beyond their elastic limit return to a position of zero strain after rupture has taken place at a fault. Faulting is a result of the sudden release of elastic energy that has slowly accumulated in the tension zone of the earth's crust, especially at a plate margin where major trans current fault occur.

elevation (1) In general, term used to describe a topographic eminence. (2) A specific altitude or height above a given level, e.g., above sea-level. (3) In surveying the term refers to the angle between the horizontal and a point at a higher level.

embayment (1) An open bay forming an indentation in a coastline. (2) In structural geology, the term refers to the way in which a large area of sedimentary rocks projects into an exposure of crystalline rocks.

emergent shoreline Any type of former coastline that once marked the juxtaposition of land and ocean but which is now either raised above present high-water mark or located at some distance inland from the present coastline.

emissivity The ratio of the total radiant energy emitted per unit time per unit area of a surface at a particular wavelength and temperature to that of a black body under the same conditions.

emplacement The actual formation of a mass of igneous rock within a body of existing country-rock but without implying its mode of formation. Thus a granite, for example, may be emplaced by the mechanism of intrusion, by stopping or by replacement.

end moraine A term synonym for terminal moraine. A mass of ice-transported debris, usually unsorted, marking the maximum limit of a glacial advance.

endogenetic A term referring to processes that originate from within the earth and result in such forces as igneous intrusions and different types of uplift and depression.

endorheic drainage A term given to an inward flowing pattern of drainage in the world's semi-arid zones.

endothermic reaction Any reaction in which absorption of heat occurs.

endotrophic The ability of an organism to receive nourishment from within.

endrumf A German term, introduced by the geomorphologist W. Penck to describe the almost flat land surface produced as the final or end product of denudation.

energy flow Energy is defined as the ability to do work and is the essence of life and controller of the ecological systems. The types of energy which animate the biological processes of ecosphere are gravitational, solar, geothermal or magnetic in origin. The radiant energy produced in the sun travels through space and only a friction of solar radiation reaches the earth to provide energy for the ecosystem. In an ecosystem, energy flow controls the entire food chain. Odum has explained the entire energy flow system in a simplified diagram as shown in figure.

energy grade line The rate of dissipation of kinetic energy along the length of a stream channel, a rate which varies according to the channel geometry.

englacial Pertaining to the environment within a flakier. It denotes melt water moving freely through the ice-body.

engulfment The collapse of the cone of a volcano owing to the extrusion of molten lava laterally through a side fissure, rather than through the central vent. Beacuse of its underlying support, the volcano may collapse inwards to form a caldera.

entrenched meander A meander cut into the underlying rock as a result of regional uplift or lowering of the regional base level.

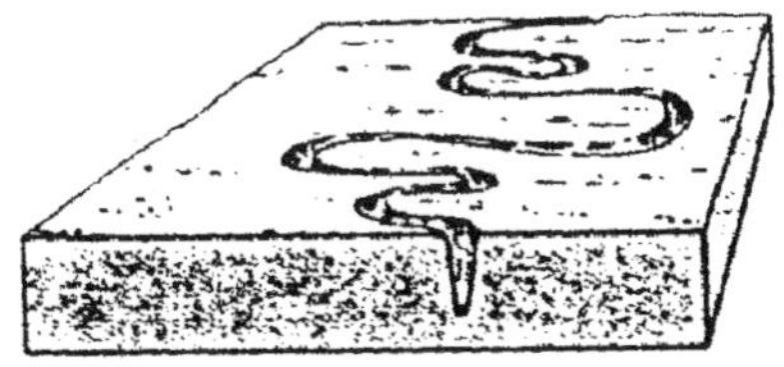

Entrenched meander

Energy Flow in a Food Chain

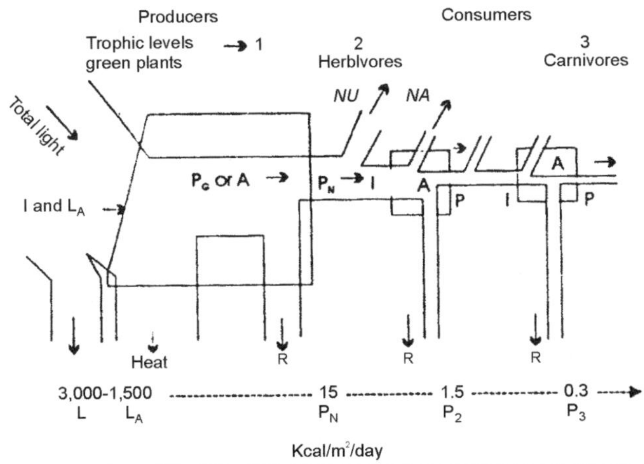

(After E. P. odum)
(I-total energy input; L_A-light absorbed by plant cover;P_G-grass primary production; A-total assimilation; P_N-net primary production; P-secondary (consumer) production; NU-energy not used (stored or exported); NA-energy not assimilated by consumers (egested); R-respiration; Bottom line in the diagram shows the order of the magnitude of energy losses expected at major transfer points, starting with a solar input of 3,000 Kcal per meter per day.)

environment The sum total of all conditions, agencies and influences which affect the development, growth, life and death of an organism, species or race.

eon The largest geologic-time unit, incorporating a number of eras.

epeiric sea Shallow sea which extends far into the interior of a continent, e.g., Hudson Bay and the Baltic Sea. The term also denotes shallow sea areas that cover the continental shelf and are partially enclosed, e.g., the North Sea.

epeirogenesis The large scale upward or downward movements of continental or oceanic areas.

ephemeral stream A stream which flows only after rain or snow-melt and has no base flow component. A desert Wadi may form an ephemeral stream.

ephemeral stream A stream which flows only after rain or snow-melt stream.

epicentral angle (Δ) The angular distance between an earthquake focus and a seismic station.

epicentre The point on the earth's surface immediately overlying an earthquake focus.

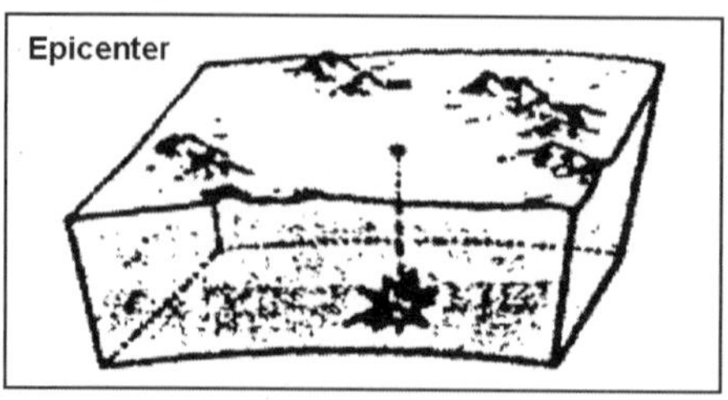

epiclast A clast produced by the chemical or mechanical weathering, or other surface processes of volcanic rock.

epigene Produced or occurring at the earth's surface. The term is used especially in relation to the processes of weathering, erosion and deposition.

epigenetic ice A type of ice lens formed under conditions of permafrost in the subsoil.

epigenetic ore Deposit layer in origin than the host rock.

epilimnion The warmer, less dense topmost layer in the water body of a lake or ocean.

epoch A term given to the third rank order in a subdivision of geological time. Several epochs form a period.

equable A term used in climatology to define a climate type exhibiting a minimum of variation throughout the year.

equation of time The difference in value between apparent time and mean solar time. If the sun arrived overhead at the meridian before 12.00 noon according to mean solar time, it is said to be fast and the equation of time is positive. Conversely, if the sun arrives late over the meridian, the equation of time is negative and the sun is said to be slow.

equator The great circle of the earth, at 0° parallel of latitude. It lies midway between the two poles in a plane at right angles to the axis of the earth. It is 40,067 km in length.

equatorial climate The type of climate which occurs in low latitudes (approx 10°N to 10°S) on low ground only. It is characterised by constantly high temperatures and humidity and by an almost equal duration of day and night. Because the midday sun is overhead or nearly overhead at all times the seasonality is reflected largely by difference in patterns of rainfall, as converging air masses are affected by movements of the inter-tropical convergence zone.

equatorial current Oceanic current which flows in an east-west direction in the equatorial regions of all the oceans. The broad (1,000–5,000 km) westward flowing currents (north and south equatorial currents) are separated by a comparatively narrow westward flowing countercurrent. The flow tends to be limited to the upper 500 m of the water column and the velocity is 0.25–1.0 m/s. The strength and position of the equatorial currents are controlled by the overlying wind system.

equatorial rain-forest The forest occurring in the zone of constant rainfall which typifies an equatorial climate.

equatorial trough Shallow low-pressure zone around the equator, where the trade wind systems tend to converge.

equatorial westerlies A name given to westerly wind at low latitudes, caused when the NE and SE trade winds are deflected on crossing the equator and develop a westerly component, thereby becoming the equatorial westerlies whenever the inter-tropical convergence zone is more than 5° away from the equator.

equilibrium line A term given to the boundary between the ablation zone and the accumulation zone of a glacier.

equinoctial A synonym for the celestial equator.

equinoctial gale A strong wind generally experienced at or around the equinox.

equinox One of the two periods during the year when the sun's apparent path intersects the plane of the earth's equator. Thus, the sun is directly overhead at the equator at noon on 21 March (the vernal equinox) and on 22 September (the autumnal equinox). The term equinox is derived from 'equal night' because at that time the length of day and night are approximately equal throughout the world.

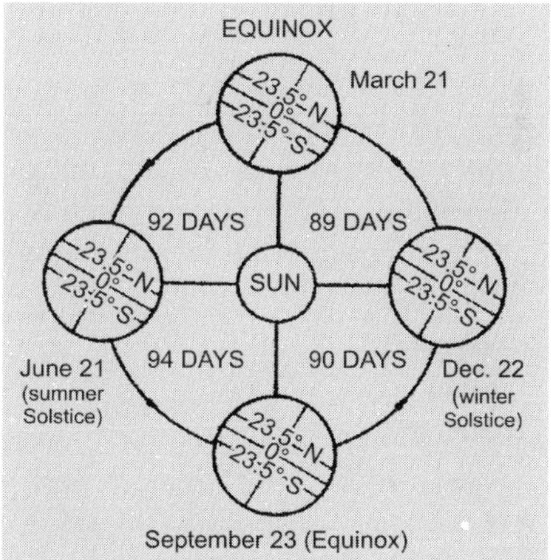

Equinox & Solstices

era First-order geologic time unit composed of several periods.

erg An Arabic world referring to a region of sand dunes, generally of very wide extent, particularly in the Sahara.

erosion A term referring to those processes of denudation which wear away the land surface by the mechanical action of the debris which is being transported by the various agents of erosion (glaciers wind, rivers, marine waves, currents).

erosion rate The rate at which geomorphological processes wear away land surfaces.

erosion surface A geomorphological expression referring to a near-level surface formed by agents of erosion, in contrast to a level surface formed by deposition—i.e., it is degradational not aggradational.

erratic A large boulder carried by glacial ice to an area far removed from its point of origin.

Erratic

eruption The release of lava and gas from the earth's interior onto the earth's surface and into the atmosphere.

escarpment A cliff or very steep slope.

esker Long, sinous, steep-sided, narrow-crested ridge which consists of cross-bedded sands and gravels. It is laid down by glacial melt water either at the retreating edge of an ice sheet, or in a subglacial or englacial ice tunnel, or in an ice-walled tunnel.

estuary The mouth of a river where it broadens into the sea and within which the tide ebbs and flows, leading to an intermixing of saline and fresh water. Estuaries are usually zones of deposition, especially if the river discharges more sediment than can be removed by tidal currents or wave action.

etang A French word referring to a small, shallow lake of brackish water located among coastal sand dunes or behind old beach-ridges. There are several examples along the landes coast of SW France.

etchplain Plain produced in a tropical or subtropical environment as a result of a phase of deep chemical weathering during tectonic stability, followed by one of erosion in which the weathered debris is stripped away. Etchplain relief reflects differences in the resistance of the bedrock to weathering and involves double placation.

etesian winds Greek name for north easterly, easterly, northerly or north-westerly winds which blow between May and September in the Aegean Sea. The equivalent Turkish term is meltemi.

eugeosyndine A type of geosyncline in which volcanic rocks are abundant formed some distance from the cration.

Euler's theorem A theorem referring to the distribution of plate margins and the variations of velocity which occur at different localities. The theorem indicates that the velocity of relative motion across a plate margin is proportional to the angular distance of the given point from the axis of rotation and to the angular velocity about the axis of rotation. The axis of rotation, for the motion of two plates in relation to each other is itself made up of a number of small circles.

eulifforal zone That part of the coastal zone which extends seawards of high-water mark down to the limit of attached plants, usually to depth between 40 and 60 m.

euphotic zone The surface layer of any body of water through which light can penetrate, thereby leading to photosynthesis.

eurasian plate One of the present-day major lithospheric plates, extending from the oceanic ridges in the Atlantic Ocean to a currently poorly identified margin with the North American Plate in the east. The southern boundary is basically the Alpine-Himalayan fold belt, involving a collage of micro plates, minor plates and the African and Indo-Australian plates.

eutasy The worldwide changing of sea level caused either by tectonic movements, or by the growth or decay of glacier.

euxinic environment A geological expression referring to an environment in which there is restricted circulation of sea water owing to the partial or total insolation of a basin from the open ocean. Thus, there are large volumes of stagnant, de-oxygenated water in which the dominating process is one of reduction, giving rise to black, pyretic carbonaceous muds. Euxinic conditions are characteristic of the deeper parts of the Black Sea, hence its name.

everglades (1) In general, a marshy tract of land with tall grasses and occasional trees, seasonally flooded by summer rainfall. (2) The proper name for a low, flat, limestone plain in Florida, USA.

evorsion A term given to the fluvial process which scours out cavities in the rocky bed of a stream, probably owing to the action of eddies.

exaration The action by which glacier ice, unladen with debris, carries out erosion of the bed rock by quarrying or joint-black removal. This process must be distinguished from that of glacial abrasion, which is carried out by the transported debris within the glacier.

exfoliation Weakening and separation of the surface layers of rock as a result of chemical or thermal weathering, or of pressure release due to erosion. Through this process shells, slabs, sheets or flankes are broken loose and stripped away from a rock mass.

Exfoliation

exhumed topography An ancient landform or landscape that has been buried beneath younger rocks or sediment and that is exposed by their subsequent erosion.

exogenetic processes A blanket term for those processes which operate on or close to the surface of the earth and which involve weathering mass movement, fluvial, eolian, glacial, periglacial and coastal processes.

exogesynclines That part of a geosyncline which borders a craton but receives its sediments from erosion of complementing highlands uplifted from the longer, narrower orthogeosynclinal belt lying away from the cration.

exosphere The outermost zone of the earth's atmosphere, within the thermosphere above about 700 km and beyond which neutral particles of atomic oxygen, ionised oxygen and hydrogen escape into space.

extending flow A concept of differential ice movement throughout the length of a glacier, introduced by J.F. Nye in 1952. It refers to the extension and related thinning of the glacier ice in those zones where the surface velocity increases.

extrusion The emission of magma from a vent or fissure on to the earth's surface where it forms a lava flow. Gas is released quietly rather than explosively from the magma at the source vent and consequently there is little associated pyroclastic activity. Gas may also be vented from a different source, or extrusion may be of lava which has been degassed previously.

'eye' of storm Central part of a tropical cyclone with light winds, generally clear skies, and a slight, horizontal pressure gradient. The diameter of the 'eye' averages 20 km but in a large cyclone can be 40 km or more. The 'eye' is an area of some air subsidence, which produces adiabatic warming.

F

fabric The physical arrangement of particles and minerals in a rock including its texture and structure, both microscopic and macroscopic.

facet (1) Any plane surface abraded on a fragment of rock by ice or wind. (2) A term used in terrain classification to denote a grouping of land elements of similar genesis.

faceted spur The truncated end of a ridge, where the smooth convex curve of the ridge crest is abruptly terminated by either a steep bevel or a cliff. This feature may be due to faulting, glaciation or river erosion.

facier A rock unit or group of units that exhibits lithological, sedimentological and faunal characteristics which enable them to be classified as distinct from another rock until or group.

Fahrenheit scale A temperature scale introduced by a German physicist, G.D. Fahrenheit, in about 1709 in which 32°F represents the freezing point of water and 212°F—the boiling point of water.

fair-weather cumulus A type of cumulus cloud associated with a summer anticyclone during which prolonged spells of dry bunny weather occur. It is characterised by a limited vertical extent because major convection is prevented by an inversion layer linked with the high pressure and subsiding air.

fall-line A line or zone marking the tract where a series of almost parallel rivers descend from a mountain region or plateau edge onto a lowland by means of waterfalls.

fan cleavage A type of cleavage in which the planes exhibit a fan-like arrangement instead of the more normal parallel alignment.

fan-folding Folding in which the limbs dip either towards each other or away from each other because the axial planes of the folds converge towards a centre. It is commonly exhibited in an anticlinorium and a synclinorium.

Farallon Plate A present-day minor lithospheric plate subducting beneath the North American Plate, the Farallon Plate is the remnant of a plate that was once large.

fast ice An extensive unbroken sheet of sea-ice formed by the *in situ* freezing of sea water when water temperature falls below –2°C in calm weather.

fathom A nautical unit of water-depth measurement used in soundings. One fathom = cft; 100 fathoms = 1 cable; 1000 fathoms = 1 nautical mile.

fault Approximately plane surface of fracture in a rock body, caused by brittle failure and along which observable relative displacement has occurred between adjacent blocks. It is a fracture in a rock, induced either by tensional or compressive forces. The main types of fault are: **normal faults**, developed by tension and lowering of rocks on the side towards which the fault-plane is dipping; **reversed fault**, resulting from compression; **tear faults**, of rocks on either side of a near vertical fault plane. Faulting is responsible for creation of rift valley and fault scrap.

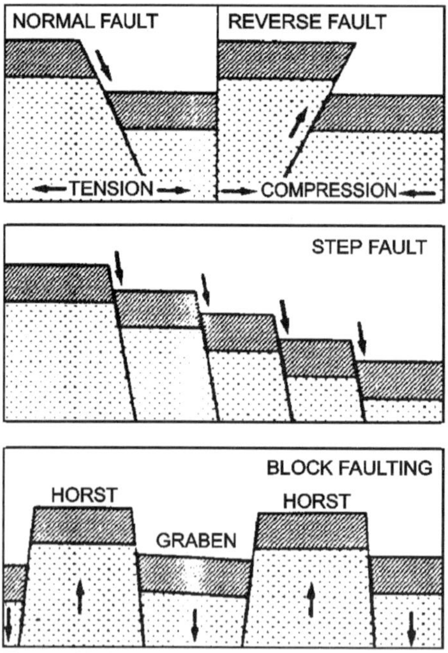

Types of fault structure

fault block A rock mass which is bound on at least two sides by fault planes. The block may be uplifted or depressed in relation to adjacent blocks.

fault creep The slow, almost imperceptible displacement of a fault, usually a tear fault, which tends to reduce the accumulation of stored strain.

fault scarp A cliff formed directly by the displacement of a recent fault, but usually on a small scale. It is a relatively transient land form since denudation will soon modify the scarp, turning it into a fault-line scarp or obliterating it altogether.

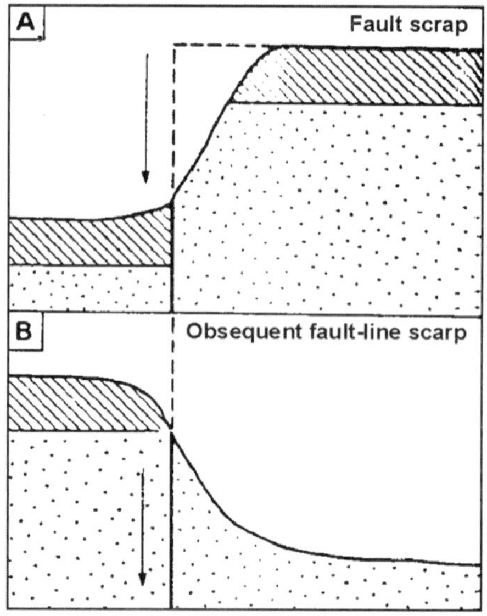

fault zone A region, from metres to kilometres in width, which is bounded by major faults within which subordinate faults may be arranged variably or systematically. Single fault zones are marked by fault gouge, breccias, or mylonites.

fault-spring A spring formed as a direct result of an impermeable bed of rock being brought into juxtaposition with a permeable bed by virtue of faulting.

faunal realms The division of the world on the basis of faunas, also known as faunal kingdoms or empires. The Sclater-Wallace division (1876) and Smiths (1983) faunal division of the world are notable. The main faunal realms are as follows: the **Nearctic**, comprising North America; the **Neotropical**, including Central and South America; the **Palearctic**, Europe and Asia north of the tropics; the **Ethiopian**—Africa South of Sabaria; the **Oriental**, covering tropical Asia; and the **Australian** including SE Asia, Oceania, Australia and New Zealand.

feet (1) A lagoon of brackish water separated from the sea by a bar or spit. The best known is the Flect, impounded by Chesil Beach, Dorset, England. (2) A small creek or inlet on a coastline.

fell A Norwegian term for an open mountainside, which occurs in many place names of the northern Pennines and the Lake district of England.

felsenmus A German term, literally meaning a 'sea of rock', which has been widely adopted to describe the mountain top spreads of angular boulders due mainly to frost action on well-jointed rocks, probably during the later cold periods of the Pleistocene.

felsic A term used to describe light coloured rocks containing an abundance of feldspar and quartz as compared with mafic rocks.

fen An area of wooded, swampy land in which the groundwater is alkaline or neutral but not acid.

fen soils A name given to those organic soils which form above fen peat in an environment of base-rich groundwater. Thus, fen soils have an alkaline or neutral reaction and when drained give rise to extremely fertile and productive agricultural soils.

Ferrel's Law A law relating to the rotation of the earth, postulated by an American scientist, W. Ferrel, in 1856. Because of the earth's rotation a body moving across the surface is subject to an apparent deflection to the right in the N. Hemisphere, although there is no actual effect at the equator. Its most important effects are shown in oceanic and air movements.

ferrisols Tropical soils similar in composition to latosols and fer(ri) siallitic soils.

fetch Length of water surface over which the wind blows in generating waves. Together with wind velocity and duration, this determines wave height. Many features of coastal deposition tend to become oriented normally to the direction of maximum fetch.

fiard (fjard) An indeed drowned coastline fringing a rocky, glaciated low land but without the deep glacial troughs of a fiord with which it should not be confused. It is generally deeper than a ria and is characterised by a large number of islands, similar to the coastal area of SE Sweden. Fizard is the Swedish spelling.

filling A term used in meteorology to describe the increasing atmospheric pressure near the centre of a depression, which ultimately leads to its disappearance. The converse term is deepening.

finger lake A narrow linear body of water occupying a glacially over-deepened valley and sometimes impounded by a morainic dam.

fiord (fjord) Long narrow, deep, U-shaped coastal inlet which usually represents the seaward end of a glaciated valley that has been partially submerged. The water depths often exceed 100m except near the mouth where a bar or sill may be present.

firn (neve) Snow that has survived a summer melting season. It is an intermediate material in the conversion of snow to glacial ice. Normally, it is granular, due to the partial melt.

firn line A line on a glacier marking the upper limit to which winter snowfall melts during the summer ablation season. It is often clearly marked, and on many glaciers separate hard, blue ice from snow above.

firn wind (glacier wind) Down-hill airflow which develops over a glacier during the day, usually in summer. The greatest air density over the glacier than over the surrounding surfaces causes this air to sink.

fissure eruption A volcanic eruption along a linear crack, fault or other structural weakness in the crust.

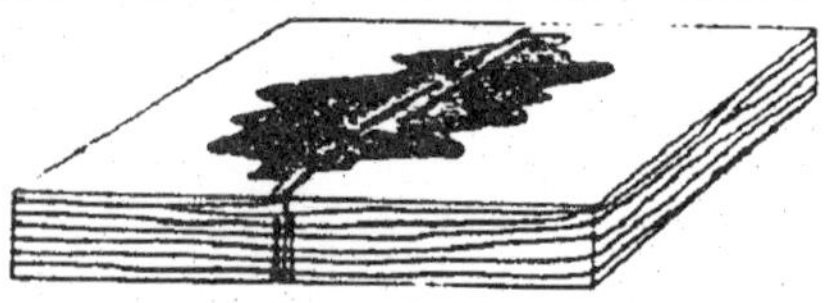
Fissure eruption

flap A geological term for a gravity-collapse structure, in which a rock stratum which was originally on the limb of an anticline has slide down its side, become overturned and finished upside down as it came to rest.

flash flood A brief but powerful surge of water either over a surface or down a normally dry stream channel. Usually it is caused by heavy convectional rainfall of short duration, and is typical of semi-arid and desert environments.

flat iron A land form of roughly triangular shape, and with one side that is both steep and uniform. It is formed between two adjacent valleys that cut through a hogback ridge roughly at right angles to its trend. It is a common land form along the eastern front of the rocky mountains, USA.

flatness index A measure of the shape of a rock fragment. It is achieved by summing the long and intermediate diameters of the rock fragment and dividing the result by twice the short diameter.

f-layer A zone at a height of about 250 km within the atmosphere that reflects high frequency radio waves.

flexure A type of fold where bedded rocks have slipped over each other owing to tensional stress during deformation in the incompetent rock layers. Sometimes referred to as flexure-slip folding.

flint Variety of chest, which occurs commonly as nodules and band in chalk. It is deposited in the porous, permeable structures of sponge, diatom and echinoid skeletons and also in burrows.

Flohn's climatic classification A classification, outlined in 1950,based on a combination of planetary wind belts and precipitation characteristics.

flood plain The part of a river valley that is made of unconsolidated river-borne sediment and periodically flooded. It is built up of relatively coarse debris left behind as a stream channel migrates laterally and of relatively fine sediment deposited when bankfull discharge is exceeded.

flood tide Rising tide: the phase of the tide between low water and the next high tide.

Florida current Part of the Gulf Stream; it extends from the southern tip of Florida to cape Hatteras, N. California. It is a fast flowing (1–3 m/s), narrow (50–75 km wide), and deep current, still evident at depths of 200 m where velocities of upto 10 cm/s have been measured. It is an example of a western boundary current.

floristic province A large geographical area containing a distinctive plant assemblage found nowhere else. In 1953, R. Good recognised thirty-seven floristic provinces in the world, ranging in size from the tiny Pacific island of Juan Fernandez upto the vast Euro-Siberian Province.

flow folding Folding which results mainly from continuous simple shear or viscous flow within layers, producing deformations analogous to lamellar flow in liquids. The term flow fold has also been used synonymously with ptygmatic fold.

flow till Sediments which flow after they have been deposited by ablation.

flowstone A smooth deposit of calcium carbonate accumulated by flowing karstic water on the floors and walls of caves.

flume The channel of an artificially constructed stream built either for industrial uses or for simulation of stream flow characteristics during hydrological or sedimentological research.

flute An intermittent sub-conical groove pattern eroded on the floor of a stream channel as a result of turbulent flow. It is one example of sole markings that are preserved on the base of a bed of sedimentary rock which itself forms a cast of the groove.

fluted moraine A ground moraine with a distinct lineation parallel to the direction of former ice movement, perhaps formed by a process similar to that which forms drumlins.

fluvial Terms pertaining to rivers and river action, respectively.

fluvio karst Land forms of karstic character formed by the combined action of fluvial and true karst processes.

focal depth A measurement referring to the shortest distance between the focus of an earthquake and the surface of the earth.

fog The obscurity exhibited by the surface layers of the atmosphere owing to the presence of water droplets formed by condensation, amplified by the suspended particles of smoke and dust which may occur.

fogbow A virtually colourless arc of brighter light, similar in form to a rainbow, seen in a jog, when the sun is behind the observer. The water droplets are too small (0.05 mm) to enable refraction and reflection to create the normal colour spectrum exhibited by a rainbow. The colours overlap and the bow appears white.

fohn wind Generic term for warm, dry winds in the lee of a mountain range. It was originally used in the European Alps. After cooling on the windward ascent at the saturated adiabatic lapse rate of 0.5°C/100 m, with the resulting condensation and precipitation, the air descending on the leeward side of the mountain range is warmed through compression at the dry adiabatic lapse rate at

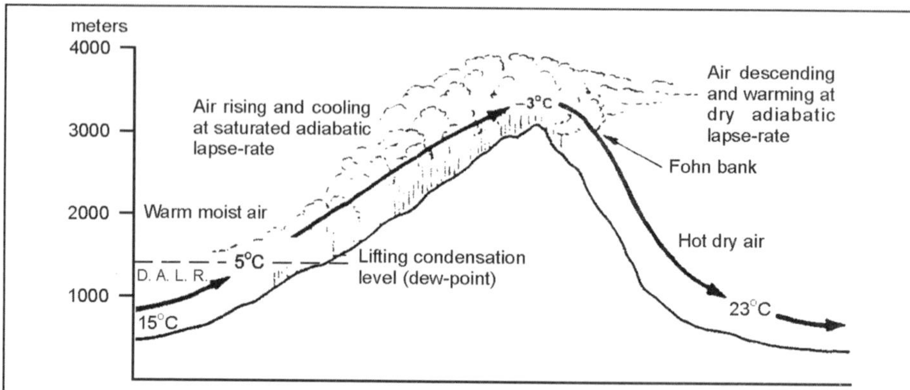

Fohn

1°C/100 m, with the resulting condensation and precipitation, the air descending on the leeward side of the mountain range is warmed through compression at the dry adiabatic lapse rate at 1°C/100 m. This produces a warming wind on the lee side with higher temperatures than occurred in the same air on the upslope side of the mountains.

fohrde A German term for the shallow estuaries of the drowned linear valleys in the boulder clay lowlands fringing the southern coast of the Baltic, e.g., Kieler Fohrda.

fold A flexure or bending of the earth's crustal rocks owing to compressional forces. In simple folding, an upfold is termed an anticline and a downfold a syncline, the various types of folds are **symmetrical**, **asymmetrical**, **monoclinal**, **isoclinal** and **recumbent**.

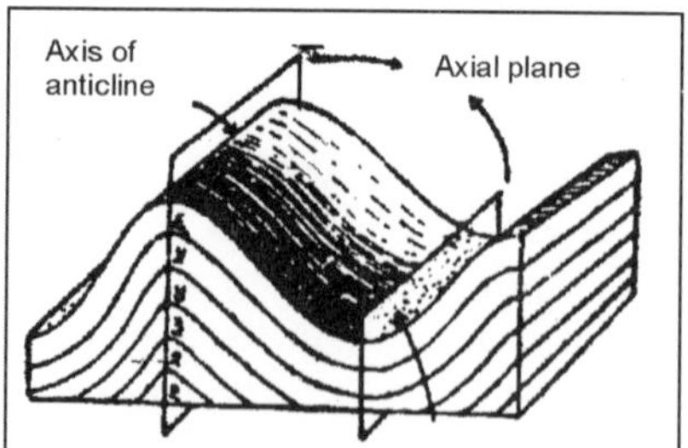

Various parts of a fold.

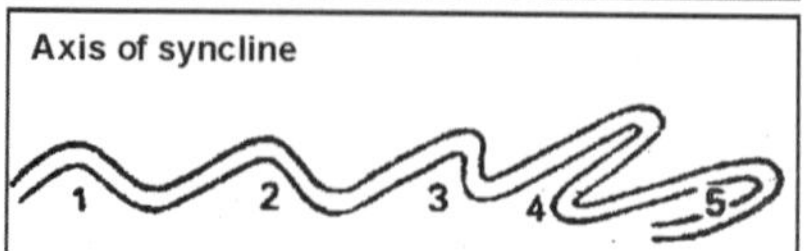

Types of fold. (1) Symmetrical, (2) Asymmetrical, (3) Monoclinal, (4) Isoclinal (5) Recumbent.

fold angle The angle between the limbs of a fold whose tightness or openness reflects the intensity with which the structure has been deformed.

fold axis A line which lies parallel to the hinge line and marks the intersection of the axial plane with the hinge zone.

fold limb The generally planar region of a fold which lies between two adjacent hinge lines and is confined between the zones of maximum curvature.

folded mountain belt A long, linear zone of earth's crust where rocks have been intensely deformed by horizontal stresses and generally intruded by igneous rocks. The great folded mountains of the world (such as the Himalayas, Appalachians, Rockies and the Alps) are believed to have been formed at convergent plate margins.

foliation A laminated or banded structure within a metamorphic rock, caused by the segregation of different minerals into parallel layers following the schist of the rock.

food chain The sequence of energy transfer in the form of food from organisms in one tropic level to those in another, achieved when one organism eats or decomposes another.

food web A series of interconnected and overlapping food chains in an ecosystem.

foot hills The lower line of hills that lie parallel with and at the foot of a higher mountain range.

fore deep A linear, narrow crustal depression in the ocean floor on the convex side of the island or a coastal range of fold mountain. Also known as trench.

fore shock A seismic tremor recorded prior to the main shock of an earthquake. It is caused by the initial slip or fracture in the vicinity of the major focus and precedes the main earthquake by a short time interval.

fore shore The lower zone of a beach extending between low-water spring-tide level and high-water spring-tide level, i.e., the part of the beach covered by spring tide.

fore-dune A term for part of a coastal dune system. It is the dune nearest the sea, occupying the same windward position, as does the advance dune in the desert dune system.

fore-set beds Those parts of the bedded layers of a delta which are situated between the top-set beds and the bottom-set beds. They are inclined at an angle, one above another as the accumulated sediment descends the steep frontal slope of the delta.

forest bed An interglacial body of sediment containing macroscopic remains of trees and other vegetation.

form-line A line drawn on a map where there is insufficient information available to enable a contour to be constructed. It is not instrumentally surveyed and is intended merely to give an impression of the terrain. It is usually depicted as a dashed rather than a solid line.

Food Web in a Grassland Ecosystem

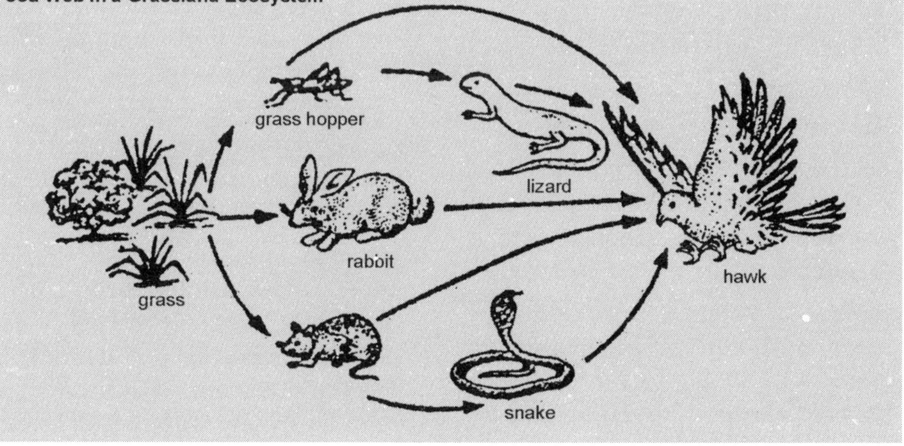

form-ratio A relationship between river width and river depth expressed as a ratio.

fortin barometer A type of mercury barometer in which the level of mercury is adjusted to zero before a reading is made.

fossil The remains of the whole or part of any formerly living organism preserved by natural causes in crustal rocks.

fossil fuels All fuels derived from the fossilisation of formerly living organisms.

fracture zone A linear fracture on the deep-sea floor across which the lithosphere changes abruptly in both age and water depth most fracture zones transect oceanic ridges and are small circles whose radii of curvature depend on the distance from the pole of rotation of the two lithospheric plates which are diverging at the ridge.

fractus A type of cloud the ragged or torn nature of which indicates storminess in the atmosphere.

fraigem A strong cold wind in Southern Brazil caused by a sudden outbreak of cold southerly air associated with a winter anticyclone.

frazil Minute needles of ice which develop into spongy accumulations in moving water below freezing-point.

free atmosphere The atmosphere above the level to which the effect of friction reaches (commonly taken as about 500 m).

free face The steepest part of a slope profile, with an arbitrary minimum slope angle of 45°, i.e., too steep for any loose material to rest upon it.

free stone A term used to describe any sedimentary rock which is so evenly textured that it can be freely worked.

free-air anomaly The gravitational acceleration remaining after correction of a measurement for the theoretical gravity, usually the International Gravity Reference Field and by the free-air correction. No correction has been made for the gravitational attraction of rocks between the observing station and the reference datum, which is usually sea level.

freezes thaw action A type of weathering process in which water freezes and the resulting ice melts as the temperatures fluctuate below and above freezing point. When the water occurs in cracks in solid rocks it causes nine per cent enlargement of these fissures during freezing and ultimately leads to frost splitting. The broken rock fragments are then moved by melt water during the period of thaw most frequently downslope.

fresh water Water with less than 0.2 per cent salinity.

freshet A surge of flood water down a small river or stream causing it to overflow its banks, resulting from heavy precipitation or sudden snowmelt around the head waters.

fret A narrow belt of sea-mist along a coastline due to differential surface heating.

frigid zone One of the three climatic zones recognised by early geographers. Now used only to describe one of the climatic types defined in Koppen's climatic classification or more loosely as a synonym for permanently snow-covered and permafrost areas.

fringing reef A coral reef which is attached to the shore, either as a continuous wave-washed erosion platform or separated from the coastline by a shallow lagoon. Beyond its seaward margin, the ocean water deepens rapidly.

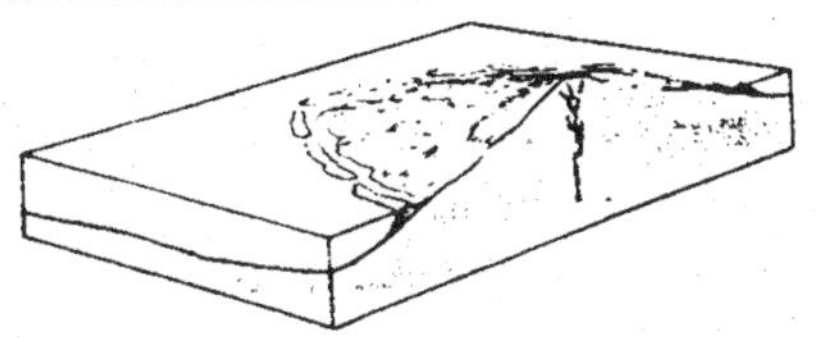

Fringing reef

front Boundary or boundary region separating air masses of different origins and characteristics. Temperature gradients in any horizontal surface are large through the front. Different types of front are distinguished according to the nature of the air masses separated by the front, the direction of the front's advance, and the stage of development.

frontal zone Transition zone, sometimes amounting to a discontinuity that separates adjacent air masses. Some turbulent mixing takes place. The sloping down separating a cold wedge under warm air typically extends about 1 km vertically and about 100 km horizontally.

frontogenesis Development and intensification of frontal boundaries between adjacent air masses.

frontolysis The gradual break-up or dissipation of a front or frontal zone. This is the antithesis of frontogenesis, and is effected mainly by horizontal divergence of air from the frontal zone, together with subsidence.

frost Condition in which the prevailing temperature is below the freezing point of water (0°C). This may lead to a deposit of ice crystals on objects, e.g., grass or trees. Such deposits result from condensation when the dew point temperature is below freezing.

frost action The mechanical weathering process caused by alternating cycles of freezing and thawing of water in pores, fissures and cracks at the surface.

frost heaving The doming or vertical lifting of the soil surface into frost hillocks owing to pressures caused by the freezing of groundwater under periglacial conditions.

frost hollow Area that is very liable to severe and frequent frosts as a result of dense, cold air moving down slope (katabatic flow) and collecting there under conditions of radiation cooling.

frost hollow, frost pocket A topographic depression susceptible to frosts while the surrounding hill slopes may remain frost free. The cold dense air which has lost heat by terrestrial radiation at night drains down slope by graving as a katabatic flow before accumulating in the hollow wherein the temperature may fall below.

frost wedging Fracturing of rock by the expansionary pressure associated with the freezing of water in planes of weakness or in pore spaces.

frozen ground That part of the earth's land surface which has temperatures at or below 0°C and generally contains groundwater in the form of ice.

fula An Arabic word referring to a shallow pool formed during the rainy season in the N. Sudan and an important source of water for pastoral tribes.

Fuller's earth (1) A fine earth similar to a clay but lacking its plasticity. (2) Capitalised Fuller's earth is the stratigraphic name of a Jurassic clay formation outcropping in Southern Britain.

fumarole A small vent in a volcanic area from which steam, gases and various acids are ejected, often with considerable force in the form of jets.

funnel cloud Cloud produced in a low pressure vortex in the centre of a spiral storm, e.g., a tornado or waterspout.

furrow (1) A channel or linear depression cut by a plough. (2) A linear depression on the ocean floor.

fusain A friable and dusty layer within a bed of coal, high in ash content.

G

gabbro An igneous rock of coarse grain, dark colour and basic composition. It is the plutonic equivalent of the volcanic basalt and the hypabyssal dolerite.

gada An Arabic term for a flat-topped table-like hill in the Atlas mountains of N. Africa. It is also used in the Sahara to describe a hamada, but its more common usage is related to an elevated landscape.

gal The unit, named after Galileo, for measuring gravitational acceleration: 1 gal = 2 cm/5.[2] The gal has been largely replaced by the gravity unit.

gale A strong wind recorded 10 m above ground level, which has been classified on the Beaufort scale as force 7, near gale to force 9 strong gale.

gallery forest A dense growth of tropical forest following the course of a river in savanna country.

gangue The non-valuable metalliferous ores or the relatively worthless non-metallic mineral material in the lode or mineral vein.

gannister An erinaceous seat earth beneath the coal seams of the coal measures in Britain. It is a very pure silica sand, the high refractory properties of which make it of great use for furnace linings because of is heat resistance.

gara An Arabic word widely used in N. Africa and the Middle East to describe an isolated rocky hill with steep slopes and usually with a tabular summit. Its plural form is gour.

GARP The Global Atmospheric Research Programme, an international project incorporating numerous studies on atmospheric circulation and atmospheric dynamics at a global scale.

garrigue An evergreen scrub with xerophyte's characteristics found on thin limestone soils in the drier parts of regions experiencing a Mediterranean climate.

gat (1) An inshore channel or strait dividing offshore islands from the mainland, e.g., the Frisian Island of N.W. Europe. (2) An opening through a line of sea cliffs allowing access to the coast from inland.

gate (1) In topography, an opening, derived from the Anglo-Saxon word geat: (a) A break in a line of hills where a river passes through a relatively narrow and restricted water gap, (b) An entrance to a natural harbour between two promontaries, e.g., the Golden Gate, San Francisco. (2) In hydrology—a barrier across a water channel built to regulate the flow of water, i.e., a spillway gate of a dam.

geanticline A broad, complex fold in the earth's crust, generally the result of compressive folding of the sediments within a geosyncline.

gelifluction A term proposed in 1956 by H. Baulig to define one type of solifluction that associated with frozen ground, covering with seasonal freezing in addition to permafrost.

gendarme A French term now used universally to describe a rocky pinnacle on a narrow mountain ridge, forming an obstacle to climbers and walkers.

general circulation The term generally used to describe the large scale circulation of the atmosphere over the globe, or over one hemisphere, with its more or less persistent features and all the transient features on various scales. Although in its nature it is a matter of the winds, the general circulation may be studied by means of barometric pressure maps because of the intimate relation between pressure and winds.

geo A linear, narrow cleft in a sea cliff worn by marine erosion along a line of weakness in the rock. It originally formed as a narrow cave but various stages of roof collapse can be seen in the many examples found on the hard rock coasts of Britain, especially in Scotland.

geocentric Relating both to the centre of the earth and to the earth as a centre.

geochemistry The study of earth's elements and its parts in order to determine the composition of the earth and the distribution of its elements.

geochronology The study of age determination in the history of the earth.

geochronometry The measurement of geological time.

geocryology The study of frozen ground (both seasonally frozen and permafrost but excluding the study of glaciers).

geode A hollow cavity in a rock, almost spherical in shape, in which inward-pointing crystals line its interior walls. It is also termed a vugh.

geodesic line A line drawn on any mathematically defined surface, representing the shortest distance between any two points on that surface.

geodesy The science related to the size and shape of the earth, together with the determination of the exact position of particular points on its surface by taking the earth's curvature into account.

Geographic information system (GIS) "It is designed for the collection, storage and analysis of objects and phenomena where geographic location is a important characteristic or critical to the analysis" (Aronoff, 1989). Using computers, a GIS stores and integrates large amounts of spatially referenced data. GIS has received considerable attention from researchers, planners, bureaucrats and others.

geoid An earth-shaped body considered either as a mean sea-level surface extended continuously through the continents or as an undulating surface somewhat lower than the surface of the spheroid beneath the oceans and rather high beneath the continents.

geomagnetic The study of the earth's magnetic field.

geomagnetic equator The great circle around the earth connecting all points of zero geomagnetic latitude. It is equidistant from each of the geomagnetic poles and its plane is at right angles to the earth's magnetic axis.

geomagnetic field The magnetic field of the earth that causes a compass needle to align north-south.

geomagnetic Poles The Poles of the earth's magnetic field, located some 6,400 km above the surface (and note, therefore, corresponding to the surface). They are situated above 73.5°N, 69°W and 78.5°S, 11°E respectively.

geomorphic region Any region defined solely by the nature of the geomorphic processes operative within it.

geomorphology The scientific study of the origin of land forms based on a cause and effect relationship. Geomorphology as a science of land form study was established largely by the pioneers efforts of scholars such as Gillant and Davis at the end of 19th century. It has undergone rapid development and change during 20th century. At first, there was emphasis on land form description and explanation in terms of cycle erosion surface and after 1960 there was much grater emphasis on the study of processes affecting land forms, climatic and applied geomorphology.

geopotential The potential energy of unit mass in the gravitational field of the earth. The unit of geopotential is the potential acquired when a mass is raised from sea-level through the unit distance in a field of force of unit strength.

Georef system A global reference system referring to the location of points on the earth's surface. It is based on the graticule, thereby facilitating the definition of a map reference in terms of letters and numbers.

geostrophic flow The term used to describe the horizontal flow of air in its theoretical movement parallel to the isobars, thus creating a geostrophic wind. The wind is resultant of two opposing forces: the pressure gradient or pressure force in one direction and the deflecting force in the other. Geostrophic flow is found only in the upper atmosphere, well clear of the friction created at the earth's surface, where winds blow obliquely across the isobars, not parallel with them.

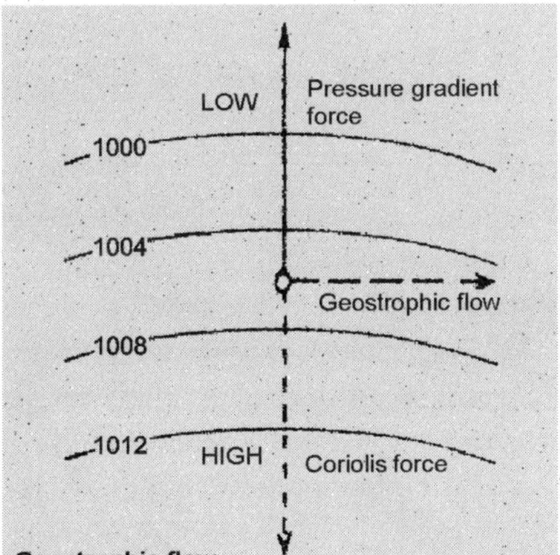

Geostrophic flow

geosynclines A major structural down fold in the earth's crust on a

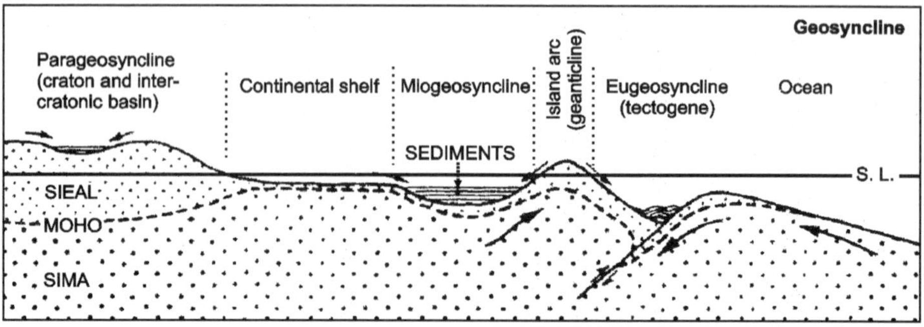

subcontinental scale. It comprises a linear basin which serves as a respitory for considerable thickness of sediment derived from neighbouring land masses.

geotectonics The scientific study of the form, structure and disposition of the rock masses which make up the crust of the earth.

geothermal energy The energy derived from hot rocks in the earth's crust, some of which is emitted at the surface as hot water or steam.

gestalt concept A term of German derivation, taken from gestalt theory, in which configurations or structures can be studied in relation to each other. In physical geography, the gestalt concept is generally utilised to denote the ways in which a number of climatic elements can combine to form a recognisable pattern on a synoptic chart.

geyser A violent ejection of steam and superheated water from an underground source through a hole in the ground. The term is derived from the icelandic geysir (= roaser or gusher) in a country with numerous geysers. High pressure forces the liquid into a gaseous phase once a critical temperature threshold has been crossed and the geyser shoots violently into the air, sometimes to heights of 60 m. Some geysers exhibit a remarkably regular periodicity of ejection, e.g., 'Old Faithful', Yellowstone National Park, USA.

geyserites A deposit of siliceous material around geysers and thermal springs. Not to be confused with travertine, which has a differential chemical composition.

gibber An Australian term for a level surface covered by a thick deposit of gravel or broken siliceous pebbles, occurring in the more arid parts of the continent.

glacial Strictly, the term refers only to a glacier, but apart from its many adjectival and adverbial uses, the term is also used as a noun to describe a cold phase during an ice age.

glacial cycle of erosion The cycle of erosion in a glacial region includes—the grooved upland of youth; the early fretted upland where smooth-topped ridge is broken by cirques; and the nature fretted upland in which the residual summits are now horns and the intervening ridges as arêtes; the old-age stage is the monumented upland. During these

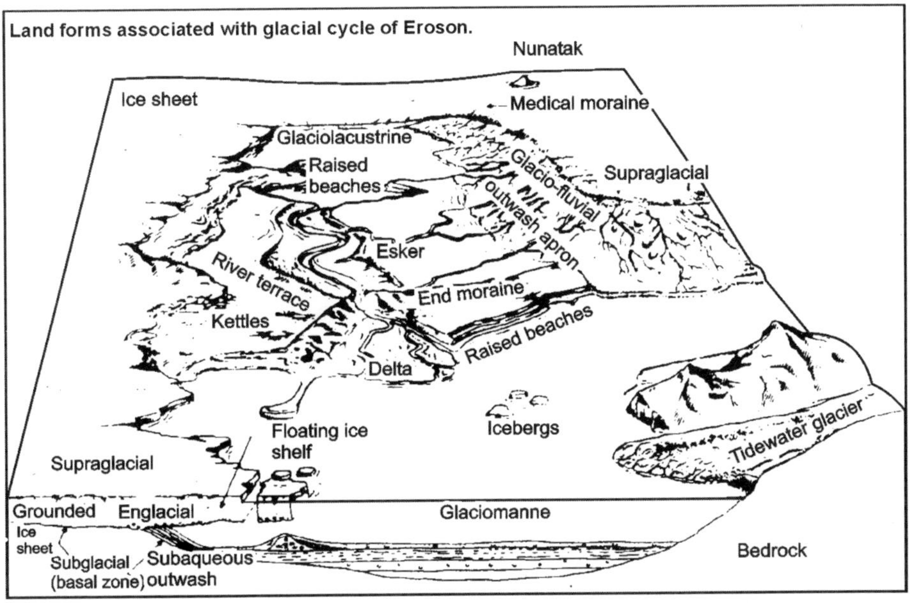

stages several landforms have been created by glacial erosion.

glacial deposit An expression which refers to any type of debris or sediment which has been derived directly or indirectly from a glacier or ice sheet. The deposit may be stratified or unstratified, homogenous or heterogenous. The glacial deposits are grouped under term 'drift' and 'till'.

glacial diffluence The local overriding of a pre-glacial subsidiary watershed by glacier ice so that ice flows through a col into a neighbouring valley. The erosive effect of ice on the col is such that it is lowered and the watershed is said to be breached.

glacial drainage channel A general expression which covers stream channels associated with glaciers and ice sheets including (i) an overflow channel and (ii) a melt water channel.

glacial erosion The destruction of the surface by the passage of a glacier or an ice sheet and the removal of the eroded material.

glacial lake A small body of water impounded between the margin of a glacier and its valley wall or associated moraire, e.g., the Marjelen sea at the edge of the Aletsch Glacier, Switzerland.

glacial limit A line marking the furthest extent of a former glacial advance. It may be identified on the ground through the recognition of features associated with glacial margins, including lateral and terminal moraines, outwash spreads, marginal melt water channel and proglacial lakes.

glacial outburst A flood caused by the sudden release of an ice-dammed or englacial lake, often with disastrous consequences. It is also known as jokulhlaup, an Icelandic term.

glacial recession A phase marked by a reduction in both the sides and the thickness of a glacier or ice-sheet.

glacial retreat The state of a glacier when its front or snout is receding although the body of the ice itself is still moving forward. This stage is reached when the rate of ablation exceeds that of accumulation, leading to a negative mass balance, glacier regime.

glacial stairway Long profile of a glacier trough It is characterised by alternating rock bars and rock basins giving the impression of a stairway. The structure is attributed to variations in the erosive power of ice or to the influence of rock jointing.

glacial trough A relatively straight, steep-sided, U-shaped valley that results from glacial erosion.

glacial-control theory A theory to explain the existence of coral at considerable depths below sea level. In 1910, R.A. Daly postulated that during the quaternary ice age the level of the oceans fell owing to the abstraction of water to form the ice sheets. Coral reef were planed own by marine waves and all living corals were destroyed because of the cooler water temperatures. As the ice sheets melted, the theory asserts that coral began to grow again on the worn-down stumps and as the released waters returned to the oceans soothe coral's upward growth kept pace with the rising sea-level.

glaciation The covering of a landscape or larger region by ice; an ice age.

glacier A larger mass of ice, resting on or adjacent to a land surface, and typically showing movement. The several types of glaciers are: cirque glacier, valley glacier, piedmount glacier, transection glacier, etc.

glacier flow It is downslope movement of the ice body. It includes sliding on the rock floor,

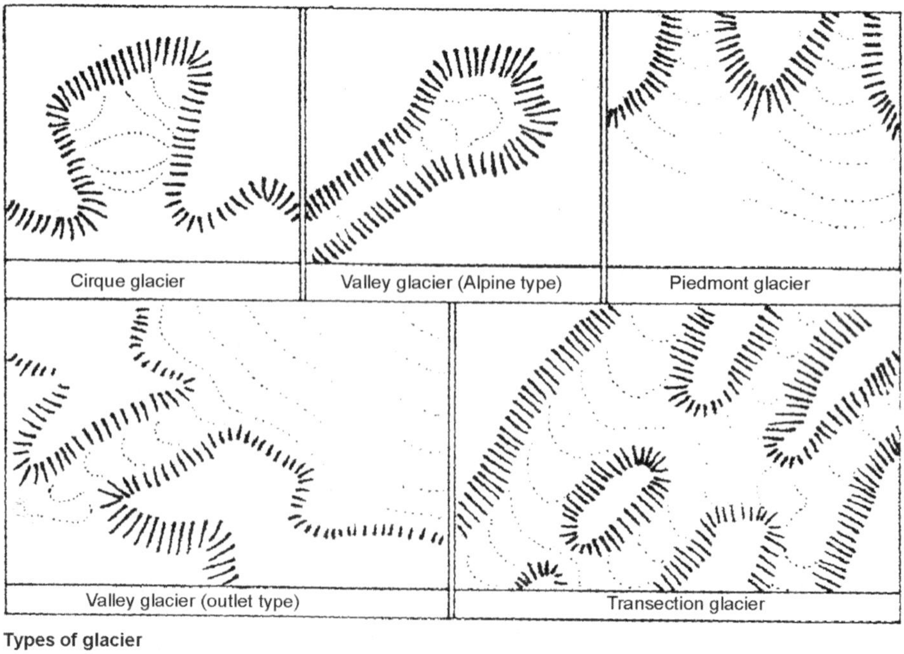

Types of glacier

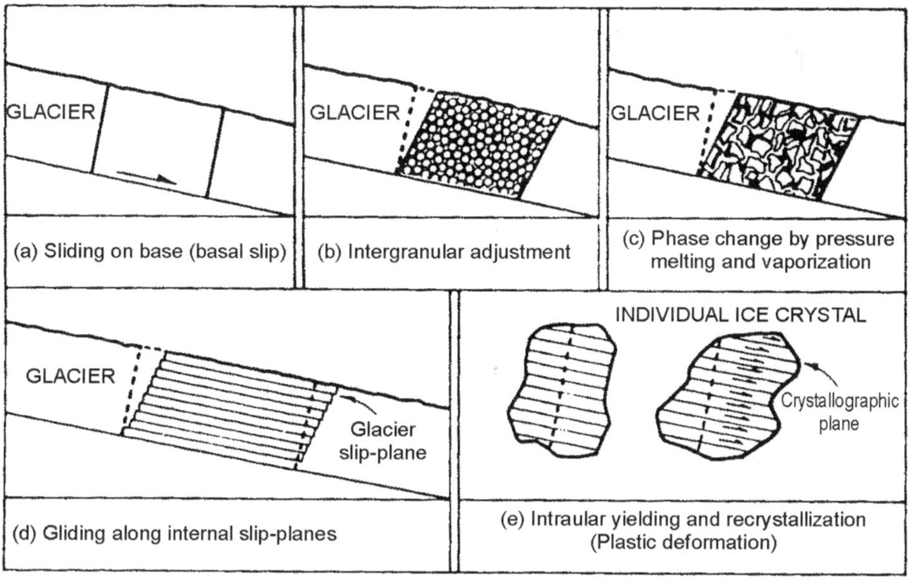

Types of glacier flow

inter-granular adjustment of ice crystals, local and temporary phase changes of sliding and inter-granular yielding.

glacier wind A cold wind blowing down a valley glacier in day time, especially in summer, as a result of cooling of the air in contact with the ice. This is a type of katabatic wind, a term which, on a larger scale, is sometimes applied to gravity winds which descend from the high ice-sheets of Antarctica and Greenland.

glacierisation The gradual advance of an existing ice-sheets or glacier over ice-free terrain. It differs from the term glaciations which refers to a former glacial advance of ice-sheets that are no longer in existence.

glacioeustasy The theory that sea levels rise and fall in response to the melting of ice during interglacial and the accumulation of ice during glaciations. A major glacioeustatic oscillation has an amplitude of some 100 m.

glaciofluvial A term referring to the processes and the land forms related to the action of glacial melt water. The fluvial transport of material and the associated mechanisms of erosion and deposition are similar to those of a river when the action is occurring beyond the edge of the glacier or ice-sheet.

glacis A French term for a gentle slope, which when used as a prefix has several different meanings: (i) glacis d accumulation = a bahada, (ii) glacis de denudation = a pediment, (iii) glacis d' ennoyage = detritus slope in a depression.

g-layer The inner solid core of the earth, extending from a depth of about 5150 km to the earth's centre at 6371 km depth.

glazed frost, glaze A coating of clear ice which forms when rain falls from an advancing cloud layer onto a ground surface, the temperature of which remains below freezing-point.

glen A Scottish term for a steep-sided valley in the highlands. It is narrower than a Strath.

global warming The increase in global temperatures brought about by the increased emission of greenhouse gases into the atmosphere (see **greenhouse effect**).

globigerina ooze A calcareous marine deposit largely comprising the shells of tiny foraminifera, especially the globigerina species in which the calcium carbonate content ranges from 30 per cent to 97 per cent.

gnammahole An Australian term for a rock hole found at the base or on the flanks of a bare granite dome in an arid environment.

goletz terraces Terraces formed on hill slopes of a combination of freez-thaw action and solifluction under periglacial conditions.

gondwanaland The name given by geologists to the super-continent thought to have existed in the S. hemisphere over 200 million years ago, and formerly part of the even larger Pre-Cambrian continent of **Pangea**. This was an ancient continental landmass that is though to have split apart during Mesozoic time to form present-day continents of South America, Africa, India, Australia and Antarctica.

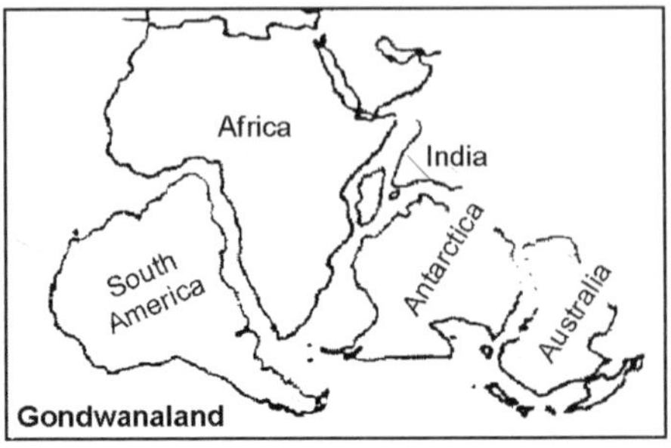

Gondwanaland

gorda plate A minor lithospheric plate which is subducting under the North American Plate to the West of Oregon.

gorge A deep and narrow chasm with precipitous rocky walls, currently occupied or formerly occupied by a river. It is more steep-sided and enclosed than a ravine, but is of the same magnitude. Any river valley of broad flood plain and gentle bluffs may pass into a gorge as it is traced up or downstream, possibly owing to the presence of a belt of hard rocks, e.g., the Avor Gorge at Bristol.

graben A German term for a structurally defined block of land downthrown between parallel faults. It may not coincide with a topographical trough, but where a fault-guided trough is occupied by a valley it is in this case synonymous with a rift valley. It contrasts with a horst.

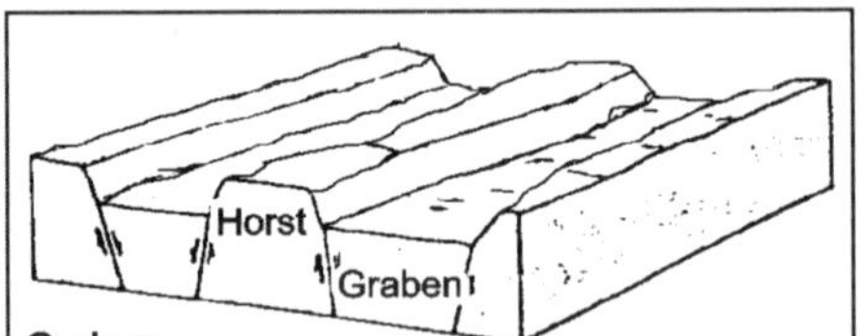

Graben

gradation (1) The bringing of a land surface to a state of uniform grade by processes of denudation. (2) Synonymous with degradation.

graded bedding Sedimentary structure in which there is an upward

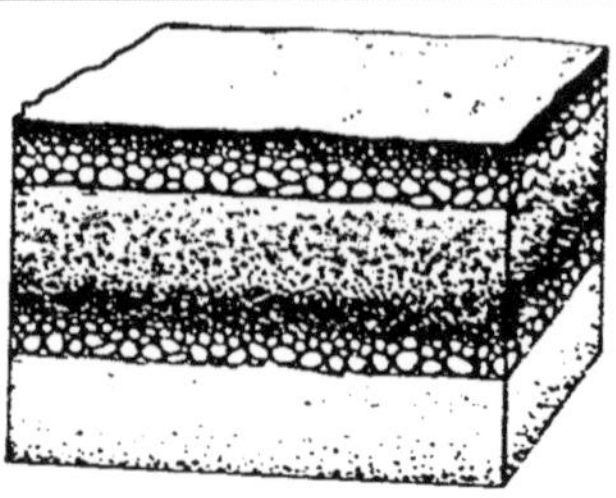
Graded bedding

gradation from coarser to finer material, caused by the deposition of a heterogeneous suspension of particles. The feature may be used to establish the way up or natural succession of strata.

graded shoreline A theoretical stage said to be achieved in a conceptual cycle of coastal development. It is represented by the cutting back of headlands by marine erosion, complemented by the enclosing of bays by the building of bars across the entrance, thereby leading to the straightening of an irregular coastline.

graded time The time-scale necessary for a stream or a slope to achieve a graded profile. It is one of the three categories in the time-scale of geomorphology devised by S.A. Schumm and R.W. Lichty.

gradient (1) The steepness of a slope expressed in degrees, percentages, or as a ratio. The latter is expressed as a proportion between the vertical interval (VI) and its horizontal equivalent (HE). (2) A change or variation in the value of a variable, especially those relating to climate, e.g., temperature gradient.

gradient wind The wind which results from a balance between the coriolis force and the centrifugal force due to any curvature displayed by the isobars. Strictly speaking, a gradient wind can only exist in the upper atmosphere for nearer the earth's surface this balance is affected by frictional forces.

grading The determination of particle size by sorting of clastic sedimentary material.

grain A description of the textural coarseness of a rock, i.e., coarse-grained or fine-grained.

grain size The size of mineral particles which make up an igneous rock. The grain size is always determined on the ground mars, with phenocrysts being ignored.

gram A unit of weight in the metric system equal to a thousandth part of cubic decimetre of water.

granite An igneous rock of coarse grain and Plutonic origins, consisting essentially of quartz (20–40%) together with alkaline and Plagioclase feldspars, and very commonly a mica.

granular disintegration One of the main types of mechanic weathering of rocks, caused either by the freezing of pore water and the dislodgement of particles, or by differential expansion and contraction due to insolation, leading to disintegration of the grains.

granule (1) A spheroidal soil aggregate. (2) A fragment of rock, the particle size of which is between 2 mm and 4 mm diameter. It is smaller than a pebble but larger than a sand grain.

graphic texture A rock texture, especially in granite, resulting from the regular intergrowth of alkali feldspar and quartz. The textural pattern resembles runic or cuneiform writing or hieroglyphics.

graphite A soft, grey-laminar form of pure carbon, commonly referred to as black lead.

grassland One of the four main subdivisions of world vegetation. It is characterised by limited precipitation, insufficient for tree growth and a season of drought. Several types may be recognised—**tropical grassland, temperate grassland (downs, pampas, prairie, steppes-veld), montane grassland.**

graticule An intersecting system of lines drawn on a map to represent meridians and parallels of latitude. Graticule intersections are the small crosses depicted on the face of a map indicating where the meridians and parallels intersect.

graupel A word of German derivation, meaning soft hail that can also be applied to pellets of partly melted and refrozen snow, each with a soft core, rarely exceeding 5 mm in diameter.

gravel In the British classification for particle sizes, grains with diameters between 2 and 60 mm.

gravel train A synonym for a valley train. It was a term used to describe the gravel deposits which are found in the higher terraces of the R. Thames where it enters the London Basin in S. England. The Thames gravel terrain, occurring at some 70 to 100 m above present river levels, is thought to be older than the Cromerian interglacial. It has between subdivided into a higher and lower gravel train.

gravimetric survey A geophysical technique for determining rock structures at depth. It is based on the variation in the gravitational field of the earth due to density differences in the crustal rocks.

gravitational gliding The downslope sliding of rock strata resulting from tectonic movements (e.g., nappe formation) that produce uplift.

gravity The force excreted by the earth and its rotation on a unit mass. Newton's law of gravity (1686) states that each body in the universe attracts every other body with a force directly proportional to the product of their masses and inversely proportional to the square of the distance between them, measured from their centres of mass along a line joining these centres.

gravity anomaly An expression used in geophysics referring to the difference between a theoretical computed gravity value and an observed gravity value. Excess observed gravity is a positive anomaly while a deficiency is a negative anomaly.

gravity collapse structure A tectonic structure developed in folded rocks where erosion of incompetent strata allows more competent strata to slide or collapse under gravity, thereby creating contorted folds.

gravity flow A type of glacier movement first described in 1751 by J.G. Altmann who realised that gravity was the cause of glacier movement on a sloping surface.

gravity slope A slope lying at the angle of rest of the eroded material. Synonymous with constant slope.

great circle A circle on the earth's surface, the plane of which passes through the centre of the earth. Thus, the shortest distance between any two surface points represents an area of a great circle.

great red spot An anticyclonic storm system in the hydrogen-helium atmosphere of the southern hemisphere of Jupiter. It covers 10° at latitude and about 12000 km across, and has been observed for more than 300 years.

green muds A fine-grained sea-floor deposit, occurring at depths of

150–300 m on the continental slope. Its colour is derived from a high proportion of glauconitic, but calcium carbonate is also present in high quantities.

greenhouse effect An expression referring to the ability of glass to allow incoming short-wave solar radiation to pass easily into a greenhouse, but which blocks some of the radiated outgoing long-wave terrestrial radiation. The process has been used to describe how short-wave radiation passes easily through the atmosphere to reach the earth's surface whereas outgoing radiation is absorbed and radiated by water vapour, droplets and carbon dioxide.

Greenhouse Effect

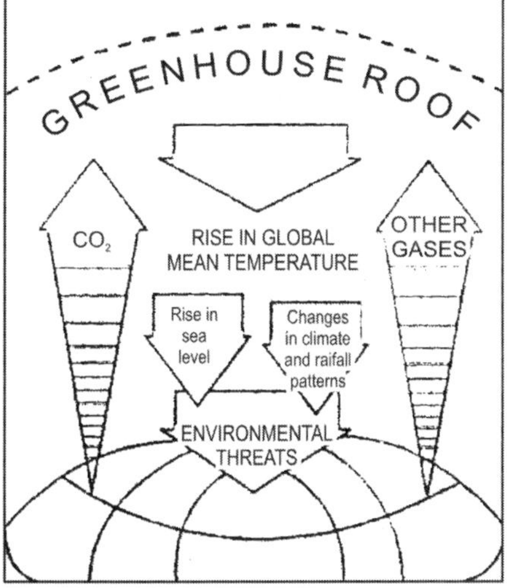

greensand (1) A sand or sandstone with high proportions of gluconite. (2) A lithostratigraphic unit of the cretaceous (upper and lower greensand).

Greenwich mean time (GMT) The local time at Greenwich, England, located on the 0° meridian (Greenwich meridian). It is the basis on which the standard times for virtually all the countries are calculated, at hourly or half-hourly intervals fast or slow of GMT. One hour is represented by 15° longitude from the Greenwich meridian.

greywether An isolated surface block lying on the down lands of S. England and named from its resemblance to a sheep. Synonymous with Sarsen.

grid A network of parallel lines intersecting at right angles and producing a series of identical squares on a map. Grid lines are numbered both east and north of an origin, thereby giving each square and any point within that square a unique location of castings and northings. Such a location is known as a grid reference.

grid north The meridian of origin of a grid system. The directions of all the south-north grid lines on a map are coincident with the meridian of origin, but only the latter coincides with true north, the remainder diverging from it by different amounts.

grike Deep, tapering cleft, normally a solution widened joint, cut into the surface of a near level area of hard limestone. In northwestern Yorkshire, England, it may be 15–16 cm wide at the surface and commonly 0.5–3 m deep. The deepest grikes form where several joints intersect, and may develop into shallow caves.

grivation The angle between grid north and magnetic north.

groove A deep scratch cut across a smoothed rock surface by a boulder held in the base of an ice-sheet, or by melt water.

ground fog, ground mist A fog caused by long wave terrestrial radiation and confined to low-lying areas or valley floors. When the ground surface cools rapidly at night by excessive terrestrial radiation loss, it lowers the temperature of the lowest air layers, which then become denser and flow into topographic hollows owing to gravitational effects.

ground frost A term which signifies that the minimum temperature at the ground surface is below 0°C (32°F), although the air temperature may remain above freezing point.

ground information A remote sensing term referring to information derived from data on the physical state of the earth. It is obtained from sources other than primary remote-sensing data sources.

ground moraine A thick sheet of till forming on undulating surface of low relief and deposited when the morainic debris carried in the base of an ice-sheet or glacier is released during a phase of melting. It tends to obscure former solid rock features.

ground swell (1) Waves in deep water exhibiting great height and length. (2) Waves that, on entering water with a depth of less than half the wave length, decrease in length and increase in height as they are affected by the sea-bed.

groundwater All the water contained in the void space within rock. The term is generally taken to exclude vadose water (water travelling between the surface and the water table). Most groundwater derives from surface sources; the remainder is either introduced by magmatic processes or is connate water.

growler A small iceberg, generally awash and therefore a particular hazard to shipping.

growth fault A fault occurring as a primary, post-depositional or syn-depositional feature commonly associated with unconsolidated sediments recently deposited in a basin which is undergoing active subsidence. Growth faults are typically normal and movement occurs at the same time as sedimentation, so the sediment is usually thicker on the down-faulted side.

groyne, groin An artificial construction built out into the sea from a shoreline to maintain material on a beach, thereby checking long shore drift and inhibiting coastal erosion.

grumsol A modern US name for a black cotton soil or tropical black soil.

guano A substance rich in phosphates and nitrogenous material accumulating as a thick deposit of bird-droppings on certain islands and sea coasts of the world.

gulf (i) An inlet of the sea of large areal proportions, more indented than a bay and generally more enclosed, (ii) used to describe in some regions, an elongated steep-sided and flat-floored valley in a karstic terrain.

gulf stream A warm current of sea water, originating in the eastern Gulf of Mexico before flowing past Florida and the eastern seaboard of the USA following the edge of the continental shelf. It leaves the coast at about 40°N and flows north-eastwards as a weaker and broader North Atlantic drift towards the British Isles and Norway.

gull, gulling A fissure which opens up in the surface rocks owing to tension as a result of camber.

gully A small but deep channel or ravine formed by fluvial erosion but not permanently occupied by a stream.

gumbo N. American term for a fine-grained alkaline clay soil which becomes very sticky when wet but dries out brick hard.

gumbo til A weathered, leached, clay-rich till in which the clay has been deoxidised.

gunz One of the earliest of the Pleistocene glacials in Central Europe. Distinguished as the earliest phase of glaciofluvial outwash to the north of the Alps by A. Panck and E. Bruckner in 1909.

gust A short-lived but rapid increase above the average wind-shaped, in contrast to the momentary decrease below average wind speed known as lull.

gut A narrow river channel or strait prior to joining an open ocean or estuary.

Gutenberg channel A layer of plasticless rigid material occurring at depths between 100 and 200 km below the lithosphere and at the upper surface of the mantle. It is thought to coincide with the asthenosphere.

Gutenberg discontinuity One of the two major discontinuities in the structure of the earth. The Gutenberg discontinuity occurs at a depth of 2,900 km and separates the mantle from the core. It has a considerable effect on seismic waves, for at this boundary between a solid and a liquid S-waves cease to be transmitted and P-waves slow down from 16.6 km/sec to 8.1 km/sec.

guyot A flat-topped variety of a seamount occurring mainly in the Pacific ocean. Their summits are almost entirely at depths of more than 1000m but rise upto 3 km from the ocean floor. The conical shape of the guyot has suggested to some scientists a volcanic origin, with the table-like summit having been planed down by marine erosion and finally submerged by a rise of sea level. Their summits are covered by sediments of various ages and of different derivation, ranging from cretaceous sediments to volcanic materials of tertiary age.

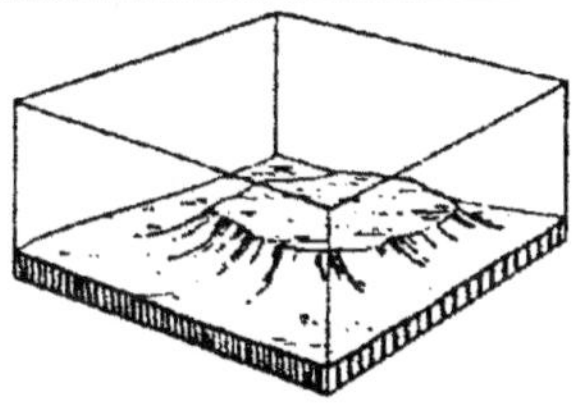

Guyot

gyre A term used in oceanography to describe the circulation of surface water in closed circulatory systems around the subtropical high-pressure system at latitudes 20°–30°N and S. In the S. Atlantic, for example, the gyre consists of a west-flowing south equatorial current, a southward-flowing Brazilian current, an eastward-following west wind drift and a north-flowing Banguela current. Double gyres occur in the S. Pacific and then Indian ocean.

gytija A Swedish term for a nutrient-rich peat or organic mud, consisting mainly of plankton.

H

haar A cold, coastal fog which drifts in from the North Sea onto the eastern coast of Scotland and N.E. England, especially in spring and early summer. Also called sea fret or sea roke. It results from the cooling effect of the North Sea on the lower layers of the atmosphere, which brings the moist air to saturation point and forms low stratus cloud.

haboob From the Arabic habb meaning 'to blow', a local term for a dust storm in northern Sudan. Typically, the storm is experienced late in the day during summer.

hachures Short lines drawn on a map to run in the direction of maximum slope, which indicate the relief by their thickness and spacing. Hachures give no information on absolute altitude. They are frequently employed to denote precipitous slopes when the contour-spacing becomes too close for clarity.

hade The angle between a fault plane, mineral vein, or lode and the vertical.

Hadley cell One of the most fundamental divisions of the global wind circulation comprising the net ascent of air over the lowest at latitudes due to convection, the compensatory downward motion in the subtropical anticyclone belt, and the resulting trade winds which blow towards the meteorological equator. It is named after G. Hadley who in 1735, seeking to account for the trade winds, proposed a single large-scale convective cell representing a thermally driven, low-latitude atmospheric circulation.

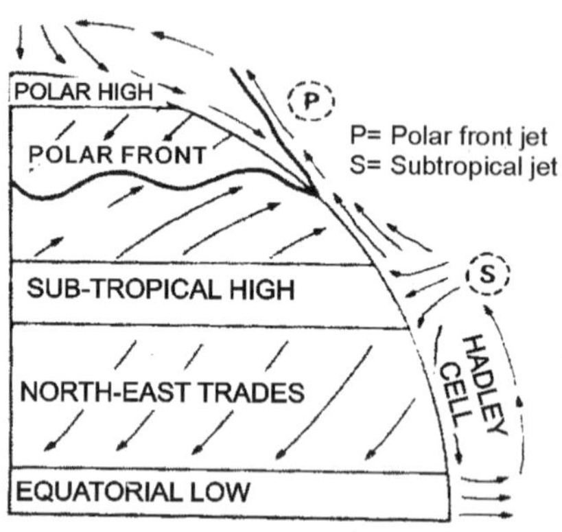

The Hadley cell and the atmospheric circulation of the northern hemisphere

haff A German term for a shallow coastal lagoon resulting from the growth of a spit across a bay or river mouth. The lagoon may be marine or brackish water with virtually no saline content. The haff is best exemplified along the southern coast of the Baltic Sea.

hag (hagg) An eroded cliff or residual of peat due to active dissection of a peat bog.

hail It is a form of precipitation comprising ice in the shape of balls or irregular particles, whose concentric structure indicates a growth by coalescence and freezing of super-cooled water drops. Hail is usually associated with cumulonimbus cloud.

hair hygrometer Instrument that indicates relative humidity, based on the expansion and contraction of a treated human hair.

hairpin A sudden change in a palaeomegnetic polar wander path, usually taken to indicate a tectonic collision between the tectonic plate concerned and some other plate.

haldenhag A German term to describe the rock-cut slope which occurs as a basal slope, generally covered with talus, below the steeper, rock face above. It is part of a scheme devised by W. Panck in 1924 to explain slope evolution.

Hale-Bopp A comet with an orbital period of more than 4000 years; perihelion date 31 March 1997; perihelion distance 0.914 AU.

half-dome A bare rock exposure, generally in coarsely crystalline rocks, rising as a prominent convex rock spheroid, often at the edge of an escapement, particularly in tropical latitudes. As a residual hill on the escarpment gets smaller by exfoliation the dome form extends between the dome face and the talus slope until its rock face meets the valley floor or plain at an abrupt angle. Since this change usually develops only on one side of the hill or escarpment, the resulting land form is a half-dome. The world's most famous example is Half Dome in Yosemite, USA but this has been truncated by glacial erosion.

half-time The length of time taken for one half of the atoms in a given amount of radioactive substance to decay into another isotope.

halo A ring of light surrounding the sun or moon at a time when the sky is partly obscured by cloud, especially cirrostratus. It results from a refraction of the light through ice-crystals within the cloud. Not to be confused with a corona, which is produced by diffraction.

haloclasty A process of rock weathering due largely to periodic wetting leading to the crystallisation of salt and the creation of stresses in the rock owing to swelling. It is one of the major weathering process in hot deserts.

hamada A rocky desert, or desert region, which does not have surfacial materials and which consists mainly of boulders and exposed bed rock. Two basic types occur: **stony hamada**, jaggedly developed across crystalline rocks; and **pebbly hamada**, cut across sedimentary material and mantled with bedrock fragments.

hanging valley Tributary valley whose floor is well above that of the adjacent main valley and where, there is therefore often a waterfall. It is typical of glaciated uplands, where it may result from glacial widening and deepening of the main valley.

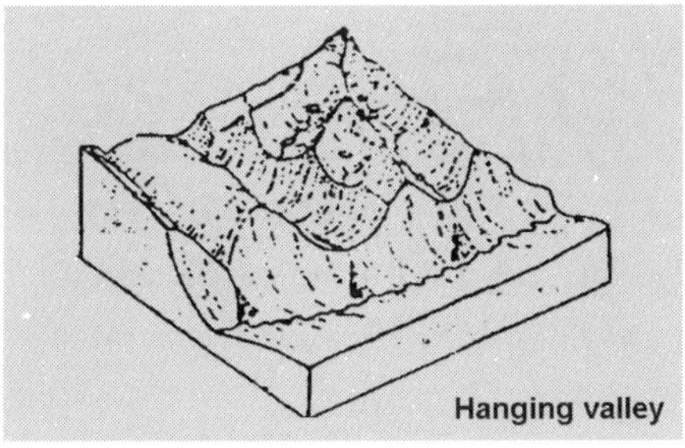

hanging wall The fault block which lies above any inclined fault surface.

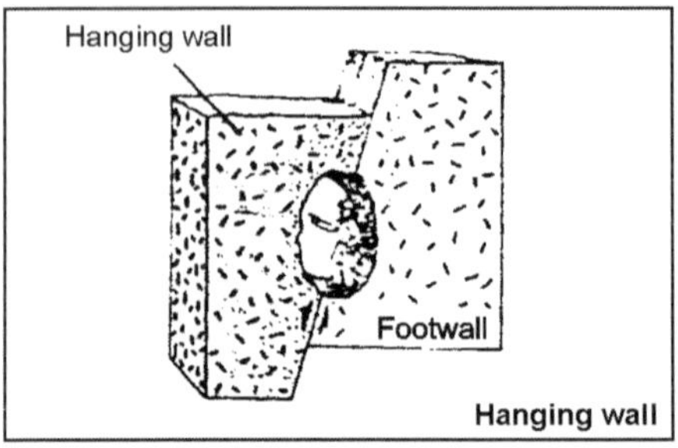

hard water Water which contains a high concentration of dissolved carbonates of calcium and magnesium derived from rocks through and over which the water has flowed.

hardpan Hardened soil horizon, usually found in the middle or lower parts of the profile, that may be indurated or cemented by a variety of possible cementing materials.

harmattan wind Dry, dusty, north-easterly or easterly wind which occurs in W. Africa north of the equator. Its effect extends from just north of the equator in January, almost to the northern tropic in July. In W. Africa is known as **the doctor** because of its invigorating dryness compared with humid tropical air. The harmattan wind stream occasionally extends south of the equator during the northern winter as an upper air wind over the south-westerly monsoon.

harmonic fold A fold which maintains its geometric form, integral wavelength and symmetry throughout a sequence of layers.

haugh A term used in N. Britain to describe a valley floor or floodplain.

Hawaiian 'high' One of the high pressure cells in the atmospheric circulation. Located in the N. Pacific this 'high' is more pronounced in the northern summer than in the northern winter.

Hawaiian eruption An eruption during which great quantities of extremely fluid basic lava flow out from a fissure or a central vent to form a typical shield volcano, exemplified by those in Hawaii (e.g., Maura Loa) the eruption is not usually explosive, so that Pyroclastic phenomena are rare.

haze A state of obscurity in the lower layers of the atmosphere owing to the presence of large numbers of condensation nuclei. There are tiny particles of dust, smoke or salt spray around which droplets of atmospheric moisture condense. It is officially classified as haze when visibility falls below 2 km but is more than 1 km. In summer, the shimmering caused by solar heating of the ground surface causes a heat haze.

head dune A variety of dune in which sand accumulates on the windward side of an obstacle in the small space of eddying air due to the uplifting effect of the obstacle on the flow of air.

head land (1) A cape or promontory jutting seawards from a coastline, usually with a significant sea cliff. (2) A strip of land left at the end of a furrow in a field in order to facilitate the turning of the plough. In England, it often became a field boundary after enclosure.

head waters The upper parts of a river drainage system; the source or sources of a river at the head of its catchment area.

headcut A term for a step in the bed of a river channel where a gentle gradient suddenly steepens down stream into a vertical drop.

headward erosion The action of a river in cutting back upstream, thereby lengthening its valley.

heat 'island' A term introduced to describe the zone of slightly increased air temperatures that occur in association with a large urban area.

heat balance It is an average condition of balance in the earth's atmosphere between incoming solar radiation and outgoing heat losses due to reflection or re-radiation from surface. Thus, heat balance prevents significant rises and falls in atmospheric temperatures.

heat wave A popular term to describe a spell of exceptionally hot weather. It is not specified in meteorological circles, either in terms of temperature or in terms of its duration.

Heaviside-Kennelly layer The lower part of the thermosphere, above the stratosphere, some 100–120 km above the earth's surface. It allows short radio-waves to penetrate to the Appleton layer, but reflects medium and long radio-waves back to earth. It is intensely ionised and is sometimes known as the E-layer.

heavy soil A soil having a high percentage of fine particles especially clay, making it difficult to cultivate.

helectite A type of stalactite in which the shape is abnormal, often in the form of a thin column reflected from the vertical by winds.

helicoidal flow A type of cork-screw flow superimposed on the primary downstream flow of a river. The flow moves across the water surface from the inside bank of a meander towards the outside bank where it descends to the channel bed before returning as a reverse flow towards the inside of the meander. The flow tends to erode meanders on the outside banks and cause deposition of material on the inside banks of the meander, resulting in a downstream migration of the meander.

heliotropism The action by which plants turn towards the light.

helm wind Local name for a type of lee wave frequent in winter and spring on the western (lee) slope of the cross fell rang in Cambria, England. Associate with this cold, gusty, north-easterly wind, the helm itself is a thick bank of cloud along the mountain range, together with a layer of narrow almost motionless cloud away from the range.

hemisphere Literally, half a sphere. Planet Earth is divided into two hemispheres by the equator. In a less precise usage, the continents of S. America and N., America are referred to as the western hemisphere.

Hercynian orogeny The mountain-building episode of carboniferous/Permian times. Its name is derived from the Harz mts. in S. Germany.

heterosphere The outer zone of the earth's atmosphere, distinguished from the underlying homosphere by virtue of its gaseous constituents which exhibit very low densities.

heterotrophic Refers to the inability of an organism to manufacture its own food. It must, therefore, obtain it from other organisms, living or dead, and it is said to depend, therefore, on external nourishment.

hiatus A gap in a stratigraphic sequence of rocks, where the missing strata either were never deposited or were destroyed by erosion prior to deposition of overlying strata.

hill An area of upland smaller than a mountain but with no specific definition of absolute elevation.

hinge fault A fault in which the amount of displacement increases from zero to a maximum amount as it traced along the strike.

hinge line (1) A line separating a region of the earth's crust undergoing uplift caused by isostatic readjustments and a region of relative stability. (2) A line of abrupt flexure in folded strata.

hinterland (1) A term given to the mobile crustal block which is reputed to move laterally towards a foreland, thereby crushing and folding the sediments in the intervening geosyncline. (2) The region lying behind a coastline.

hoar frost An ice deposit with a feathery and needle-like appearance which forms on moist ground, surface objects that have been cooled by terrestrial radiation on windless clear nights so that any condensation which occurs will freeze once the temperature below freezing point.

hog's-back ridge, hogback A steeply dipping cuesta, forming a narrow-crested ridge in which both slopes are of comparable steepness, e.g., the Hog's Back near Guildford, Surrey, England.

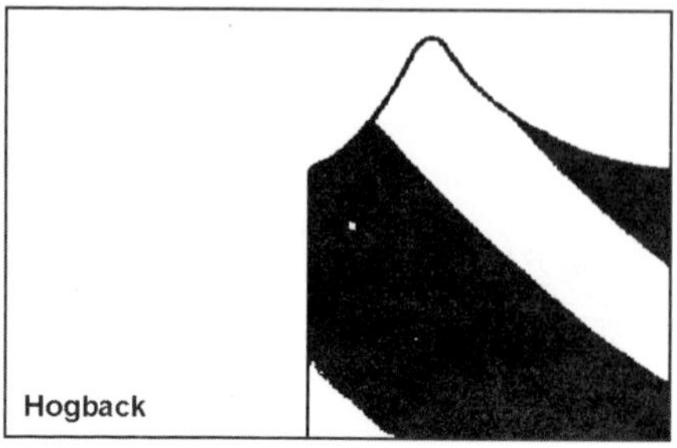

holm (holme) (1) A small island, e.g., Stockholm, Dyfed, Wales. (2) A floodplain in N. Britain.

holo karst A term introduced by J. Cvijic in 1918 to denote the total entity of karstic landforms in a region upon which the karstic cycle is said to operate.

Holocene A term meaning 'the whole of recent life' and referring to all of the time which has elapsed since the Pleistocene. It is one of the two epochs which make up the quaternary.

Holstein The interglacial which separated the Elster and Saale cold stages in the Pleistocene sequence of NW Europe.

homoclinal terrain Landforms created from rock strata which are tilted by the same degree of dip in the same direction, resulting in a type of scarp-and-vale topography, especially when referring to one of the eroded flanks of an anticlinal fold. Its use is most common in the USA, whereas in Britain the term uniclinal is preferred.

homoscismal line, homoseism A line connecting all points on a map that are affected simultaneously by a seismic shock. It is synonymous with isoseist.

homosphere The inner zone of the earth's atmosphere beneath the heterosphere and distinguished from it by the higher densities of its gaseous constituents. Its chemical composition is virtually uniform upto an altitude of 90 km, beyond which the heterosphere begins. The homosphere can be subdivided into the troposphere, stratosphere and mesosphere.

homotaxis A term referring to those rock strata which occupy the same position in the stratigraphic sequence although they occur in widely separated regions.

honey comb weathering A type of rock weathering in which the rock surface assumes the appearance of honey comb.

hoodoo An American term for a weirdly shaped pillar of solid rock in a semi-arid environment similar in form to an earth pillar. It is the result of undercutting the wind.

horizon (1) A planar surface separating two beds of rock in a sequence and assumed to be a time plane of no thickness. (2) A soil layer of any thickness generally running parallel to the ground surface and exhibiting a relatively uniform character in its material.

horn A German term for a pyramidal peak. It is a sharp peak formed at the intersection of the headwalls of three or more cirques.

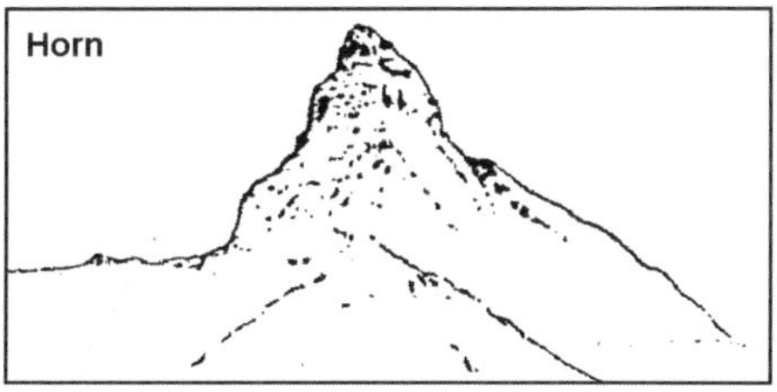

horn blende A mineral of the amphibolite group, common in both igneous and metamorphic rocks.

horn stone A fine-grained, compact volcanic ash often with irregular but indistinct layering.

horse latitudes The belts of high pressure, characterised by calm and light winds, which occur in the sub-tropics at 35°N to 30°N and at 35°S to 30°S. In the pattern of atmospheric circulation, they lie between the trade wind belts and the westerlies of higher latitudes. They are zones of mainly descending air at the junction of the Ferrel cell and the Hadley cell from which surface air masses move equatorwards and polewards as the trade winds and westerlies respectively the belts move N and S seasonally.

horst A fault-defined block which has been left upstanding by the sinking of the adjoining land along normal faults or has been uplifted along parallel faults.

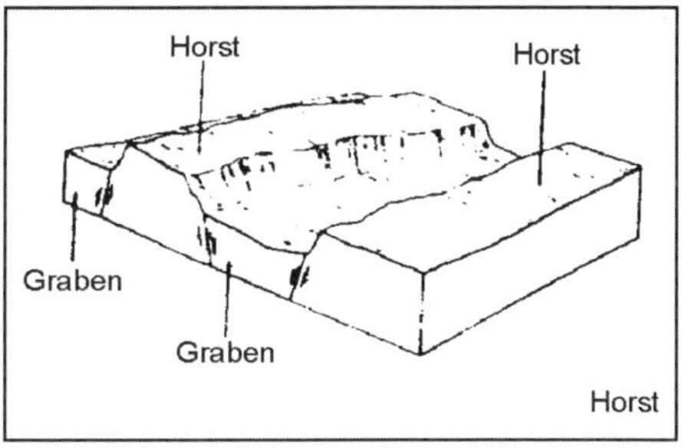

hot dry rock Rocks, usually granite which have abnormally high heat production as a result of the decay of radiogenic elements rather than merely residual heat.

hot spots Localities on the earth's surface where greater than average thermal activity takes place, often leading to volcanic outbursts.

hot spring A thermal spring with temperatures above 37°C, where water flows out of the ground in a continuous and non-explosive way, in contrast to the flow of a geyser.

hot tower A colloquial term used to describe the vertically rising current of unstable air which rises at the centre of a tropical cumulonimbus cloud owing to intense heating. It also refers to the primary energy cell of a tropical cyclone, where rising condensing air releases vast amounts of latent heat to fuel the storm.

how A Scandinavian-derived term for a low hill in N. England.

howe A Scottish term for a low place or depression.

Hoxnian (1) An interglacial period. (2) A series of temperature-climate deposits, named after Hoxne, Suffolk, England, with a characteristic vegetational sequence that occurs in tills of the earliest glacial stages, sometimes filling deep channels. They may be equivalent to the Holsteinian deposits of continental Europe. Sometimes during the Hoxnian interglacial, the sea rose to well above its present level

hum An isolated hill, left as a residual on the floor of a Polje, named from the village of Hun in the Yugoslavian.

humic layer, humose layer A layer of highly decomposed organic soil in which only a little fibrous material has survived.

humidity The state of the atmosphere in relation to its water vapour content, and normally referring to relative humidity unless otherwise stated.

humilis From the Latin humis meaning 'low' a species of shallow cumulus cloud which typically has a flattened appearance.

hummock (1) In general, a low mound of earth, rock or ice. (2) More specifically, a mound of earth in a Periglacial environment, having either a core of mineral soil or a core of stones, upto 70cm in height and upto 3 m in diameter. Active hummocks are closely related to the presence of permafrost.

hummocky moraine A type of moraine characterised by considerably undulating terrain owing to the presence of a large number of kettle holes and crevasse fillings. It is often thought to be the result of an ice-sheet stagnation *in situ*.

humus A black or dark-brown organic substance produced by the processes of mineralisation and humification within the soil.

huronian The stratigraphic name of part of the Pre-Cambrian in Canada, during which there is evidence of an ice age.

hurricane (1) The name given to a tropical cyclone that develops over the N. Atlantic and Caribbean. Hurricane move westward then swings north, on tracks that often carry them across inhabited islands and coastal areas of Mexico and the USA. (2) A wind blowing at more than 120 km/h which is force 12 on the Beaufort scale.

hydraulic action The force exerted by moving water on rocks without resource to the use of its load. The erosive force of fast flowing water is alone sufficient to remove loose material from river banks, while the air forced into cracks in solid rocks by breaking waves is capable of causing their disintegration by expanding the fissures.

hydraulic friction The resistance to the flow of a stream caused by roughness of the channel bed.

hydraulic gradient The slope of the water table. It can be expressed as the difference in water levels between two points divided by the length of the shortest flow between those points.

hydraulic radius An alternative expression for hydraulic mean depth expressed as the cross-section of the water flowing through a channel divided by the wetted perimeter of the channel. The greater the hydraulic radius, the larger the flow for a given channel and gradient.

hydraulic tidal current A local current created in a relatively narrow stretch of sea water to compensate for high-tide height differences resulting from contrasting times of high water at either end of the strait, e.g., Menai Strait, N. Wales.

hydric layer A layer of water in organic soils, extending from a depth of not less than 40 cm to more than 160 cm.

hydrogenic soil A soil formed in an environment of prolonged water logging, mainly in cold humid regions.

hydrograph A graph depicting the discharge of a stream as a function of time at a given station.

hydrolaccolith A type of frost-blister in the soil formed under permafrost conditions by the same mechanisms as those operative in Pingo formation.

hydrological balance budget A term referring to the relationship between evaporation (E), precipitation (P), run-off (R) and ground surface storage (ÄS) for a given station over a given period of time. It is expressed as P = E + R + ÄS, where ÄS may be positive or negative.

hydrological cycle The cycle of water movement through the earth atmosphere system initiated through the acquisition of water vapour by evaporation and transpiration from water and land surfaces released into the atmosphere by condensation and deposited on a land and water surface by precipitation. At the earth's surface, the precipitation is stored on the surface (lakes, glaciers) or at depth as groundwater, or is evaporated or transpired with the balance returned to sea through flow.

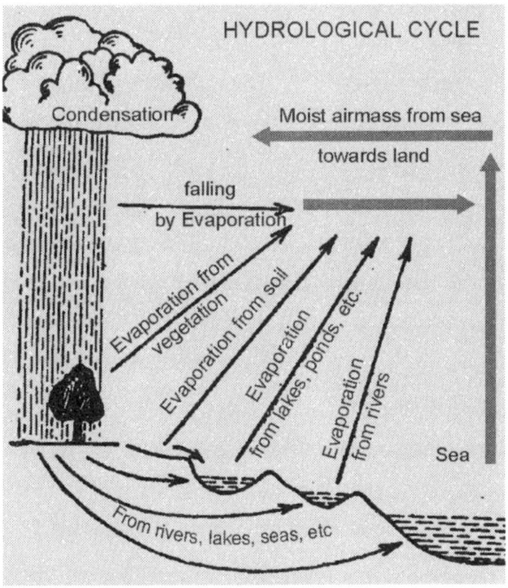

hydrolysis The main type of chemical weathering, in which water combines with rock minerals to form an insoluble precipitate and insoluble components, except in the special case of carbonation where only soluble products are created.

hydrosphere One of the three components of the earth's planetary system, namely atmosphere, lithosphere and hydrosphere, with the latter relating to the surface waters. The hydrosphere and the atmosphere are interlinked by the hydrological cycle.

hydrostatic equation An equation used by meteorologists to define the relationships which exist between altitude, gravity, air pressure and density. A solution to the equation produces a barometer-height formula.

hydrothermal activity Those processes associated with igneous activity which involve heated or superheated water.

hygrograph An instrument which records the changing relative humidity of the atmosphere. A continuous record is obtained on a graph affixed to a rotating drum, by means of a pen, which is linked to a hygrometer. The most common type is the hair hygrograph, which uses strands of human hair that increase in length with increasing relative humidity.

hygrometer An instrument which measures the relative humidity of the atmosphere at any point in time.

hygroscopic nuclei Particles of dust, smoke, sulphur dioxide, salts or similar microscopic substances occurring in the atmosphere around which condensation occurs. In general, because of their solubility, condensation commences on hygroscopic nuclei before the air is saturated. They range in size from a radius of 0.001 μm to 10 μm and are more prolific over continents than over oceans.

hygroscopic water Water that is adsorbed onto a surface from the atmosphere. It adheres in very thin films to soil particles and is unavailable for plant growth.

hypa byssal rocks The smaller-scale igneous intrusions, such as a sill or a dyke, which are intermediate in size between the large deep-seated plutonic rocks and the extrusive volcanic rocks. The term is rarely used today.

hypocentre The actual location, usually within the earth, of the first motion of an earthquake, i.e., the location of the focus of the earthquake. The point at the earth's surface overlying the hypocenter is the epicentre.

hypogene Mineral deposit formed by generally ascending solutions in or from below the earth's crust, or professes such as volcanicity operating within the crust.

hypothermal Applied to a mineral deposit originating at great depth and at temperatures between 300 and 500°C; below the mosothermal zone.

hypsithermal A term used in the USA to describe those warmer climatic conditions of the holocene which existed between about 9000 BP and 2600 BP. In Britain, it is sometimes referred to as the climatic optimum but is more specifically confined to the Atlantic period between 7500 BP and 5200 BP.

hypsographic curve Sometimes termed as hypsometric curve. A graph denoting the proportion of a land mass which stands above a given datum. These are plotted in terms of

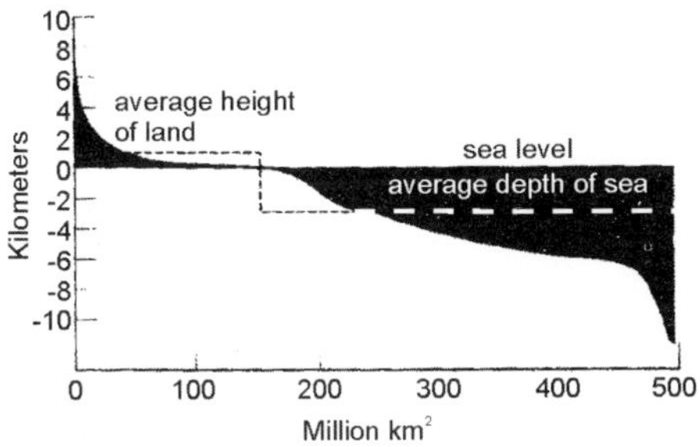

Hypsographic curve

percentage of total area or absolute areas on the horizontal axis and with altitude on the vertical axis.

hypsometre An instrument used in the determination of absolute height above sea level, which has to be used in conjunction with a set of statistical tables. Its uses have largely been superseded by the aneroid barometer. It is based on the differences exhibited by the boiling point of water according to the changes of atmospheric pressure with altitude.

hythergraph A type of climograph, in which two climatic elements (temperature/humidity or temperature/rainfall) are plotted in diagrammatic form to show broad relationships in relation to human behavioural responses.

I

ice Water in its solid form, produced by: (i) the normal freezing process of water, (ii) the sublimation process in which ice crystals are formed from water vapour when the air temperature falls below freezing point.

ice age Any period in the earth's history during which ice-sheets expand considerably and surface temperatures in the temperate latitudes are significantly lowered. The term was first used by K. Schimper in 1837.

ice apron A thin sheet of ice or snow which adheres to a mountain slope, often at a steep angle.

ice barchan A crescentic dune or barchan made not by sand but by wind-blown ice crystals at very low temperatures.

ice barrier (1) The outer edge of the Antarctic ice-sheet, namely the Ross Ice Barrier. (2) A type of ice dam blocking the melt waters of a glacier.

ice blink Whitish appearance seen above the horizon, caused by light reflection over an ice or pack-ice surface.

ice dome A main component of an ice sheet or ice-cap. It has a convex surface form of parabolic shape and tends to develop symmetrically over a land mass. Its thickness often exceeds 3000 m.

Ice edge The interface between a mass of floating sea-ice and the open water beyond.

ice field A nearly level field of ice, whose area may range from about 5 km^2 to near continental size, formed when the land surface is sufficiently high or uniform for ice to accumulate. It differs from an ice-cap in that it lacks a domed form and its flow is controlled by the underlying relief.

ice front Strictly speaking, the term ice front is applicable only to the vertical cliff, which forms the seaward limit of a floating glacier or an ice-shelf. The term is also used commonly to describe the distal limit of an ice-sheet on a land surface, although the term ice wall has been coined to describe the latter usage.

ice house period A time during which glaciers reached their maximum extent, ocean waters were well mixed and oxygenated and sea levels were low.

ice pan A large piece of flat, floating sea ice, rising less than 1m above the surface of the sea.

ice piedmont An ice-field located on a low-lying strip of land between the ocean and a range of coasted mountains. It frequently terminates in an ice shag.

ice plucking A process of block removal or quarrying carried out by glacier ice whereby well-joined blocks are mechanically pulled from the rock floor when the basal ice freezes to them generally on the downstream side of a rocky protuberance.

ice segregation The formation of lenses or bands of clear ice in frost susceptible materials, especially silt, partly by freezing of groundwater drawn upwards by capillary action. Such formation is conductive to uplift of the soil layers.

ice storm A period of heavy rainfall in which the rain freezes as soon as it comes into contact with an object the temperature of which is below freezing point.

ice stream, ice streaming That part of an ice sheet in which there occurs a corridor or stream of faster moving ice surrounded by large areas of slow-moving or stagnant ice.

ice wedge Tapering, vertically layered mass of ice, about a metre wide at the top and extending downward for some 3.7 m. It results from contraction cracking of the ground during extreme cold, followed by water penetration from the active layer and subsequent freezing. It is characteristic of uniform sediment, such as river alluvium, under periglacial conditions.

iceberg A mass of floating ice in a lake or in the sea, which has broken away from an ice-shelf or a glacier. Distinctions are sometimes made between Arctic icebergs, which are castellated and ragged in character and Antarctic icebergs, which are mainly flat-topped or tabular, having broken away from the edge of an ice barrier.

ice-cap Ice mass less than 50,000 km^2 in area, but large enough to submerge the underlying topography and whose flow behaviour is a consequence of its size and shape. It consists of a central ice dome together with outlet glaciers radiating from the periphery.

ice-contact slope The steeply slope of a kame or other glaciofluvial deposit interpreted as the slope against which the former ice body stood in juxtaposition.

ice-cored moraine Any morainic mound or ridge marginal to an ice-sheet or glacier which encloses buried ice. It is due either to a considerable accumulation of moraine in the glacier's terminal or lateral area, particularly in a state ablation or to the burial of ice avalanches or marginal snow banks by morainic debris.

ice-fall (1) Cataract of ice where a glacier moves down a steep slope gradient. It is characterised by numerous crevasses or ice-chasms, and by unstable ice pinnacles known as scracs. An example is the Khumb ice-fall beneath Everest. (2) The collapse of a mass of ice from an iceberg or a mountain ice-cap, which in the latter case leads to an ice avalanche.

ice-floe A discrete, tabular mass of ice floating on a water body; formed after the break up, usually by spring gales, of the winter fast ice. It is flatter and thinner than an iceberg and larger than an ice cake. An accumulation of a large number of individual floes represents the pack-ice.

ice-fog A surface fog associated with ice-sheets and ice-shelves where the lowest layers of the atmosphere are cooled by the ice surface sufficiently for minute ice crystals to form in the air. Here they will remain suspended only when calm conditions prevail.

ice-foot A narrow strip of sea-ice which forms along the shore in Arctic latitudes between high and low water marks and is unmoved by the tides.

Iceland low The region of the N. Atlantic in which the average value of atmospheric pressure is low, owing to the frequency of low-pressure systems moving into and occupying the region. Anyone of these systems, when present on an individual day, may be called an Iceland low, but the term is used mostly to describe the statistical or climatic feature.

ice-rafting The action by which rocks and other glacially-eroded debris are transported by floating ice, either ice-floes or icebergs. They may be transported great distance, eventually being deposited on beaches or being dropped into soft sediments on the sea-floor as the floating ice melts. The ice-rafted debris is known to be widely deposited on the sea-floor where the Arctic and Antarctic ice-floes meet the warmer currents of temperate latitudes.

ice-rind A thin crust of sea-ice formed by the initial freezing of the surface of a calm body of water. Sometimes termed slush or sludge.

ice-scoured lowland An area of low relief thought to have reached this condition largely owing to the erosive effects of an ice-sheet.

ice-sheet A large ice mass with an area usually greater than about 50,000 km^2 made up of ice domes and outlet glaciers. The Antarctic ice-sheet is the world's largest with an area of about $11.5 \times 10^6 km^2$ and a mean thickness of about 2000 m.

ice-shelf The outer part of an ice-cap or ice-sheet that extends into and over the sea. It typically ends in a cliff that may be 30 m high, and the total ice thickness may be 200 m. Ice wastage is by calving and by bottom melt.

ichor A term proposed in 1933 to describe the magnetic fluid which penetrates the country rock around a granitic intrusive body and leads to granitisation.

icicle A pendant of clear ice, generally in the form of a spike, formed by the freezing of dripping water.

igneous rock Rocks which have been formed by the crystallisation and solidification of magma. They include hypabyasal, plutonic, pyroclastic and volcanic rocks and are classified according to their feldspar character, grain size texture and chemical composition. The various forms of igneous intrusive rocks

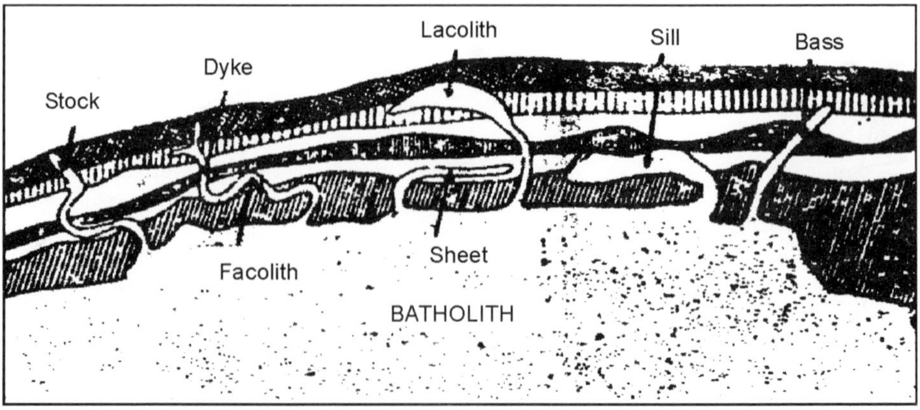

Various forms of Igneous Intrusive rocks.

are—**stock dyke, lacolith, sill, bass, facolith, sheet,** etc.

illumination, circle of A term referring to the great circle on the surface of the earth which distinguishes day from night.

illuvation The process of depositing soil substances and hydrous oxides of aluminium and in the illuvial horizon, usually from an upper horizon.

illuvial horizon A soil horizon into which material is precipitated or redeposited by the process of illuvation.

image enhancement A remote sensing term referring the process of improving the visual quality of a remotely sensed image by optical or digital means. For example, (i) contrast stretching involves the expansion of the original range of digital values in order to use the full contrast range of the recording film or display device, (ii) edger enhancement involves the emphasising of rapid spatial changes within an image.

imbricate structure (1) It is a geological structure associated with a thrust fault in which slices or wedges of crustal rocks are pushed one above another without folding. (2) A type of sedimentary fabric in a febbly rock in which the long axes of the pebbles overlap each other as they are inclined in the direction of the former current which deposited the sediment.

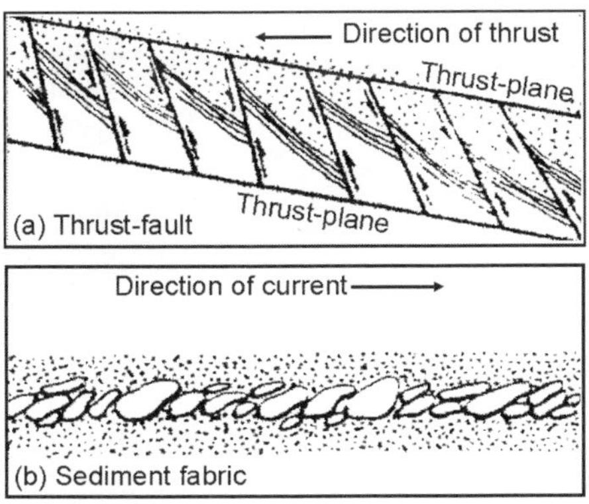

Imbricate structure

impeded drainage The restriction of the downward gravitational movement of water through the soil.

in-and-out channel An expression used to describe a type of marginal melt water channel developed in solid rock along lateral ice margins. It may have been formed by superimposition of a meandering superglacial or englacial stream which formerly wandered from the glacier onto a hillside and back again.

incised meander A general expression for river meander which is cut down deeply into bed rock owing to rejuvenation of the drainage. It includes both an ingrown meander and an entrenched meander.

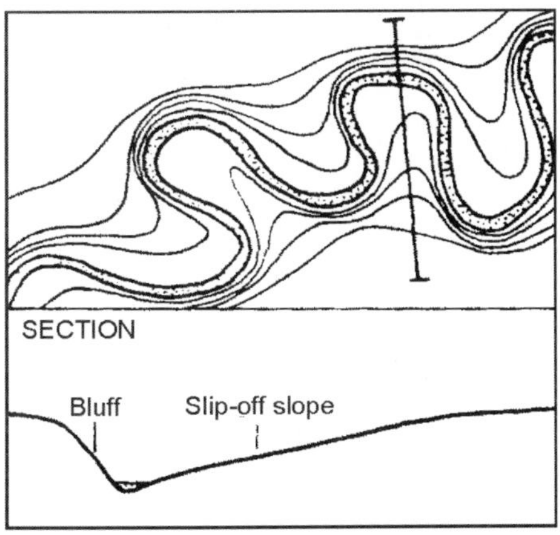

Incised meanders

inclination The angle between the horizontal and a magnetic vector conventionally, a vector with a magnetic north pole dipping below the horizontal is considered positive, and an upward vector is negative.

inclined fold A fold in which the angle of dip of the axial plane between 10° and 80° and the highest and lowest points on the fold surfaces do not necessarily coincide with the hinge points.

inclusion The enclosure of an older rock fragment within a younger rock or of crystals of one mineral within another.

incompetent bed A bed of rock which is insufficiently strong to resist stresses incurred during folding such that it deforms plastically rather than bends, thereby leading to flowage or the development of a balty cleavage. It contrasts with a competent bed.

inconsequent drainage A drainage network that shows no relationship with the underlying structure, i.e., it is not consequent upon it, because the drainage either has been superimposed or is antecedent. It is a type of discordant drainage.

incurs From the Latin incus meaning 'anvil' a supplementary cloud feature comprising the flattened, anvil-like shape of the top of cumulonimbus cloud.

index cycle A periodic development of a general wave motion in the atmosphere, resulting from the interaction between polar and tropical air. Over a six week period, especially in winter, large wave amplitudes develop associated with pronounced oscillations of the jet stream itself occurring along the deformed boundary of the cold polar air stream and the warm tropical air. The cycle terminates when cut off low pressure systems develop in the cold air. These destroy the waves and lead to a redevelopment of small amplitude waves (this state is termed a high zonal index).

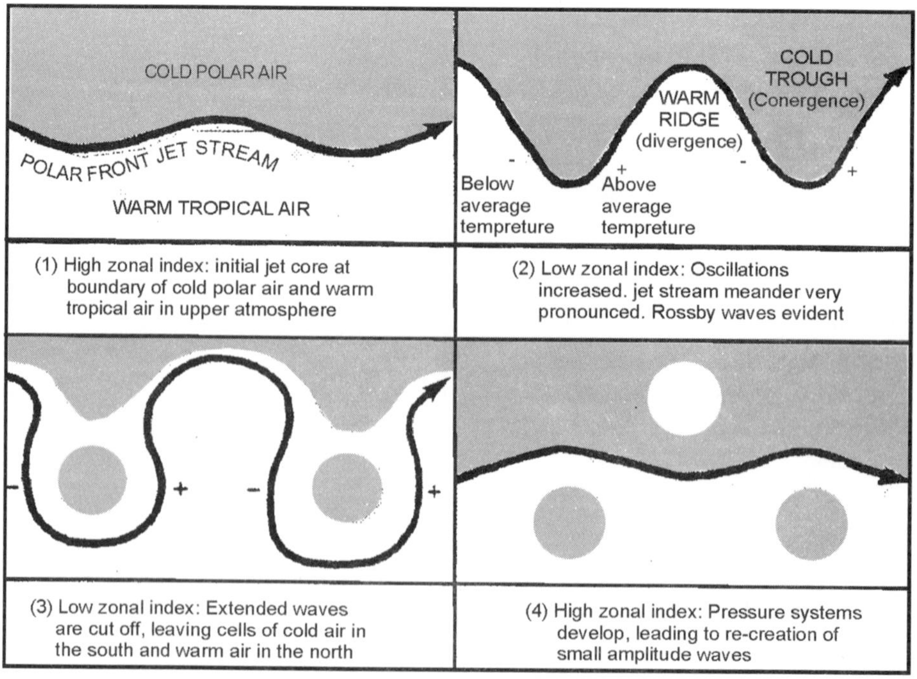

Index cycle

index fossil A fossil which is restricted to a particular assembly zone and whose name becomes indicative of a particular biostratigraphic zone.

Indian Ocean One of the world's major oceans, lying between Africa, India and Australia. It has a surface area of 77 million/km^2 and an average depth of 3872 m. The ocean receives a great deal of sediment from three of the world's major rivers (the Ganges, the Indus, and the Brahmaputra).

Indo-Australian plate One of the present-day major lithospheric plates, which is having new material added to its south and south-west along the Carlsberg Ridge and the south-east Indian Rise, but its other margins are the collision zone of the Himalayan orogenic belt, subduction zones or transform faults. It is thought this plate may now be breaking into separate plates along the line of the 90°E ridge.

induration Process of forming indurated horizons or hardpans that have a high bulk density and are hard or brittle cementing materials may be present and responsible for the induration.

inface The steeper slope of a cuesta used especially in a description of coastal lowlands in the eastern USA where the inface is on the landward side of a cuesta.

infilteration rate The maximum rate at which water can enter the soil under specified condition. It was formerly termed infiltration capacity.

inflexion point (1) The point of no curvature in a fold. (2) The point at which the curvature of a fold changes sense between one closure and the next.

influent stream (1) In general usage, a tributary stream. (2) More specifically, a stream from which water is lost by seepage to the water table that lies at a lower level than the stream channel. Such streams are characteristics of arid lands and of limestone and chalk terrain where the surface flow may be seen only in irregular stretches.

infrared radiation Electromagnetic radiation which has a wave-length between 0.7 μm and 100 μm.

ingrown meander A type of incised meander with asymmetric valley sides. A gentle slope on one flank is faced by a steep slope on the other, indicating that incision due to rejuvenation occurred more slowly than the development of the meanders so that lateral erosion of the river banks for a time kept pace with the vertical corrosion of the channel floor, in contrast to the formation of an intrenched meander.

inland sea A large body of saline or brackish water isolated from and not connected to the open ocean, e.g., the Caspian Sea.

inlet A short narrow opening or waterway which connects a smaller body of water with a larger body either of fresh water or sea water.

inlet cave A cave occurring below a swallow hole where surface drainage in a karst terrain first passes underground.

inlier An exposure of rocks which is completely surrounded by younger strata, resulting particularly from the erosion and breaching of an anticlinal structure, but which can also be formed by faulting. It contrasts with an outlier.

inner cover The probably solid centre of the earth's core, below a depth of 5,000 km.

inselberg A German term (literally, island mountain) that is widely adopted to describe a prominent steep-sided hill of solid rock, rising abruptly from a plain of low relief. It may have a narrow pediment at its base. It is best developed under a savannah climate.

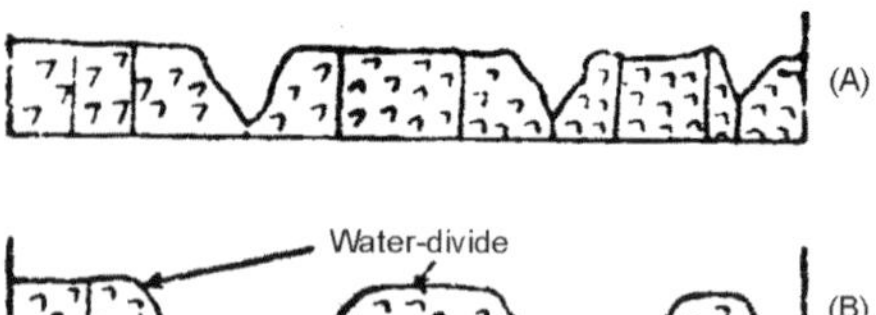

The development of inselberg

insequent drainage (1) A drainage pattern which wanders irregularly across the surface and is not influenced by topographic features. (2) In the USA, the usage is synonymous with inconsequent drainage.

insolation The amount of diffuse and direct solar radiation which reaches the earth's surface. It plays an important role in distinguishing different climatic zones, since its annual amount is 2.5 times greater at the equator than at the poles.

instability A physical condition of the atmosphere in which a body of air rises and expands because it is warmer and therefore less dense then the air immediately above it. As a result of this upward movement, the body of air will cool by adiabatic mechanisms.

instantaneous field of view (IFOV) In remote sensing, the angle through which a detector is receiving electromagnetic radiation. It is often expressed as a function of the group area visible at any one time, which is dependent on height and the angle of radiation reception.

instrumented catchments The catchment area of a stream in which a number of recording instruments have been set up in order to measure the hydrology of the stream.

insular shelf The zone of relatively shallow water surrounding a large island, analogous with a continental shelf.

intensity of an earthquake Measure of the effects of an earthquake at a particular place on the earth's surface. It is measured either on Mercalli scale or the Rossi-Forel scale. Its magnitude is measured on the Richter scale.

interbedded That which occurs between two layers or beds. A sandstone, for example, may be interbedded between two beds of shale or a lava may be interbedded between beds of sedimentary rocks. It is synonymous with intercalation.

interbolate moraine A term given to a terminal moraine in which two lobes merge together and curve back at right angles to the ice-frontal margin as a type of medial moraine.

intercepted moisture That part of the precipitation which is prevented from reaching the soil by any plant surface.

interfluve The area of higher ground separating two rivers which flow into the same drainage system.

interglacial The time period between two glacial stages, during which there is an amelioration of climate with temperatures in the temperate zone rising to about the level of those of the present day. In the quarternary ice age in Britain at least five interglacials have been recognised. These are (with increasing age)—the Ipswichian, the Hoxnian, the Cromerian, the Pastonian and the Bramertonian.

interlocking spurs The interdigitation of the valley slopes as the stream follows a winding course down a V-shaped valley whereby each projecting spur overlaps its neighbour and obscurer in view up and downstream. They are unlikely to survive the passages of a valley glacier, in which the spurs would be truncated and the valley converted to a U-shaped valley.

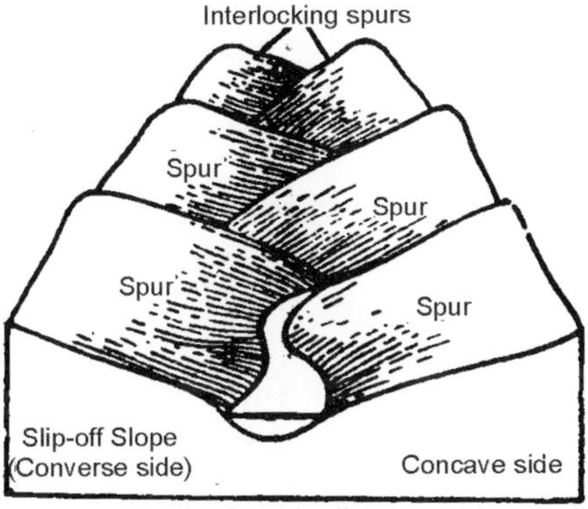

V-shaped valley with interlocking spurs

intermediate contour A contour that occurs between index contours.

intermittent rocks Any igneous rock which contains between 52 per cent and 66 per cent silica, by weight, e.g., diorite.

intermittent stream A stream which ceases to flow in very dry periods, such streams tend to have permeable beds and during periods of flow water leaking through their beds is added to the local groundwater. The chalk bournes of Southern England provide typical examples of intermittent streams.

international date line An internationally agreed time-change line drawn approximately along the 180° meridian but deviating to both W and

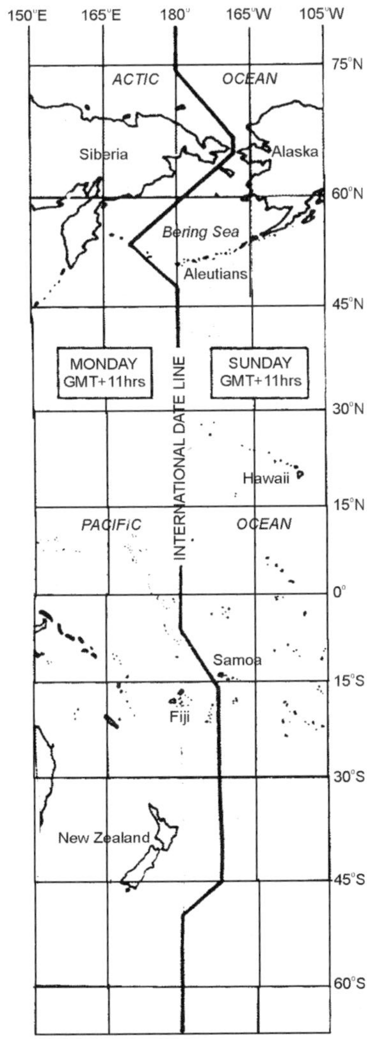

International date line

E to avoid land areas in the Pacific Ocean. A crossing of the date-line entails repeating one day when travelling eastwards and omitting one day when travelling westwards. This is to compensate for the accumulated time change of one hour for every 15° of longitude.

interplavial A stage of quaternary time in low latitudes during which rainfall is thought to have decreased from the maxima recorded during the pluvial stages. It was once believed that there was a direct correlation between the interglacial stages of higher latitudes and the interplavials of low latitudes, but this correlation is no longer accepted.

interrupted stream A stream which flows at some places but not at others, probably owing to the varying height of the water table at various points along the valley. It can be compared with an intermittent stream which flows at sometimes, but not at others.

interstadial A short phase of climatic amelioriation within a glacial stage. It is a shorter duration than an interglacial and may represent only a minor recession of an ice-sheet not a complete disappearance.

intertropical convergence zone (ITCZ) This is a low-latitude zone of convergence between air masses coming from either hemisphere at the boundary between north-easterly and south-easterly trade winds. Low-latitude depressions often form along the zone, which move latitudinally with the seasons, their occurrence being mainly in the ocean sectors and sometimes leading to tropical hurricanes when the zone is displaced relatively far from the equator. Over land, continental wind systems, e.g., the south-westerly monsoon and the hot-dry desert winds in Africa.

intertropical front Meeting point of air brought by the trade winds from the circulation of winds of the northern and southern hemispheres. It is not always a sharp front, but is always a convergence zone. The equatorial rains are associated with this convergence.

intrusion A body of igneous rock which has been injected in a molten state into existing crustal rocks and which on cooling becomes an igneous intrusive rock or igneous intrusion. Intrusions can occur along the bedding plans as **concordant intrusions** (sill) or across them as **discordant intrusions** (dike).

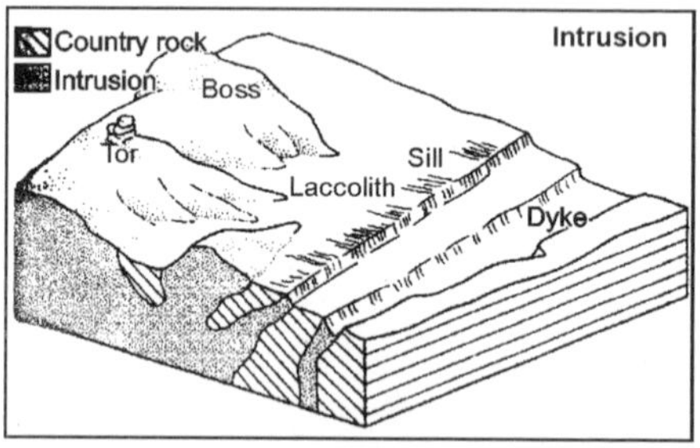

inversion of temperature Normally, air temperature decreases with increasing elevation but under certain weather conditions the converse may be true. Thus, contrary to the normal environmental lapse rate over a limited height range, air temperature increases with height so that a layer of warmer air overlies a colder air.

inverted relief A topography created by prolonged period of denudation in which anticlines from the valleys or low ground and the synclines coincide with the peaks or the high

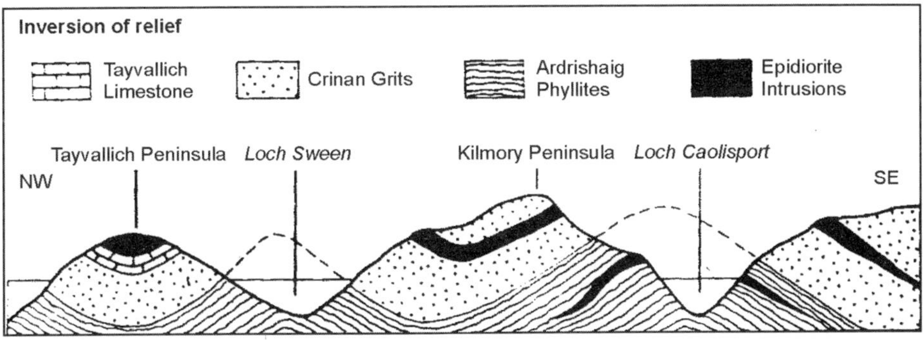

ground. This is because the tensional joints along the anticlinal crest are structurally less resistant to denudation that are the tightly compressed joints along the sunclinal axis. Thus, there is a tendency for the anticlines to be eroded more rapidly than the synclines.

ionosphere The part of the atmosphere that lies above 80 km altitude, with the height concentrations of ions and free electrons. The most intense concentration is at 100–300 km altitude. Long distance radio communications use waves that are reflected by certain regions of the ionosphere where there are particular concentrations of ions and free electrons. This allows radio waves to be transmitted around the curved surface of the earth. Communications satellites make it easier to transmit higher frequency waves (e.g., television transmissions) around the earth, but reflection from the ionosphere continues to be used for radio transmission, being cheaper.

iridescent clouds High clouds, as usually observed within 30° of the sun with red, green yellow, blue, or violet tints along their edges. Tiny cloud particles cause diffraction of light rays, producing the coloured effect.

isallobar A line on a weather chart connecting places where the same atmospheric pressure changes have taken place over a given time period. It is used to give the forecaster a picture of the pressure tendency, i.e., how pressure systems, especially depressions are moving and developing.

isanomalous line A line on a map connecting places which exhibit an equal deviation or difference from the mean or normal of any climate element; thus a line of equal anomaly.

isentropic line A line drawn to join up points having equal potential temperature.

island An area of land completely surrounded by water. Greenland (2.2 million km^2) is the largest island that is not a continent. Other big islands of the world are: New Guinea (8,21,030 km^2), Borneo (7,44,360 km^2), Madagascar (5,87,040 km^2), Baffin Is. (5,08,000 km^2), Sumatra (4,73,600 km^2), Honshu (2,30,500 km^2), Great Britain (2,29,880 km^2), etc.

island arc A lengthy chain, usually curvilinear, of oceanic islands which frequently exhibit a fore deep or ocean trench on the convex side. Many island arcs are volcanic and are generally the locating of severe seismic activity, related to their position along plate margins in the subduction zone.

isobar A line on a weather chart joining places with equal barometric pressure.

isobare A line drawn on a map joining up places having the same amount of uplift or subsidence of land during a specified time period.

isobath A line joining points on the sea bed or on a lake bed, situated at an equal vertical distance beneath the surface.

isobathy therm A line joining points having the same temperature at a given depth below sea-level, or below the ground surface.

isobront A line on a map connecting places which have thunderstorms at the same time.

isoceraunic line A line joining places having the same percentage frequency of days per annum on which thunder is heard.

isocheim A line on a map connecting places which have the same mean winter temperature.

isochional line (1) A line joining points on the same altitude of snow limit. (2) A line joining places having the same frequency of snowfall, duration of snow cover or depth of snow.

isochronous surface A surface within a body of sediments equivalent to a time plane.

isoclinal fold A fold of any type in which the limbs have parallel dips.

isofrequency line A line joining places having the same number of tremors following an earthquake.

isogonic line A line drawn on a map joining points of compass declination.

isograd Line joining points where rocks exhibit the same grade of metamorphism, and therefore areas of the crust which experienced equal pressures and temperature during metamorphism.

isohaline A line joining points in the ocean having the same degree of salinity.

isohel A line on a map joining places which have the same amount of recorded sunshine hours.

isohyet A line on a map joining places of equal rainfall amount. It can be used for any time interval from a single period upto mean annual rainfall totals.

isohypsometric line A line joining points in the crust which exhibit the same height of groundwater table.

isokinetic A line on a map joining places which exhibit equal degrees of wind velocity. Also referred to as an isotach.

isomer A line on a map joining places which have the same mean monthly rainfall expressed as a percentage of the mean rainfall.

isoneph A line on a map connecting places of equal cloudiness.

isoniph A line on a map connecting places of equal snow depth.

isopach A geological term referring to points having the same thickness of a rock stratum.

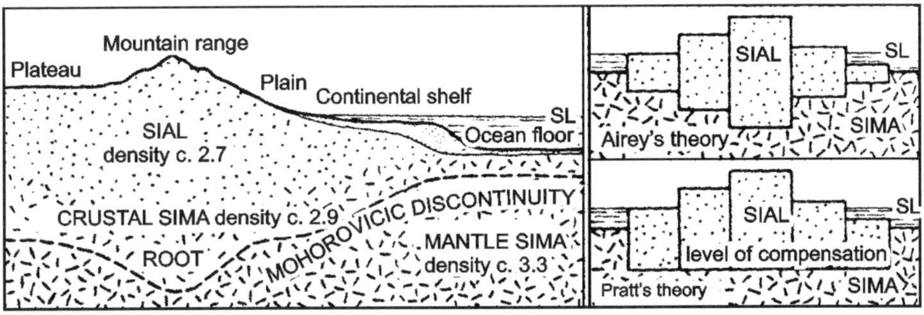

Isostasy

isopleth A line drawn on a map connecting points assumed to be of equal value.

isopycnic A line joining up points which exhibit the same density on certain meteorological and oceanographic charts.

isoryme A line drawn on a map joining places of equal frost intensity.

isoseismal line A line joining places having identical degrees of earthquake intensity during a specific earthquake.

isoseist A line joining places at which the earthquake-wave arrival time from a particular epicentre is identical. It is synonymous with homoseismal line.

isostasy The state of balance which the earth's crust tends to maintain or to return to by isostatic compensation if anything occurs to upset that balance.

isosteric surface A surface in the atmosphere which exhibits constant density.

isotachyte A line drawn on a map of a glacier joining points which show the same rate of movement.

isothere A line on a map connecting places which exhibit the same mean summer temperature.

isothermal layer (1) Formerly, a term used for the stratosphere but this use has now been abandoned. (2) Modern usage refers to any vertical section of the atmosphere in which temperature remains constant with height, i.e., exhibits a zero lapse-rate.

isothmus A narrow tract of land separating two bodies of water and connecting two larger bodies of land, e.g., the Isthmus of Panama.

isotropic Having the same properties in all directions. If the properties are unequal the term anisotropic is used.

iteration A repetitive process of successive approximations in a procedure. It is used in numerical solving of complex equations.

J

jebel An Arabic name for a mountain or hill in a hot desert environment.

jet A hard, black and lustrous form of lignite, found in isolated masses within organic rich shales.

jet stream A narrow belt of high-altitude (above 12000 m) westerly winds in the troposphere. Their speed varies from a mean of 110 km/h in summer to about 184 km/h in winter, although velocities of over 370 km/h have been recorded occasionally. A number of separate streams have been identified: (i) the most constant that of the subtropics (between 30° and 20°N and S of the equator); (ii) that associated with the polar front of mid-latitudes, which is greatly variable in its location, partly because of the transitory thermal contrasts of the depressions of these latitudes; (iii) an Arctic jet stream; (iv) a polar-night jet stream above the troposphere. Surface depressions and anticyclones occur in zones which exhibit large surface-temperature gradients. These differences in turn cause marked thermal contrasts in the troposphere, thereby creating divergence and convergence of air streams in the tropospheric circulation.

Easterly tropical jet stream

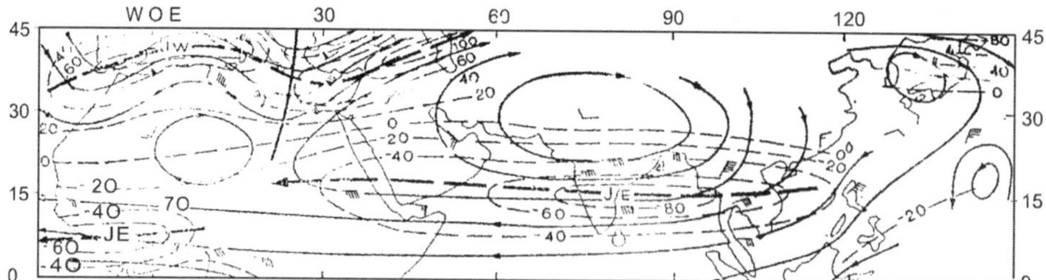

The location of the easterly jet streams at 200 mb. Streamlines are shown in solid lines and isotachs (wind speed) dashed. Wind speeds are given in knots (westerly components positive, easterly negative).

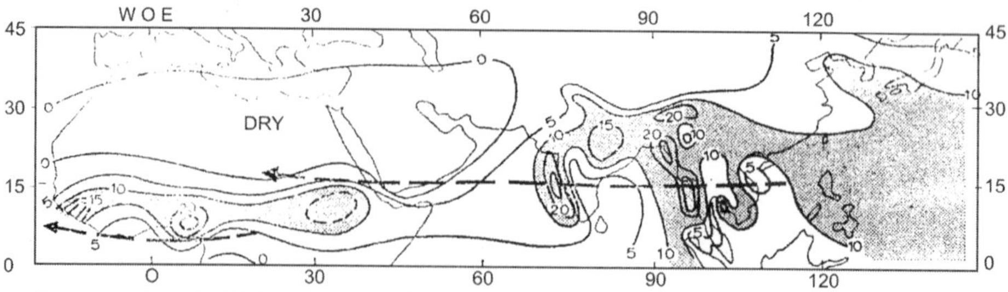

The average July rainfall (shaded areas receive more than 10 inches or 25 cm) in relation to the location of the easterly jet streams.

joint A fracture in a rock but one which exhibits no differential movement, in contrast to a fault. Joints can be formed in three ways: (1) By folding and thrusting. (2) By shrinkage due either to cooling of an igneous rock or to drying out of a sedimentary rock. (3) By unloading of a rock mass after the removal of its cover rocks, giving rise to sheet joints. The types of joints include—**longitudinal joints** parallel to the strike and the axis of folding; **cross joints** parallel with dip and at right angles to the fall; and **oblique joints** at approximately 45° to the direction of tectonic movement.

joint-block removal The process of loosening and transporting of jointed blocks of bedrock by the action of a glacier. It is a significant mechanism operating in the steepening of a cirque's back wall and in the creation of the steeper end of a roche moutonnee.

jokull An Icelandic term for a small ice-cap.

jotian A stage of the middle-upper Proterozoic, from about 1600 to 650 Ma ago, of the Baltic shield region, underlain by the Gothian and overlain by the Varegian.

jotnian orogeny An episode of mountain building that involved the proterozoic sedimentary rocks of the Baltic Shield in what is now Scandinavia.

Juan de Fuca Plate The small lithospheric plate in the northeast Pacific ocean which is being subducted slowly under the North American Plate, giving rise to the generally andesitic volcanic chain from northern California to southern British Columbia. The Juan de Fuca Plate is a remnant of the Farallon Plate. The Juan de Fuca ridge is offset from the East Pacific Rise by the San Andreas Fault.

junction (1) The contact-plane between two adjacent rock bodies. It may be completely conformable or in the character of an unconformity. (2) The meeting point of two streams.

Jurassic One of the three Mesozoic periods; it lasted from 208 to 145.6 Ma, following the Triassic and preceding the Cretaceous.

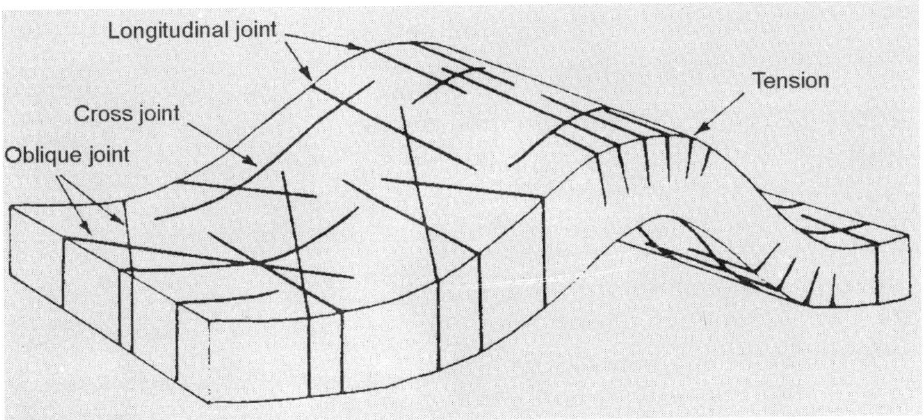

Types of joint

juvenila The term applied to volcanic material derived directly from a magma.

juvenile water Original water, formed as a result of magmatic processes. Juvenile water has never been in the atmosphere. Magmatic water can form in very large quantities. A magma body with a density of 2.5, an assumed water content of 5 per cent by weight, a thickness of 1 km, and an area of 10 km^3 contains some 1.25×10^9 m^3 of water.

K

kame Steep-sided mound composed of bedded sand and gravel which often shows signs of marginal slumping. It is a land form of glacial deposition, associated with stagnant ice whose removal by melting causes the collapse.

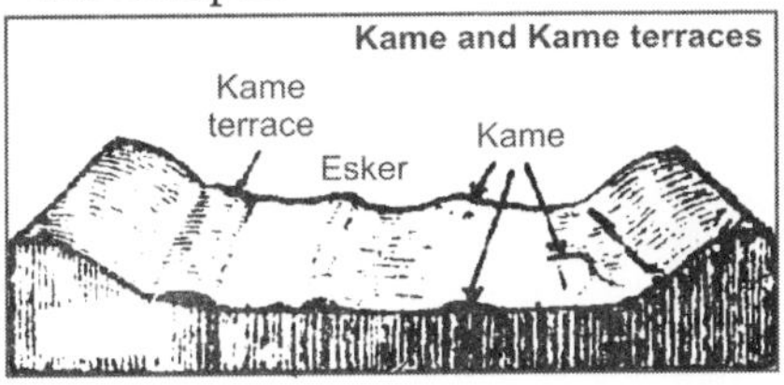

kame delta Flat-topped mound of stratified sand and gravel laid down in standing water at an ice margin. Subsequent ice melt leading to loss of support brings about collapse at the ice-contact margin.

kame terrace Continuous valley-side land form consisting of stratified sand and gravel whose outer edge typically shows collapse features. It is laid down by melt water at the junction between an ice mass and the valley wall.

kame-and-kettle terrain An undulating landscape composed of groups of kames and/or kame terraces interspersed or pitted with kettle holes. This type of land form is sometimes termed a kame complex and is created when glacio-fluvial sediments are lowered on to the sub-ice surface as the glacier or ice-sheet decays, especially when it can be established that the kame sediments were formed in separate basins of accumulation on the ice surface.

karren German term describing the group of solutional features developed at the surface of an outcrop of hard limestone. Forms range from runnels (shallow troughs) a few millimetres deep to fissures extending several metres into the limestone. The word is sometimes used as a synonym of lapies.

karrenohern A German term for a rounded solution hollow formed at the intersection of two grikes (Kluft Karren) often upto 5 m in depth. It is synonymous with lapies well.

karst (1) Area of the Dinaric Alps of Yugoslavia underlain by limestone. (2) Any region underlain by limestone and characterised by a set of land forms resulting largely from the action of carbonation. (3) By exten-

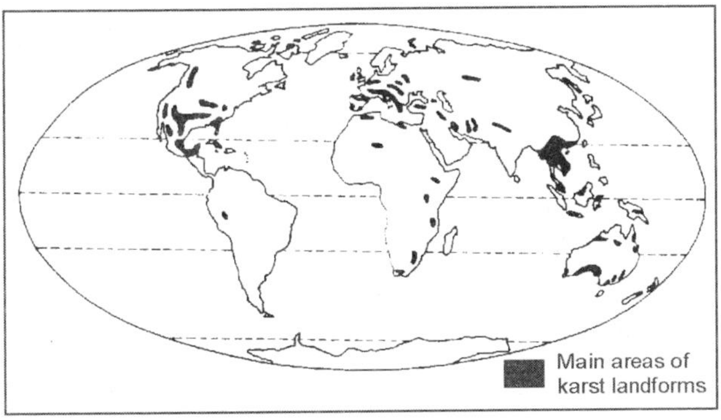

World distribution of karst landforms

sion, regions in which other processes produce similar types of land forms, e.g., thermokarst in the periglacial environment.

karst topography A landscape characterised by sinks, solution valleys, caves and other features produced by groundwater activity, it commonly develops in limestone.

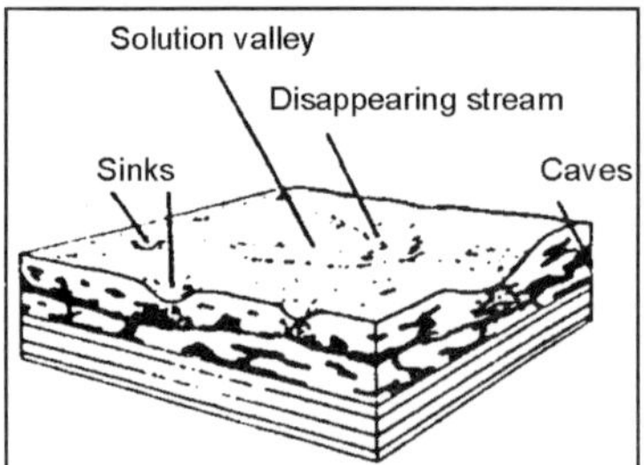

Karst Topography

karst water Water which issues from karst springs and whose chemical composition indicates that it has passed across and through kart limestone.

karst window That part of an underground river in a karst area revealed by the collapse of a section of the roof, e.g., cradle hold in the marble arch area of Ireland.

katabatic wind (Stroph) A downhill wind, usually cold, which blows down valleys at night and outwards from large ice-caps, such as Antarctica and Greenland. It is caused by ground surface cooling (often nocturnally) as

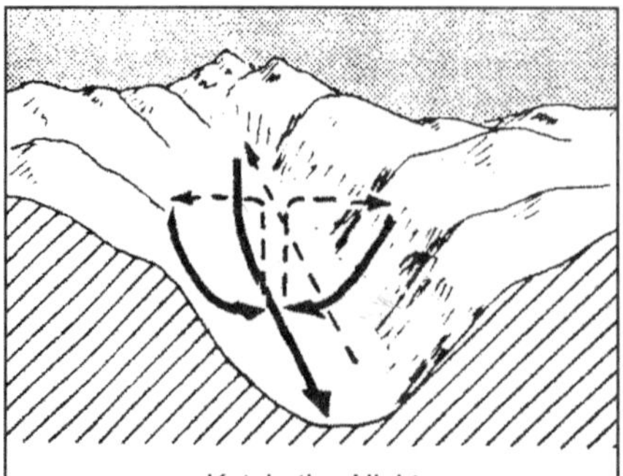

Katabatic ; Night

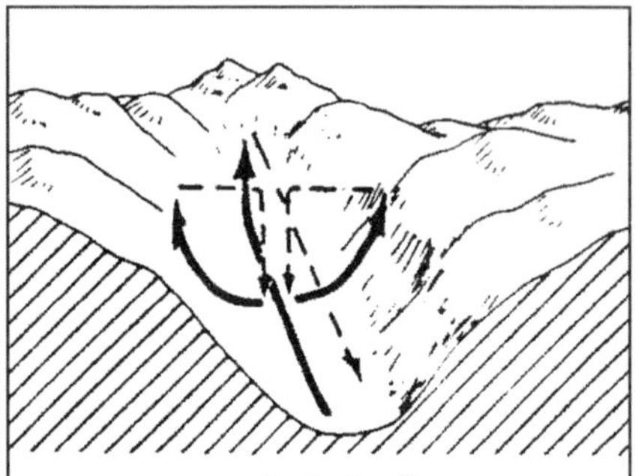

Anabatic : Day

Katabatic and Anabatic winds

a result of radiation, which in turn cools the lower air layers. The falling temperature increases the density of the air which then moves downhill by gravity flow.

katafront A depressed frontal zone in which the warm air is given very little opportunity to rise over the cold air because of the prevailing subsidence at high level. It is often associated with the perimeter of a depression.

kegel karst A type of tropical karst terrain in which the characteristic

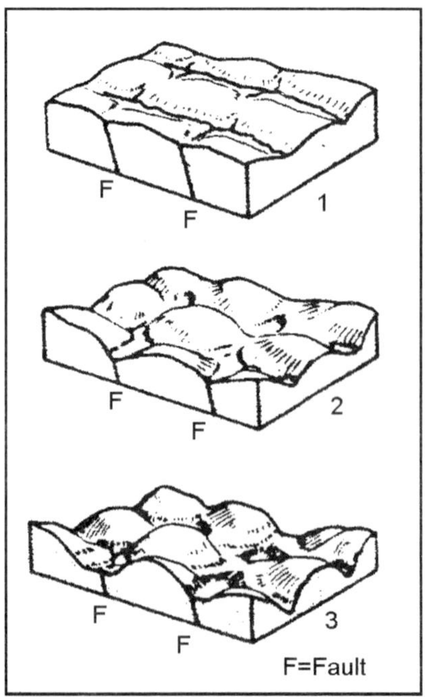

Kegel karst

land forms are numerous cone like hills, sometimes separated by cockpits, which are enclosed depressions interconnected by sinous corridors between the cones. The cockpits generally occur in lines following joint or fault patterns. Key karst is well-developed in Indonesia, Jamaica, Puerto Rico, Parts of Central America, Vietnam, S. China and Malaysia.

kelvin wave An oceanographic term which related to a tidal system in which the tidal range of a rectangular sea area increases to the right of the direction of the progressive wave and decreases to the left. This is exemplified in the English Channel where the north coast of France exhibits a tidal range of 11–13 m while that of the south coast of England has no more than 4 m.

kennlose winter A term to describe the singular temperature conditions which characterise the long winter night of Antarctica. Although temperatures decrease as winter approaches, it has been found that at the end of the winter they are not very much lower than they were at its commencement. This is thought to be the result of the balance achieved between surface radiational cooling and atmospheric counter-radiation from above the strong surface temperature inversion.

kettle hole A depression in the surface of glacial drift (especially ablation or kettle moraine), resulting from the melting of an included ice mass. It may be filled with water to form a small lake (kettle lake).

kettle moraine A morainic area characterised by an extremely undulating terrain of kames and kettle holes.

kettled sandur (pitted sadur) An outwash plain of glaciofluvial deposits the proximal parts of which are pitted with kettle holes. The sandur was created by ice melt streams flowing in superglacial, englacial or subglacial locations.

khamsin Hot, dry, dusty wind originating in N. Africa and blowing across Egypt, usually between April and June, ahead of depressions which move eastward or north-eastward in the Mediterranean Sea or across N. Africa, with high pressure to the east. The term is also applied to very strong southerly or south-westerly winds in the Red Sea.

kilometre A metric length of 1000 metres, equivalent to 0.621 of a mile.

klint (1) A hill or ridge formed from a reef-limestone that has proved more resistant than the surrounding rocks. (2) A Scandinavian term for a steep sea cliff in the Baltic, especially in Denmark and Sweden.

klippe (klippen) A German term referring to an isolated overthrust mass of folded rocks, usually a nappe, cut off from the main fold structure by erosion. Best examples can be seen in the Pre-Alps of Switzerland but, in Britain some of the isolated mountains in Assynt, Sutherland and N.W. Scotland, are klippen.

kloof An Afrikaans term for a rocky gorge or a mountain pass.

knee fold A type of gravity-collapse structure resulting from the slipping of competent rocks after the erosion of incompetent beds.

knick Term introduced in geomorphology by L.C. King to describe the sharp freak of slope between the inselberg and its surrounding pediment. It may be the result of chemical weathering.

knick point Abrupt change of gradient in the generally smooth long profile of a stream, and typically separating two concave up segments. It is often attributed to a fall in base level: this, it is said, initiates a knick-point which then slowly travels upstream. It may alternatively be due to a change in rock type or load size or to tributary entry.

knob and kettle (say and swell topography) The landscape sometimes found on a recent terminal moraine complex and consisting of a hummocky mound (the 'knob') alternating with a depression (the kettle).

knock-and-lochan A type of landscape, found particularly in NW Scotland, resulting from the erosion of a relatively low-lying area of hard rock by an ice sheet.

knoll a small rounded hill (Koppen's climatic classification) A system of climatic classification based upon temperatures and precipitation linked to vegetation zones. It was devised by German biologist, W. Koppen in 1918 and revised many times, especially by Koppen and R. Geiger. Five major categories subdivide the earth's climates: **tropical forest**, **dry**, **warm temperate rainy**, **cold forest** and **polar**. Further, subdivisions are made on the basis of rainfall regime, temperature characteristics and other special features.

koppie (also kopji) A small rocky hill in South Africa.

Kuroshio A warm ocean current which fed by the north equatorial current runs from the Philippines to Japan.

L

Labrador current A cold current found in the North Atlantic Ocean.

laccolith An igneous intrusion, forming a small hill on the surface. Laccoliths are formed by magma. It forms a pod-shaped or lens-shaped body with a generally horizontal floor.

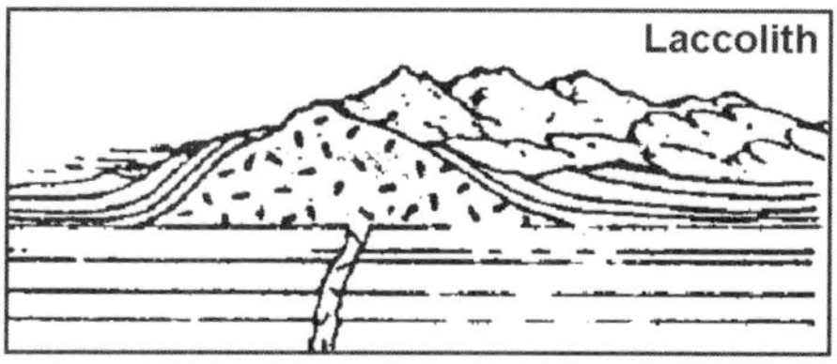

lacustrine are the deposits which have been laid down on the bed of a lake.

lagoon A stretch of shallow water partially or completely cut off from the sea. Lagoons are often found behind spits, bars and coral reefs.

lahar (mud flow) A mass movement feature on the flank of a volcanic cone. The volcanic ash may move either as a mudslide when saturated with water or as a dry landslide as a result of earth tremors.

lake An enclosed body of water, usually but not necessarily fresh water, from which the sea is excluded.

lake breeze A local breeze blowing during the afternoon off a large enclosed body of water during periods of calm weather.

Lamb, Hubert Horace (1913–97) A British climatologist, who was among the first scientists to draw attention to variability of climates and their effects.

Lamb's dust veil index An index, devised by H.H. Lamb, of the amount of finally divided material suspended in the atmosphere after great volcanic eruptions.

land The solid part of the earth's surface/a specified geographical tract of the earth's surface comprising its geology, topography, hydrology, soils, flora and fauna, etc.

land and sea breezes Circulation of air common along coasts, caused by a low-level pressure gradient due to the differential heating of land and sea. The land cools down more quickly at night than the sea. Air starts to rise from relatively low pressure of the

Land and Sea Breezes

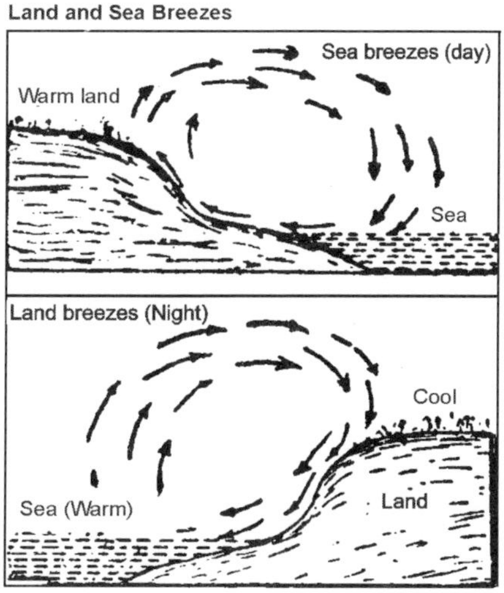

warm sea, and cool air drifts out from land to sea until the sun has warmed up the land in the morning. During the day, a reverse flow, or sea breeze will develop.

land-bridge An isthmus or other connection between two land masses across which animals and plants move to colonise a new environment. The Bering land bridge linking Alaska and Siberia across the Bering strait is the best example.

landform Any feature of earth's surface having a distinct shape and origin. Landforms include major features (such as continents, ocean basins, plains, plateaus, and mountain ranges) and minor features (such as hills, valleys, slopes, dunes, drumlins, etc.). Collectively, the landforms of earth constitute the entire surface configuration of the planate. The study of landforms is called **geomorphology.**

landsat One of the unmanned, earth orbiting, NASA series of earth resource satellites. The satellite records images of the earth's surface that are relayed to ground receiving stations located around the world.

landscape The secondary, including the geomorphological features as well as the man-made features.

landscape ecology A term introduced by German geographer Carl Troll who also later used the term geo-ecology. It is the science, which investigates the relationships between the biosphere and authoposphere and either the earth's surface or the abiotic components.

landslide The downslope gravitational movement of a body of rock. It may be induced by natural agencies, e.g., heavy rain, earthquake or it may be caused by human interference with the slope stability. Landslide includes **slump**, **planar slide**, **wedge failure** and **block slide**.

langley A unit of energy used as a measure of solar radiation, named after S.P. Langley (1942). It is equivalent to one calorie per square centimetre per minute.

lapies A feature of limestone topography, consisting of shallow, straight

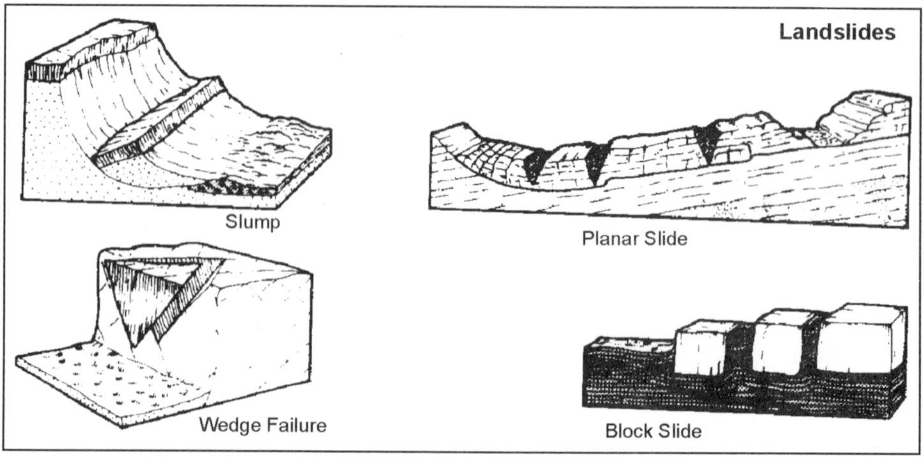

Landslides

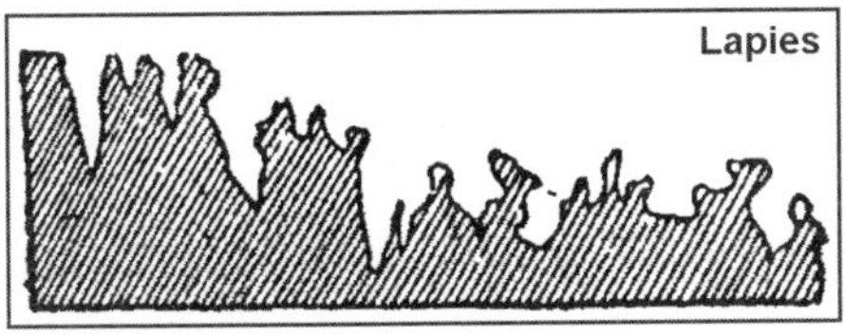

grooves incised by a solution into a sloping surface of limestone.

Laplace, Pierre Simon, Marquis de (1796–1827) A French physicist, best known for his nebular hypothesis of the formation of the solar system.

lapse rate The rate of temperature change in the atmosphere with height, sometimes referred to as the vertical temperature gradient. The average lapse rate is about 0.6°C per 100 m. (1.08°F per 330 ft), and this is called the environmental lapse rate. The rate will vary according to the water vapour content of the atmosphere. If dry air rising, it will cool more quickly, at the dry **adiabatic lapse rate** (about 1°C per 100 m). If the air is wet, the cooling rate will be slower, at the **saturated adiabatic lapse rate** (about 0.5°C per 100 m).

lateral dune A small dune alongside a large dune in a dessert topography.

lateral erosion The action performed by a river when it erodes banks in a horizontal direction. Such action is most common in a meander zone, where river impinges more strongly on the outside of a meander, thereby undercutting the banks and causing bank-caving and repeated action resulted in channel migration.

lateral moraine A linear pile of rocks deposited by a glacier. As the glacier moved, rock fragments were pushed out the sides where they left ice melted.

laterite A reddish hard-baked soil which is found in tropical parts of the world. It results when hot wet conditions wash away most of the nutrient content of the soil. Laterites usually form when the soil has become exposed by forest clearance.

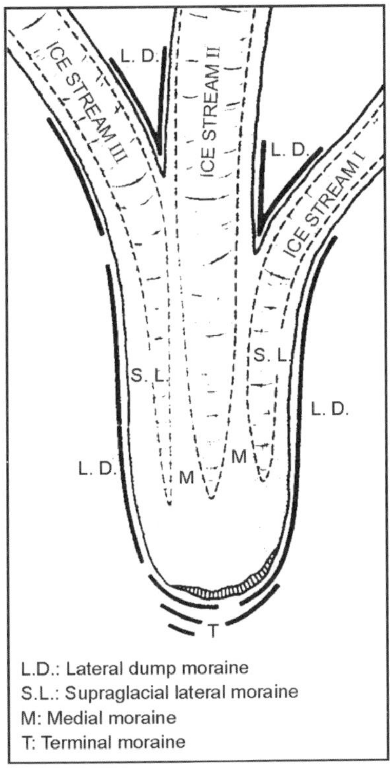

Lateral , Medial and Terminal Moraines

Laurasia The northern part of **Pangaea**, a supercontinent thought to have been broken up by continental drift. It was separated from southern part by the Tethys ocean.

lava Molten rock or magma which has reached the earth's surface as a result of volcanic activity. Its types are acid lava, basic lava and viscous lava.

lava dome A massive protuberance of lava forming a mountain with convex slopes, resulting from the accumulation of lava.

lava flow Movement of lava from a volcano or fissure.

lava fountain A forceful vertical expulsion of a jet of molten lava from a vent or a fissure.

lava lake Created by deposition of lava at creator, e.g., that of Kilauea in Hawaii.

lava plateau A large levelled area which has been built up by a series of volcanic eruptions. A lava plateau covers several thousand sq km in Deccan Plateau of India and in the Snake/Colombia area of the NW USA.

lava tunnel An open-ended tube within a lava flow caused where the lava surface cools to form a crust, thereby creating a roof below which the more mobile lava may drain away to leave a lava-enclosed tunnel.

law of cross-cutting relationships An igneous rock fault or other geologic feature must be younger than any rock across which it cuts.

law of original horizontality The law was first proposed in the mid-17th century by Nicolaus Steno, stating that sedimentary rocks are always deposited as horizontal strata.

layer cloud A principal form of cloud with flattened sheet-like appearance and of limited vertical extent.

leaching The downward movement of water through the soil zone which results in the removal of water-soluble minerals from upper layer and accumulation in the lower soil zone.

lee depression A low pressure region found downward of a mountain chain, representing the large scale part of the drag of the ground on the air. The cyclonic circulation is due to the interplay of the vorticity and divergence of the air to the lee of the obstacle.

lee waves Waves in the atmosphere of about 6 km wavelength extending downwind of an obstacle.

lee, leeward Downwind side of the side protected from the wind or the opposite slope to windward.

lentic environment A term referring to the ecology of standing water.

lenticular cloud A type of almond-shaped cloud. The long white cloud of Southern Alps of New Zealand.

lessivage The process by which clay minerals are moved mechanically downwards through a soil by percolating water, leading to a degree of decalcification.

levanter A Spanish term for an occasional easterly wind which bellows through straits of Gibraltar and between the Balearic Island.

levee A broad, long-crested ridge running alongside a floodplain stream or intertidal inlet, composed of coarse sand to silt sediment deposited by flood waters.

levenche A Spanish term for a hot, dry southerly or south-easterly wind blowing from the Sahara in SE Spain.

Formation of natural levees

Formation of levees

lichenometry A technique for dating Holocene events that was developed in 1950. It is especially useful for dating glacial fluctuations over past 5000 or so years.

limestone A sedimentary rock, dominantly composed of calcium carbonate and also referred to as calcareous rock. Their main types are: shelly limestone: comprising masses of whole or broken shells; coral limestone: remains of coral reefs; oolitic limestone: consists of ooliths or grains of calcareous material; magnesian limestone or dolomite and chalk: a soft limestone.

limestone pavement A horizontal or gently inclined bare limestone, broken by numerous small-scale solutional features.

lineament A large-scale liner feature on the landscape, such as a trough or ridge, that is the product of the structural geology.d.

lithification The process of the formation of a consolidated rock from originally unconsolidated sediments through cementation or other digenetic processes.

lithology The macroscopic physical characteristics of a rock.

lithosol Surficial deposits which do not exhibit soil horizons.

lithosphere The rocks of the earth's crust. Also used more specifically to describe the layers of sial and sima above Mohorovicic discontinuity.

litter The surface deposit or layer of vegetation above the soil from which **humus** is ultimately formed.

littoral deposits Materials that accumulate between high-and-low water marks, generally comprising sand, shells, etc.

littoral zone That area of a shoreline that is in juxtaposition to a water body. In other words, the zone between high-and-low water spring-tide mark.

Llaxfordian orogeny A mountain building episode that occurred about 1800–1600 Ma. ago; which is recorded in north-west trending fold in the Lewisian gneisses of NW Scotlan

load The material transported by a river in the form of dissolved load, suspended load, saltation, land load or traction load.

Loam A medium-textured soil, comprising mixtures of sand, silt and clay. Different types of loan can be identified according to the precise content of sand, silt and clay, these are **sandy loam**, **silty loam**, and **clay loam**.

local winds The winds which differ from the general winds expected from the pressure pattern due to topographical or urban or other effects. Four main types are: (i) winds intensified by topographical features; (ii) winds that blow along pressure gradient such as land and sea breezes, mountain and valley winds, and on larger scale fohn, chinook, bora, and mistral winds; (iii) wind associated with vertical instability such as those accompanying thunderstorms; (iv) strong winds due to flow over a level surface with a strong pressure gradient, such as sirocco and blizzard.

loess A fine-grained yellow soil, considered to be a wind blown genesis. Thick loess deposits are found in China, Central Europe and Russia.

long profile of a river A longitudinal section of the course of a river drawn along the river from source to mouth. It is expressed graphically as a curve. The profile is usually concave upward, is graded to a local or regional base level, and may be punctuated by knickpoints where the river cuts through former valley floors or river terraces.

longitudinal dune An elongated sand dune oriented in the direction of the prevailing wind.

longshore drift The movement of sand, mud and shingle along a coastline. The direction of drift can change, as it is caused by waves, which in turn are caused by wind. Most coastlines have a prevailing direction for long shore drift.

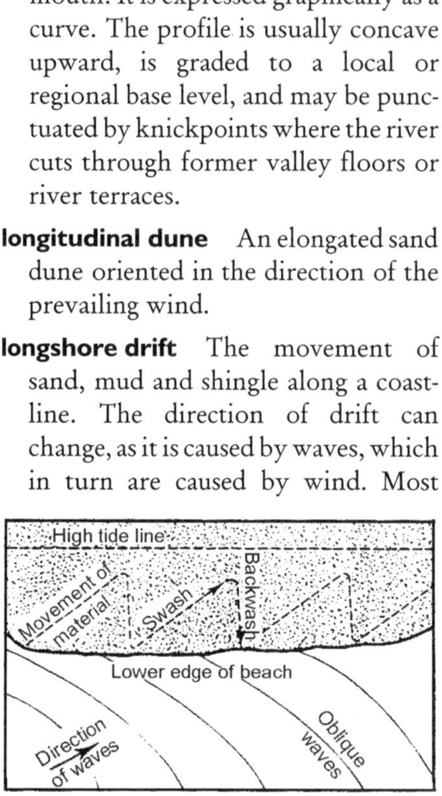

Longshore drift

lopolith A saucer-shaped igneous intrusion which is concave upwards. The **mega-monoliths** are larger and exhibit a banded structure. The largest of them cover 200,000 km^2 (Duluth gabbro, Canada) and make up some of the largest exposures of igneous rocks at the earth's surface.

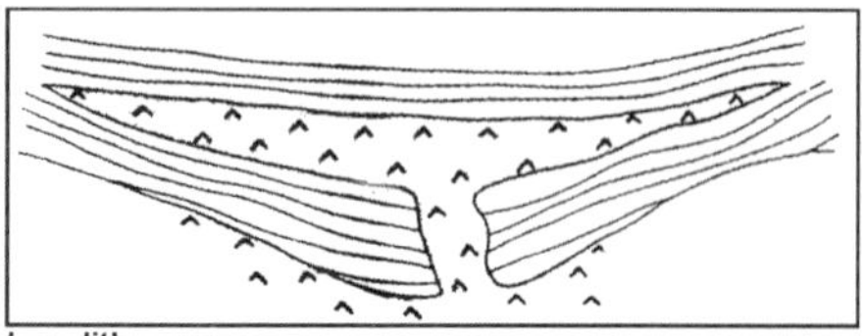
Lopolith

lotic environment A term referring to the ecology of running waters, e.g., rivers.

lowestoft till An extensive glacial deposit of Anglian age, occurring in East Anglia, England.

lunette An asymmetrical crescentic ridge of Aeolian origin on the leeward side of certain Australian lakes and swamps.

lysimeter A piece of equipment for measuring the rate and amount of water storage and percolation through a soil.

M

maar An old volcanic crater, especially in the Eifel region of Germany. A pond or lake formed in such depression.

machair A Scottish term applied to landform/vegetation systems of many dune pasture areas of parts of the highlands.

macroclimate The general climate of a large area (e.g., humid temperature, warm temperate, equatorial climates) in which controlling factors are latitude, the distribution of large land masses and oceans, and the general atmospheric circulation.

magma Molten rock material, highly changed with gases, derived from considerable depths within the earth which forms igneous rocks when it solidifies. Large masses of magma are in **batholiths**, and thin seams of magma creates **sills** and **dykes**. If magma reaches surface, it flows out as **lava**.

magnetic declination The acute angle between magnetic north and true north expressed in degrees W and E of true north. Also termed the **magnetic variation**.

magnetic equator A line joining points on the earth's surface where magnetic needle exhibits no magnetic inclination. i.e., where it remains horizontal.

magnetic interval The time period which occurs between magnetic reversals in the polarity of the earth.

magnetic meridian A line joining position of the magnetic poles on the earth's surface.

magnetic poles The two specific points of the earth's magnetic field at which the lines of magnetic force are vertical. The **magnetic north pole** is located currently on Prince of Wales Is in North Canada (approx. 70°N lat. and 100°W long.),while the **magnetic south pole** occurs in South Victoria Land, Antarctica (approx. 68°S lat. and 143°E long.).

magnetic storm A high level of magnetic disturbance produced by particles of solar origin, causing rapid field variations over the earth. Such storms disturb ionosphere causing anomalous radio propagation.

magnetic strips The approximately parallel bands of differing polarity which occur among the solid rocks of the ocean floor.

magnetic strips Parallel bends of igneous rock, occurring on the ocean floor, which are characterised by differing magnetic polarity.

magnetic survey A technique for mapping the distortion of the earth's magnetic field caused by the presence of igneous and magnetic field caused by the presence of igneous and metamorphic rocks at depth.

magnetism The earth can be compared to a huge bipolar magnet. Compasses point towards the North

Pole or the South Pole, depending on which hemisphere you are in. The magnetic poles are not quite located at true poles, and their locations changes slightly each year.

magnetite A magnetic iron ore (Fe_3O_4) common in the Urals and N. Sweden.

mangrove Plant communities dominated by the genus Rhizophora, Bruguieria and Avicemia which colonise tidal flats, estuaries and muddy coasts in the tropical and sub-tropical areas.

mantle A layer of rocks having high density (upto 3.3) and considerable thickness (over 2500 km), lying between the crust and core of the earth. The uppermost rigid section of the mantle forms the lower part of the lithosphere. In geomorphology, mantle is an alternative name for **regolith.**

mantle plume A convectional flow of hot rock that rises through mantle to the base of the lithosphere and which give rise to a hot spot on the surface. In the oceans, mantle plumes create lines of large volcanic, such as the Hawaiian chain, if the overlying plate is moving with respect to the plume.

marginal basin A basin developed along passive margin, synonym for back are basin.

marginal channel A melt water stream flowing at the margin of a glacier.

marginal deep A deep-sea trench lying on the seaward side of an island arc.

marginal depression A linear hallow surrounding the base of an inselberg or at the base of an escarpment where it meets the pediment. It is found in tropical regions.

marginal sea A semi-enclosed body of water adjacent to and widely open to the ocean, e.g., Gulf of Mexico, Caribbean Sea, Gulf of California.

marine abrasion The eroding action of sea waves and the shingle and sand by which they move load, whereby the bedrock surface along coastline is abraded into a wave-cut platform.

marine cycle of erosion A theoretical scheme outlined in 1919 by D.W. Johnson to explain the sequential development of emergent and submergent shorelines, based on the cyclic concept of W.M. Davis.

marine pollution The pollution of oceans and seas. The oceans have now become dumping grounds of the waste material, which is responsible of marine pollution. Similarly lumps of tar, flotsam and jetsam as well as other waste products and also oil seepage now has become a threat to oceans.

marine-built terrace An accumulation of materials removed during the cutting of a marine abrasion platform.

maritime air mass A large mass of air which has come from an area of sea.

maritime climate A climate which is strongly influenced by proximity to the sea. The climatic conditions near the ocean tend to be much warmer in winter and slightly cooler in summer than inland areas at the same latitude. Britain, Norway, N.W. France, British Columbia in Canada, Washington and Oregon in USA, Southern Chile, Tasmania, New Zealand are some of the examples of the areas of maritime climate.

maritime polar air mass An air mass originating from a source region in the oceanic tracts of polar latitudes.

maritime tropical air An air mass originating from source region in the oceanic tract of the tropical zone. It is characteristically warm and moist and is associated with warm fronts.

marsh A tract of low, ill-drained ground with patches of open water in which reeds, rushes and sedges abound in temperate latitudes.

mass balance The relationship between annual accumulation of snow and ice on a glacier or ice sheet and the losses—mainly in the form of melt water resulting from ablation, it is also referred as **glacier budget**. The two types of mass balance are positive and negative, while the total budget (annual accumulation plus ablation) helps to determine the activity of a glacier.

mass movement The downhill movement of soil and rock due to the effects of gravity. There are various types of mass movement. Slow movements are called soil creep or rock creep and faster movements are called **earth flow** or earth slide, and rock fall or rock slide.

massif A distinct area of higher ground or mountain with fairly similar features. A good example is the Massif of Central in France.

massive A term used to describe a rock in which there are few bedding planes, joint or other features.

maturity The stage in the cycle of erosion between Youth and Old. The maturity stage is marked by the achievement of grade by rivers, the decline in slope steepness and the resultant lowering of interfluves summits.

mean sea-level The average levels of the surface of the sea, determined by averaging all tidal levels recorded at hourly intervals.

meander A bend or curve in the course of a river, resulting from long-continued bank erosion at alternating points along the channel margins. Once a meander has been created it will continue, becoming accentuated by the erosive action of the river. On the outside of a bend there will be lateral corrosion which will gradually workout sideways. On the inside of the bend there is likely to be some deposition, which will build

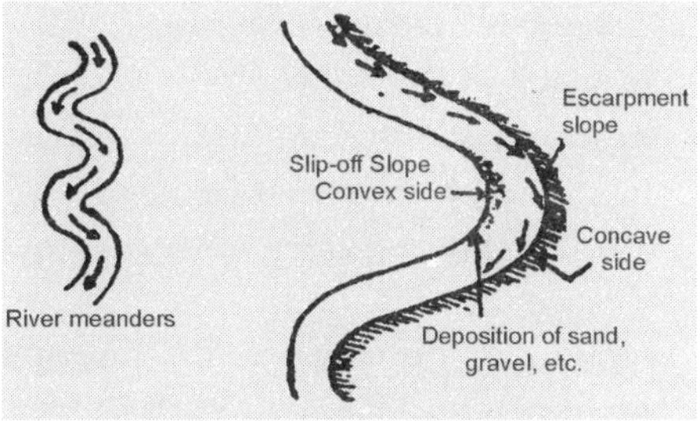

River-meanders, Escarpment slope and Slip-off slopes.

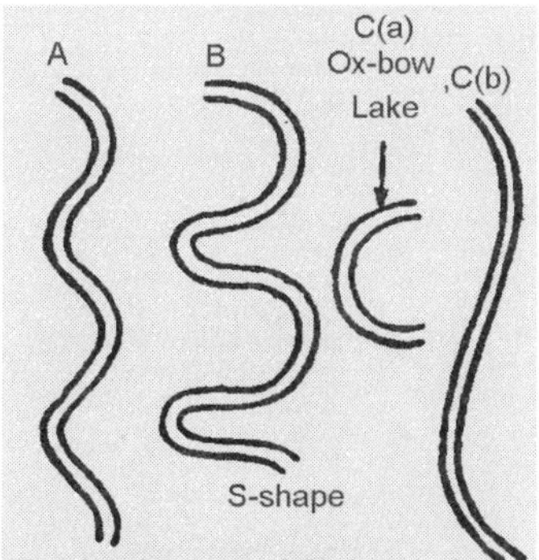

The development of Ox-bow Lake
A=meander
B=The S-shape of meander
C(a)=Oxbow Lake
C(b)=The course of the stream is straightened.

up a flat **flood plain**. Meanders tend to be quite small in the more mountainous upper course of a river, but they become larger further downstream. Down in the lower course, the meanders may spread over very large distances and some on the Mississippi River in the USA are upto 60 km. Meanders may eventually give rise to **ox-bow lakes**.

meander terrace A type of terrace formed where a river meandering freely and incising its channel at the same time.

meander valley A river valley, which displays a meandering pattern. The river valley comprising meanders of large magnitude cut into solid rock but on the flood plain of which a set of smaller scale meanders relating to the present stream are found.

mechanical weathering (also physical weathering) The physical break up of rocks owing to internal and external stresses caused by weathering agents. Disintegration is caused by exfoliation, freeze-thaw action, granular disintegration, sheeting, spalling, unloading, wetting and drying action.

medial moraine A morainic ridge, developed on the surface of a valley glacier formed by the joining of two separate ice-streams. The debris covering the ridge may be derived from weathering and collapse of the slopes above the ice streams but some time due to release of the glacier surface by ablation.

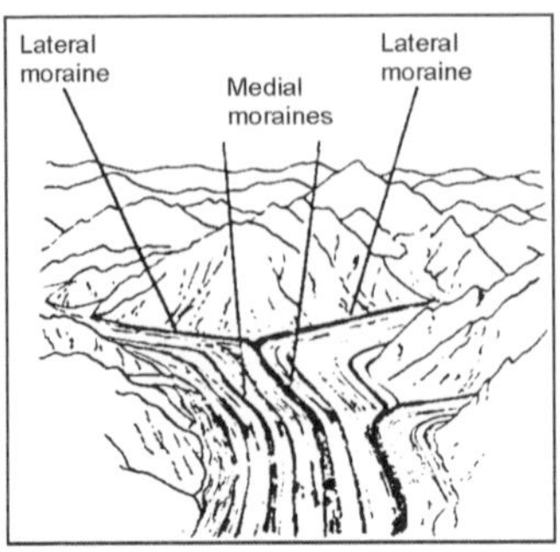

Medial moraines

mediterranean climate A type of climate associated with the countries surrounding the Mediterranean Sea, also found in Southern California in the USA, Central Chile, near Cape Town in S. Africa and near Perth and Adelaide in Australia. In this climate, summers are hot and dry, while winters are mild and wet.

meltwater erosion Erosion effected by running water from melting of a glacier ice and snow patches and more extensive snow accumulations. The most spectacular example of meltwater erosion is the large deep **pot-holes** and gorge-like channels. It is believed that many dry valley in English Chalk country are the product of past melt water erosion. The formation of melt water channels, having very steep sides are also created.

meridian A line of longitude that joins the North and South Poles as part of a great circle or semi-great circle but terminates at the geographic poles. Meridians are numbered from 0° to 180° both east and west from the Greenwich meridian (0°), also known as prime meridian or central meridian.

meridian day The term used for the day on which the International Date Line is crossed by a traveller.

meridianal flow, circulation Any large-scale air movement from north to south or vice versa, i.e., along the meridians.

merokarst A partially developed karst. It occurs where limestones are relatively thin and in areas of impure limestone.

mesa A steep-sided plateau of rocks, often in horizontally bedded rocks, surrounded by a plain. It is a Spanish term widely used to describe a table land or isolated flat-topped hill with steep slope at least on one side. A mesa represents a stage in the denudation of a plateau in a region experiencing an arid or semi-arid climate and is commonly found in the SW of the USA and Mexico.

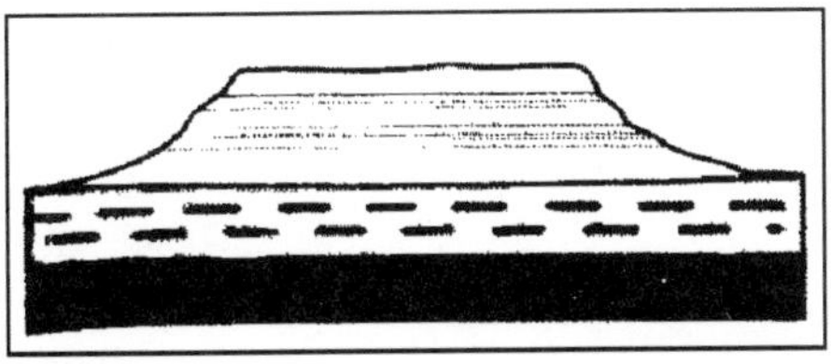

Mesa

meseta Similar to a mesa, this term refers specifically to the high plateau area in the Central Spain, which tilts towards Portugal.

mesopause The boundary between the mesosphere and the thermosphere, at about 80 km above the earth's surface.

mesophyte A plant requiring a moderate amount of moisture for successful growth. Most trees are **mesophytic**.

mesosphere The zone of atmosphere which extends from the stratopause (at about 50 km above the earth's surface) to the mesopause.

mesozoic The middle of three eras that constitute the Phanerozic period of time. The Mesozoic (literally middle life) was preceded by the Paleozoic Era and followed by the Cenozoic Era. It began with the Triassic approximately 245 Ma ago and ended around 65 Ma at start of Teritary.

metamorphism The process by which existing rocks are changed (metamorphosed) in terms of texture and structure, as a result of: (i) intensive compressive stresses (dynamic metamorphism), (ii) increased temperatures (thermal metamorphism), and (iii) penetration by active gases and liquids escaping from magma (contact metamorphism). Common types of metamorphic rock include marble, slate, schist, granite.

meteor A body of matte travelling at great speed through space which becomes luminous when it enters the atmosphere, at about 150 km above earth's surface.

meteoric dust (cosmic dust) Fine dust in atmosphere which is derived from extraterrestrial sources, but becomes trapped within the earth's gravitational field.

meteorite A solid body that manages to reach the earth's surface from an extraterrestrial source, despite frictional effect of the atmosphere.

meteorite crater A depression in the earth's surface created by the impact of a meteorite.

meteorology The scientific study of the phenomena and processes of the atmosphere or the study of weather.

methane A colourless, odourless gas, often referred to as marsh gas. It is a major constituent of natural gas, burns with a clean flame, and has high calorific value.

mica Rock forming mineral made up of silicates, which is characterised by a platy habit, easily splitting into thin sheets.

micro fauna Animal organisms too small to be identified without the aid of a microscope.

micro organisms (in the soil) The tiny organisms which help to incorporate plant residues into the soil in the form of humus.

microclimatology The study of weather conditions in a small area and a short distance above the ground.

micro-erosion Small-scale processes of denudation which occur to depths of only a few centimetres.

microflora Plants that are too small to be identified without the aid of a microscope.

microfossil A fossil too small to be identified without the aid of a microscope.

micrometeorology The study of the atmospheric processes which in combination make the different types of macro-climate to be found throughout the world.

microwave A remote sensing term referring to the region of electromagnetic radiation in the wavelength range 0-1 mm to beyond 1 m.

Mid-Atlantic Ridge (mid-oceanic ridge) The ocean ridge that runs for 14,000 km along the middle of the Atlantic ocean. It is a large-scale linear elevation rising 1–3 km from ocean floor. It is formed at a constructive margin, where two plates are diverging and new magma is flowing out to form new ocean crust. Another mid-oceanic ridge is situated in Indian ocean between South Africa and Antarctica.

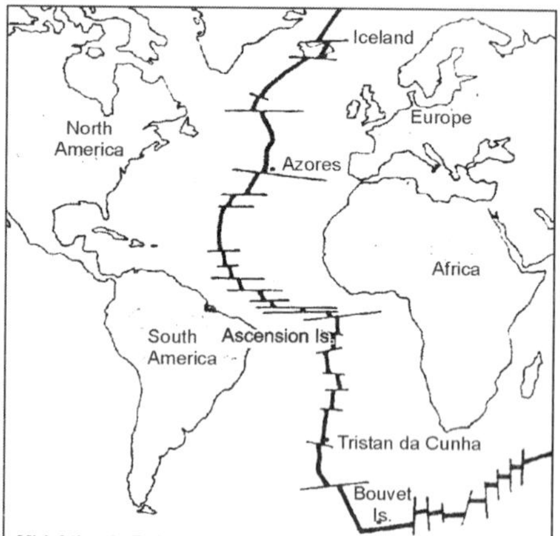

Mid-Atlantic Ridge

mid-latitude zone Also referred a temperate zone of the world. It refers to the zone lying between 23.5° and 66.5°N and S of the equator.

midnight sun Term given to the phenomena which occurs between May and July in northern hemisphere and between November and January in the southern hemisphere, during which sun remains above the horizon when viewed between latitudes 63.5 to 90°N and S of equator. The regions of northern Scandinavia are often referred to as the **Lands of the Midnight Sun**.

migmatite A coarse-grained, heterogenerous mixed rock consisting of a high-grade metamorphic component.

Milankovich cycles The cyclical changes in the rotation and orbit of the earth that Milankovich correlated to climatic effects. There are three cycles: changes in the ice intercity of the earth's orbit, altering the distance between earth and the sun and variations in the tilt of earth's rotational axis.

mima mound A small, rounded hillock generally 1.0 to 2.0 m in height and 10 m in diameter.

mineral A natural inorganic substance possessing a definite chemical composition and almost always in a crystalline form.

mineral deposit A body of naturally occurring minerals of commercial value.

mirage An optical illusion produced by refraction in the lower atmosphere as a result of differential heating of the air objects beyond the horizon may become visible.

misfit stream Stream that appears to be too small to have formed the valley though which it flows. Misfit streams are the result of either change of climate or by river capture as well as also due to underground absteraction of water. An example of misfit stream is the R. Meuse in Lorraine E. France.

mist Suspension of very small water drops in the air, causing a reduction in the horizontal visibility at the earth's surface.

mistral A cold dry north or north-west wind affecting the Rhone Valley. It is strong and most violent in winter and spring. Average frequency of mistral in Rhone delta is over 100 days per year.

mogote A large residual limestone hill which is a remnant of limestone solution and erosional processes. It is also known as bay stack hill. Mogotes are usually found in South China, Vietnam, France (Central Massif), Philippines, Yugoslavia and Java.

Mohorovicic discontinuity The boundary between the lithosphere and mantle named after A. Mohorovicic who discovered it in 1909.

Mohs' scale A scale of mineral hardness, named after its inventor F. Mohs. The scale is from one to ten: (i) talc, (ii) gypsum, (iii) calcite, (iv) flurite, (v) apatite, (vi) orthoclase feldspar, (vii) quartz, (viii) topaz, (ix) corundum, (x) diamond.

monadnock An isolated hill, the product of long-continued subarial denudation, standing above a peneplain. According to W.M. Davis monadnocks are the end-product of divide wasting, through normal cycle of erosion.

monocline It is a one-limbed fold with a very steep angle of dip on one side and virtually horizontal strata on the other.

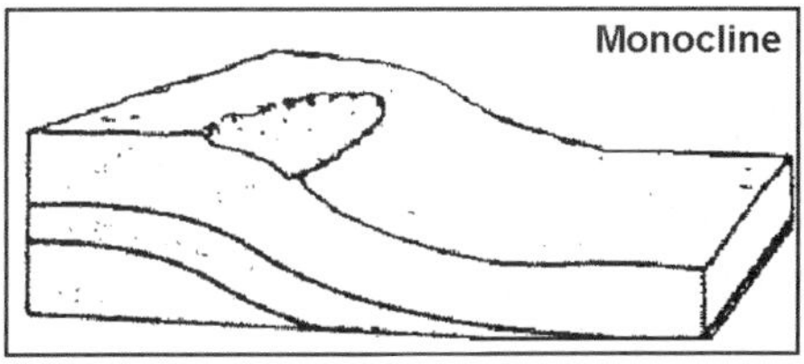

monocyclic landscape A topography which has been created by a single cycle of erosion, in contrast to a polycyclic landscape.

monolith A vertical section of a soil profile removed from the ground and mounted for display.

monsoon A term derived from Arabic word, **mausim**, meaning season. Thus monsoon are the seasonal winds over land masses and adjacent oceans. The Indian monsoon is the famous example, which is

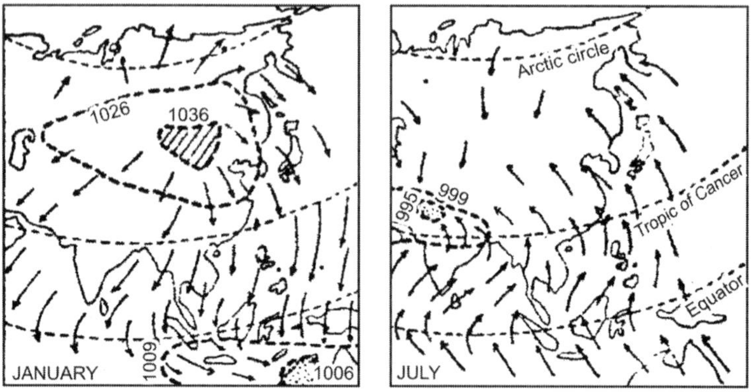

Monsoon winds in Asia—Winter (January) and Summer (July).

related to the alternating development of high and low pressure systems, the movement of the intertropical convergence zone, and the rearrangement of the westerly jet streams. The major monsoon areas are in Asia, where the seasonal reversal of wind is greatest. This is because the largest continent, Asia, is adjacent to the largest ocean the Pacific as well as the Indian Ocean.

montane Meaning mountainous. This is used to describe the montane 'cloud' forest found in upland tropical and equatorial regions.

moor A tract of bleak open land, generally on a plateau or mountain. The vegetation is likely to be heather or bracken, with some course grasses.

moraine An accumulation of debris comprising rounded, subangular and angular boulders, gravels, sand and silt, being transported on, within and

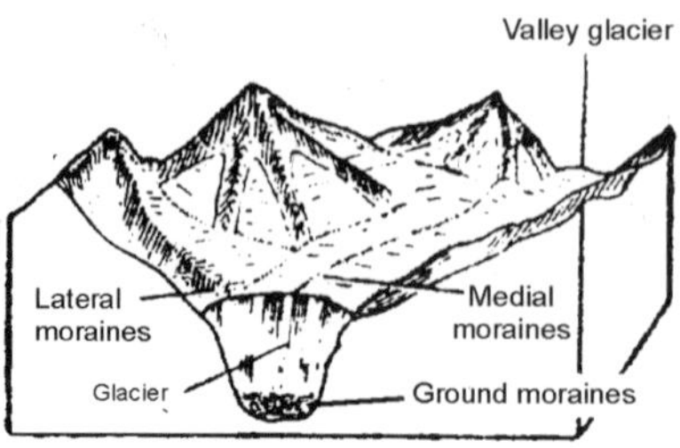

Types of moraines

beneath a glacier. Accumulations at the side of a glacier are **called lateral moraines**; at the end of glacier they are called **terminal moraines**, while the accumulation at the surface of valley is known as **medial moraine**. Moraines are also identified as: **super glacial moraine**, **englacial moraine**, **ablation moraine** and **sub-glacial moraine**.

morphogenetic regions Those regions in which it is claimed that certain geomorphological processes result from a particular set of climatic conditions, thereby giving distinctive regional landscapes.

morphological map A map designed to depict detailed surface forms, usually of a small area. Morphological maps are useful in the interpretation of relationships between landforms and soil and vegetation distributions.

morphology In context of geomorphology, it is the study of form, in which the study of size, shape and origin of landforms is done.

morphometry The quantitative description of forms. Evans (1990) has distinguished general geomorphometry which is based upon an analysis of the entire land surface as a continuous, rough surface described by the attributes at a sample points, and specific geomorphometry which relates to specific landforms. Morphometry is based on detailed measurements, taken either in the field or from topographical maps. The ultimate aim is to assist in the explanation of landforms.

mottled zone An important horizon in a tropical weathering profile, lying between an upper zone of induration and a lower pallid zone. The horizon is referred to as the mottled zone because of its 'motted' red and yellow appearance when exposed in section.

moulin A **sink-hole** in a glacier, caused by the melting out of a crevasse or ice feature, by way of which a supraglacial stream enters the ice and nourishes the englacial drainage system.

mountain A portion of land surface rising considerably above the surrounding country either as a single eminence or in a range or chain. Some authorities regard eminence above 600 m (2000 ft) as mountains, those below being referred to as hills.

muck A dark-coloured waterlogged organic soil consisting of a mixture of decomposed material and mineral soil.

mud A mixture of clay and/or silt with water to form a plastic mass with a particle size below 0.06 mm diameter.

mud lumps Small-scale landforms found in the Mississippi delta region of the USA.

mud volcano A mount built up of mud carried to the surface by geysers or gap eruptions in volcanically active regions.

mud-flat A tract of low lying shoreline submerged at high tide and composed of silt or clay to varying depths.

mud-flow A type of mass-movement on an unstable slope, similar to an earth-flow but of greater velocity because of the high percentage of water present in the mixture.

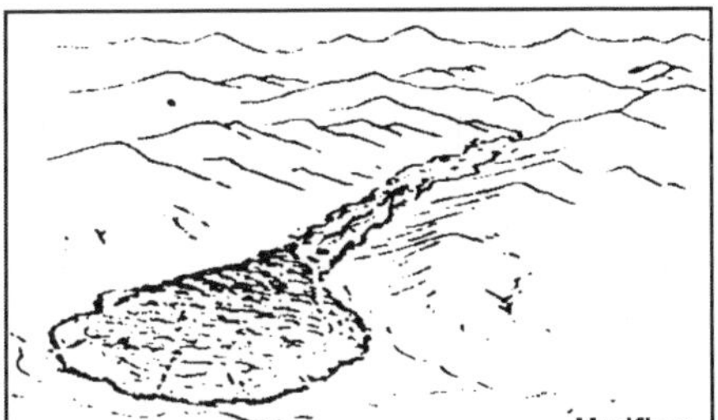

Mudflow

mushroom rocks These are formed by wind erosion, when the upper part of rocks are eroded less than lower ones. Due to greater erosion of lower portions, the rocks assume the form of mushroom. Such a rock is known as **gara** in Sahara and **pitzfelsen** in Germany.

Mushroom rocks

muskeg A boggy area which contains sphagnum moss, and other mosses and lichens. It is found in the Tundra regions of Canada.

mylonite Rock produced in zones of tectonic dislocation, e.g., fault and shear zones. Mylonite is well-laminated and often hard and splintery.

N

nacreous clouds A type of cloud seen occasionally at great altitude (approximately 22–4 km) just before sunrise or after sunset. It is also known as **mother of pearl cloud**.

nagssugtoqidian orogeny A late Archaean and Proterozoic phases of mountain-building affecting a belt 240–300 km wide in what is now western Greenland.

nappe A gigantic recumbent fold, broken by a low-angled thrust fault. The rocks on the upper limb, above the fault, are displaced forwards by many kilometres. A complex series of nappes has been formed in the western Alps, as a result of intense compression of geosynclinal sediments by the movement of the African Foreland northward against the rigid block of the European Foreland.

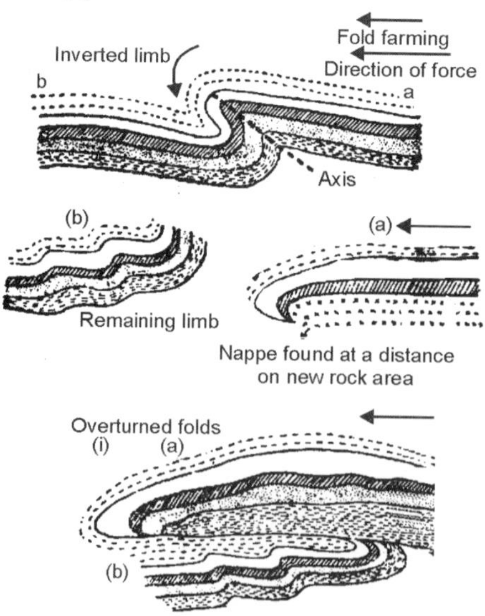

Development of a Nappe

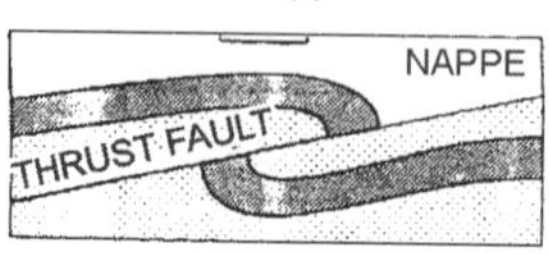

natural bridge A natural rock that has been converted into a bridge spanning an erosional valley, due to collapse of adjacent rock strata up and downstream. Also in karst regions where subterranean drainage is revealed with bridges still intact. The term natural bridge sometimes used as a synonym for a coastal arch created by marine erosion.

Natural Bridge

natural event system A system which incorporates the magnitude, frequency, duration and temporal spacing of a natural event.

natural gas A mixture of hydrocarbon gases occurring in subterranean rock reservoirs. It is composed of methane (84.6%), ethane (6.4%), propane (5.3%) and several other minor hydrocarbons.

natural region An area of the earth's surface having relatively uniform natural characteristics (e.g., relief, climate, vegetation, soil) which enable it to be classified as natural region.

natural resources Those commodities which exit in the natural state and are useful to people; e.g., minerals, soil, plants, animals, air, water, etc.

natural siphon rainfall recorder A self-recording rain gauge (hyetograph) which records both duration and intensity of a rain event.

natural vegetation The vegetation which has been unaffected by human and their activities. In other words, its growth is natural.

neap tide A tide occurring near the lunar quadrature (i.e., at right angles, with the earth at the apex) during which the tidal producing forces do not supplement each other, thereby causing relatively small tidal ranges. Neap tides occur every 14.75 days, and coincide with the first and last quarters of the moon (see tide).

near shore current system A marine current system in the vicinity of the breaker zone along shoreline, which are induced by wave activity.

nebka An Arabic term for a small arrow-shaped hummock of sand which has collected leeward of a rocky obstacle or clump of vegetation.

nebular hypothesis It was propped by Marquis de Laplace in 1796, suggesting that the matter which forms sun and planets originated as a disc-shaped cloud of gas or nebula which contracted into discrete bodies.

neck (volcanic neck) An eroded remnant of solidified lava occupying the pipe of a former active volcano, but which has now been exposed by denudation of the surrounding cone. Examples of necks are: Castle Rock, Edinburgh, Scotland and the Rocher Saint Michel and Rocher Corneille, Le Puy, Central Massif of France.

needle (1) A pinnacle or pointed mass of rock detached from a sea cliff or a mountain rock face, e.g., chalk sea stacks of the Needles, Isle of Wight, Napes Needle, Great Gable, Cumbria, England. (2) Abrasion by winds bearing sand goes on at the mountain tops which are converted into needle-like forms.

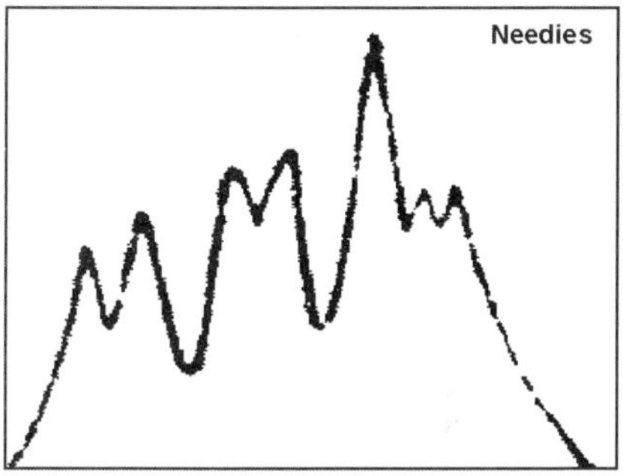

needle ice A small ice crystal, formed when freezing affects the topmost layer of the soil. Needle ice grows normal to the ground surface, and is concentrated beneath individual stones or patches of soil which are good conductors of heat.

nehrung A sand or shingle spit which separates a haff from the open sea. A bar which isolate on estuary or lagoon from the sea.

nekton The actively swimming marine organisms near the ocean surface which have the ability to move in opposition to tides and currents.

neoglacial A period of relatively cold climate, associated with several minor glacial advances. This term often used in N. America to describe the increased glacial activity in Alaska and in the western cordillera of Canada and USA following the climatic optimum hypsithermal.

neolithic An archaeological term adopted from Greek, meaning 'new stone' age and used to describe the culture achieved in the middle Post-Glacial, when mankind had extended the ability to make stone tools.

neoticetonic The study of the processes and effects of movements of the earth's crust that have occurred during Late Cianozoic (Neogene).

nephanalysis The term used to cover analysis and interpretation of spatially organised cloud data.

nephoscope An instrument measuring the height, direction of movement of cloud from a point on the ground.

neptunist A devotee of the Neptunean theory, expounded by German geologist A.G. Werner in the late 18th century, which proposed the all rocks were formed by water or in aqueous environments.

neritic zone That past of the ocean which includes both the estuarine zone and the continental shelf. The neritic zone covers about 10 per cent of total ocean floor and is affected by tidal currents.

net radiation The net exchange between all wave outgoing and incoming radiation. The net radiation is considered positive if the flux downwards exceeds that upwards, and in this case will add energy to the surface. Net radiation is also known as radiation balance.

neutral fold Where the fold axis and axial plane are inclined vertically, a neutral fold is known as 'vertical fold'.

neutral soil Soil with a pH value of 6.6–7.3.

nevados The name given to a type of katabatic wind, characterised by its extreme coldness, blowing from the Andean snow peaks down into the valleys of Equator, South America.

neve A French term, used to describe freshly deposited snow in the accumulation area of a glacier.

niche A term used in ecology to describe the environmental conditions that are necessary for an organism to maintain a viable population and the amounts of each of the resources that it requires to do so.

niche glacier A small glacier occurring in a small hollow or gully on a steepy sloping mountain wall.

nife It is based on the symbols for nickel + iron (Ni + Fe). A term used for the material forming the core of the earth.

nimbus A type of cloud from which rain is falling. It also forms combination with other cloud type, such as **cumulonimbus** and **nimbostrastus**—a uniform, low, dark grey cloud, giving continuous rain or snow.

nip A notch or low cliff cut at the edge of a coastal plain during the initial stage of wave attack on an emergent shoreline.

nitrificationbio Chemical oxidation of ammonia to nitrite and thence to nitrite by action of bacteria, thereby making it available in a form usable by plants. When soils become acid and poorly aerated, nitrite may be converted back to unusable nitrogen gas, a process termed gentrifications.

nivation Nivation is a complex process involving freeze-thaw action. It is localised erosion of a hillside by frost action, mass wasting and the sheet flow or rill work of melt water at edges of and beneath, lingering snow paths. The main effect of nivation is to produce nivation hollows, which as they grow in depth, trap more snow and thereby enhance the process deepening.

nivation cirque It is a transitional form between a nivation hollow and glacial cirque. Nivation cirques possess steep, frost shattered headwalls, sometimes more than 100 m in height and in plan are semi-circular.

niveo-aeolian These are mixed deposits of snow and wind-driven sand and other detritus in addition to the forms and microforms generated by them.

noctilucent cloud A rare cloud-form found near a height of 80 km where there is a minimum in the temperature. They are so extremely tenuous that they can be seen only against the light scattered by air molecules on summer nights.

non-conformity An unconformity in which stratified rocks rest on eroded granitic or metamorphic rocks.

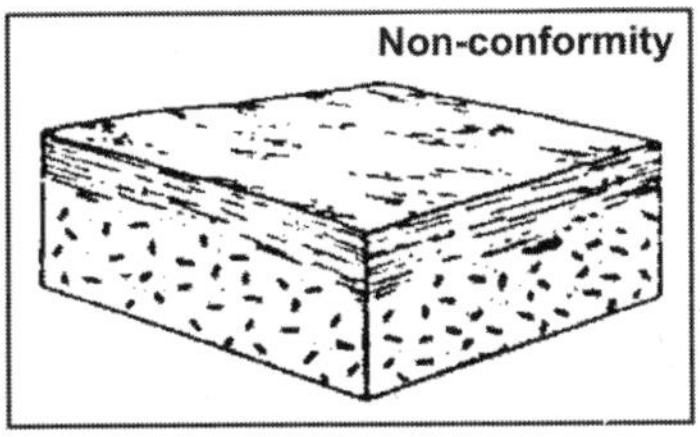

non-relict slopes Slopes whose form is related entirely to processes that are similar to those of today, implying that there has not been a climatic change over a long period to operate any earlier forms.

nor'wester A convective storm, usually from the north-west, which affects Assam and Bengal between March and May.

normal cycle See cycle of erosion.

normal fault A fault with the fault plane inclined towards the side which has been downthrown.

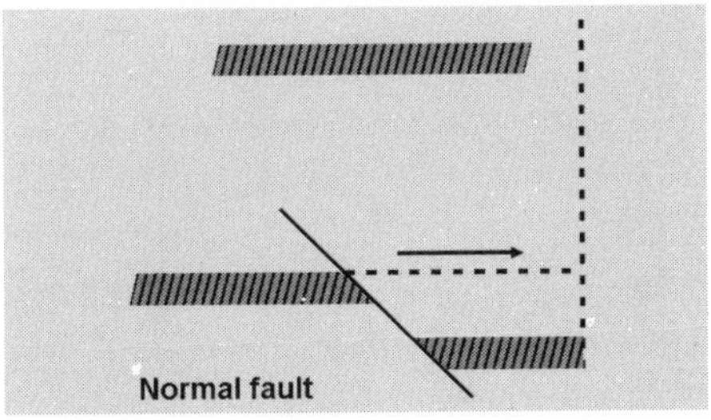

normal stress The stress or load applied to the surface of an object. It is perpendicular to the shear stresses which act parallel to the surface.

North American Plate Major lithosphere plates, the North American Plate extends from mid-Atlantic ridge in the east to a complex pattern of subduction zones and transform faults which form its boundaries with Pacific, Juan de Fuca, Gorda, Cocos, Caribbean and South American Plates in the west and south.

North Atlantic Drift The name given to the north eastward flowing extension of the Gulf Stream which swings away from the eastern coast of the USA and flows across the Atlantic ocean as a number of shifting bands of water which bifurcate as they progress eastward.

notch A land form that develops at the base of a cliff, platform or reef plant, especially in limestone and on tropical coasts.

nubbin (1) A small rounded lump of earth thought to be created by needle ice action at the ground surface in periglacial climates. (2) A small granite dome, created by subsurface weathering.

nunatak A term adopted to describe an isolated rocky peak which protrudes above an ice sheet and may never have been glaciated.

nutation A periodic minor shift in the position of the earth's axis, resulting from the gravitational attraction by the Sun and moon on the earth's equatorial bulge.

nutrient cycle The process whereby mineral elements necessary for plant growth (such as carbon, hydrogen, oxygen, nitrogen, phosphorous, and potassium) are taken up from the soil, and then returned when the plants shed vegetal matter or die.

O

oasis A place in a desert where water is available by way of springs or wells. The water may reach the surface naturally or may have to be raised by deep-well boring. In some instances, the water table may be near enough to the surface for surface springs to occur, especially in erosional hollows created by deflation.

obsequent stream A stream or river which is the tributary of a subsequent stream and flows in a direction opposite to the regional dip of the land surface.

occlusion A complex frontal zone associated with the later states of the life cycle of an extratropical cyclone.

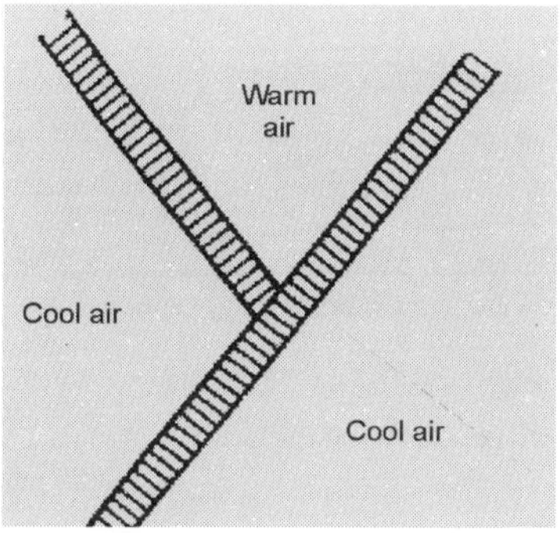

Warm occlusion

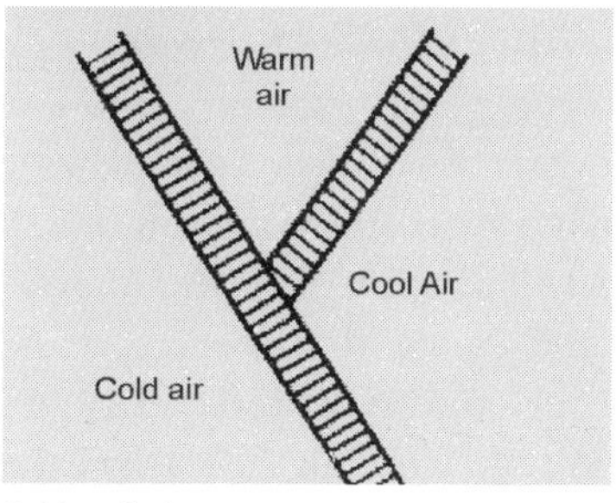

Cold occlusion

The word is used to describe an occluded front. An occlusion may be a **warm occlusion** such that warmer air follows the frontal system or a **cold occlusion** such that colder air follows in the rear. The detailed structure of the occlusion will be determined by the temperature difference of these two air masses.

ocean The mass of water occupying about 70 per cent of the earth's surface. More specifically, the mass of water which fills the ocean basins. The oceans are: the Atlantic, Pacific, Indian and Arctic.

ocean basins Those areas of the sea-floor which lie seaward of the continental shelf but which include the continental slope. The ocean basins has been divided into: the continental slope, the **deep-sea plain**, the very deep parts of the abyssal zone (below 5,500 m) known as **ocean deeps** and the **mid-ocean ridges**.

ocean currents Large-scale water movement in ocean, arising due to (i) prevailing winds (ii) tidal motion and (iii) density differences in sea water caused by differential heating and cooling, salinity difference or variations in the suspended sediments. The major ocean currents are: north and south equatorial currents in Atlantic, Pacific and Indian Ocean (only south equatorial in Indian ocean),Gulf Stream, Labrador, North Atlantic drift, Brazil current, Californian

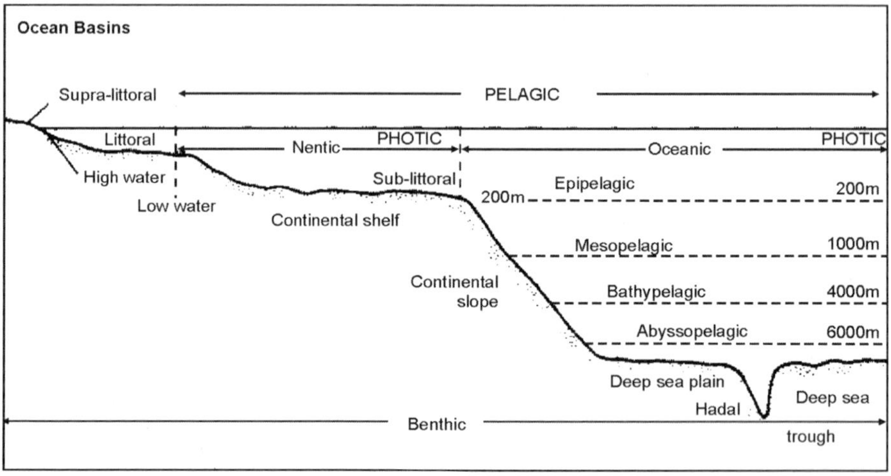

Ocean Currents

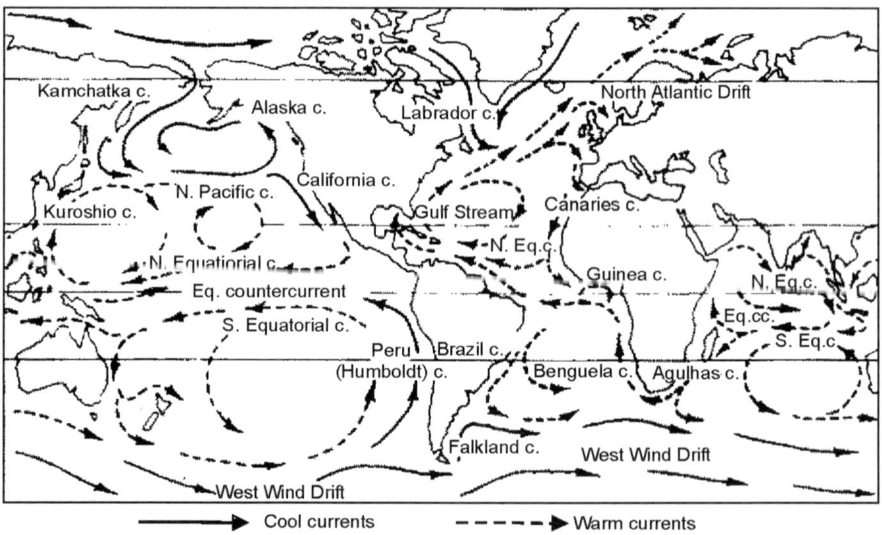

current, Peru (Humboldt) current, Bengula current, Kuroshio current, North Pacific current, Kamchatka current. In Northern Indian ocean, monsoon winds determined the direction of currents.

ocean floor The ocean floor includes the continental shelf and all the sub-divisions of the ocean-basin floor. It refers to the entire sea-bed below low water mark.

ocean waves These are different types of waves generated in ocean waters such as boundary waves, breaker, surf, tsunami.

ocean-floor spreading/sea-floor spreading A movement of the sea-bed associated with plate tectonics. For example, along the Mid-Atlantic Ridge, two plates are moving apart and as a result of a split is developing through the middle of Iceland. In the Mesozoic era, western Europe and North America were close together, but have since been drifting apart for millions of years because of the movements of the ocean floor.

oceanic crust The oceanic rocks which forms 65 per cent of the earth's surface. The oceanic crust comprises four seismic layers, commencing at an average depth of 4.5 km below sea-level.

oceanicity A measure of the degree to which a maritime climate prevails at a selected point on the earth's surface.

oceanography The study of the various aspect of oceans including the study of sea floor, the physics and chemistry of the seas as well as marine biology.

offshore A term applied to any phenomenon that moves away from or is located in an area away from the shoreline.

offshore bar A bank of sand developed some distance offshore on a very gently shelving coastline. The bar may be lengthened by longshore drift and stabilised by the formation of sand dunes. Sometimes shallow lagoon, filled with mud and silt also formed.

offshore zone Sea area extending from the outer limit of the breaker zone to the zone of the shelf edge.

oil shale Dark grey or black shale containing substances that yield liquid hydrocarbons on distillation produce petroleum.

okta A measurement of cloud cover equal to one eighth: eight oktas means that the sky is completely covered, and four oktas means that it is half covered.

old age The last stage of the cycle of erosion, when river gradients are much reduced and slopes have undergone decline to the extent that interfluves are very subdued. The landscape is now approaching the peneplain form.

old fold mountain Those fold mountains which were created by earth movements prior to the alpine orogeny. The structural features, such as folds, nappes, faults, etc. of old mountains are similar to those of young fold mountains. But due to their great age, the former have been subjected to immense amounts of erosion.

oligocene Geological era of the Tertiary following the Eocene and preceding the Miocene, extending from about 38 million years BP to 22.5 million years BP.

oligomict A type of sedimentary rock comprising fragment of a single type of material.

oligotrophic State of water body when it has a low nutrient content and is therefore unable to support a large aquatic flora and fauna.

oligotrophic lake A lake, characterised by relative lack of dissolved minerals and a generally low nutrient status. Many deep lakes in upland granite regions in parts of Scotland and Norway are the examples of oligotrophic.

olivine The chief member of a group of rock forming silicate material.

omnivore Organism which eats both plants and animals.

onion weathering (exfoliation) The destruction of a rock through the peeling off of the surface layers.

ontogeny The stages of development during the whole life history of an organism.

oolite A sedimentary rock, which is composed of concentrically layered spheres—which have formed by accretion on the surface of a grain. It largely consists of spherical grains of calcium carbonate in concentric spherical layers.

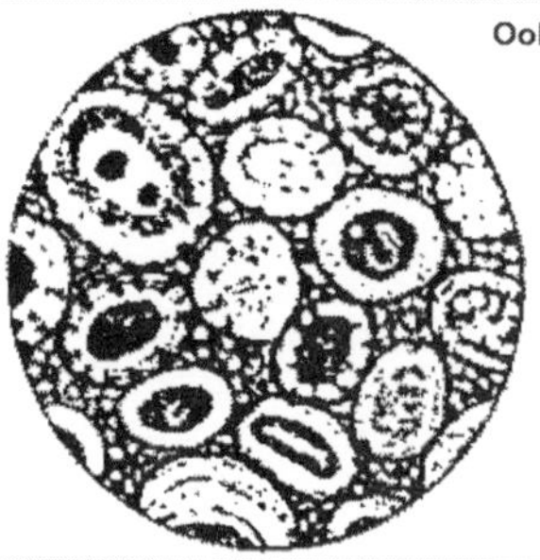
Oolite

ooze A fine-textural sediments formed in ocean basin at depths greater than 2000 m in the abyssal zone and deeper parts of the bathyal zone. The various types of ooze are: **calcareous oozes**, **globigerina ooze**, **pteropod ooze**, **siliceous ooze**, **radiolarian ooze** and **red clay**.

open fold As defined by M.J. Fleuty (1964), a fold whose inter-limb angle is between 90° and 170°.

opencast mining A method of extraction of minerals, where the overburden is limited and easily stripped and waste has to be transported to external dumps.

open-channel flow The movement of water in a river channel, influenced by the following physical properties: density, mass weight, viscosity, temperature, etc. All of these affect the channel geometry and help to explain sediment transport.

optical density A term referring to the measurement of image tone in remote sensing, ranging from 0= white to 4 = black.

orbit of the earth The counter-clockwise path followed by planet earth as it revolves around the sun in one sidereal year. The orbit is not circular but elliptical.

orbit of the moon The counter-clockwise path followed by the moon as it revolves around the earth. This orbit is not circular but elliptical and considerably more flattened than the orbit of the earth.

orbital motion It is the circular or near-circular movement of water particles beneath a wave. The path is not a completely closed loop because a water particle moves slightly in the direction of wave propagation during each wave period and setting up an orbital current. When the wave enters shallower water the wave orbit becomes elliptical in shape.

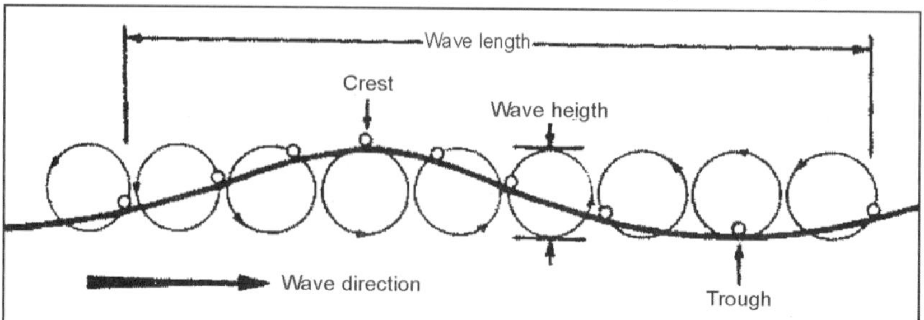

Orbital motion

order, stream (or stream order) The system of the identification of the links in a stream network by the process of ordering. Several geomorphologists have devised methods of stream ordering. R.E. Horton (1945) was the first and led to the development of 'Horton's Laws'. In Horton ordering, all unbranched streams were designated first order, two first combined to make a second order and so on. A.N. Strahler (1952) suggested that the second stage should not be effected and Strahler ordering become widely used. Statistical methods have also been used in stream ordering. A.E. Scheidegger, R.L. Shreve, K.J. Gregory and D.E. Walling, etc. have also suggested methods to stream orders. The objective of all ordering systems is to describe a link in a drainage network and also to provide an ordering system that can readily provide an identification of the discharge from a network.

organic Pertaining to organism, which are characterised by the possession of organs.

organic deposits All the sediments formed by organisms and their remains, such as chalk, coal, lignite, ooze, organic soil, etc.

organic fertiliser The fertiliser derived from organic material, such as manure, bone meal, seaweed or fish meal.

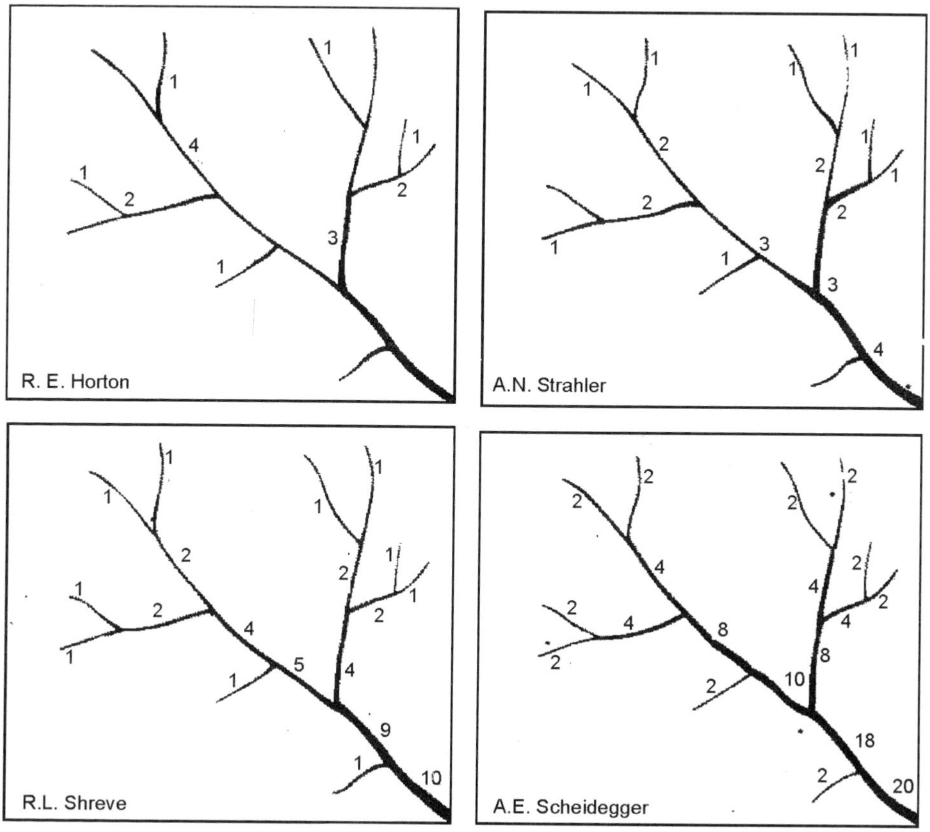

Four methods of stream and segment ordering

organic soil A soil composed predominantly of organic matter.

organic weathering The disintegration or destruction of a rock by living organisms or organic process.

oriented lakes In geomorphology, it refers to **Thaw Lake** which possesses a preferred long axis orientation. Such lakes occur widely in Alaska, Siberia and Canada. Such lakes are generally developed by thermokarst subsistence of ice rich permafrost terrain.

orocline A structural area or any structural mountain chain or orogenic belt, the line of which has been sharply displaced by subsequent horizontal movement to create a marked change in trend.

orogeny A major period of fold-mountain formation, during which the process of **orogenesis** occur. It involves powerful compressive folding and faulting of previously formed sediments, followed by uplift and erosion of these structures. The mountain chains of which the Andes and the Himalayas are contemporary examples, are known as orogenic belt. The main orogenesis to affect Europe since the Pre-Cambrian have been **caledonian**, **hercynian** and **alpine.**

orographic rainfall Rainfall which occurs as a result of winds rising to pass over high ground or topographic barrier. As the winds rise, the air is cooled, causing condensation and clouds begin to form. If air continues to rise, rain will fall. Such rains occur mostly in windward side of the mountain, while the leeward side remains as rain-shadow area.

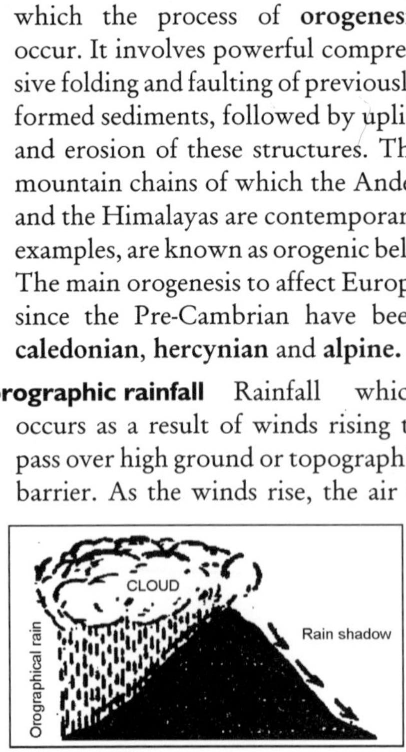

Orographical Rain

orography An uncommon word referring to the study of mountain systems and depiction of their relief.

orophyte A plant that will grow only in a mountainous environment.

orthoclase A mineral of the feldspar group, with a common constituent of granite rocks.

oscillation current A type of wave-induced movement of sea water on the floor of a shallow sea.

oscillation theory of tides The theory postulates that each ocean can be divided into tidal units each centred around a nodal point and each government by the depth, size and shape of the particular water body within the tidal unit. The body of water is set into rhythmic motion or oscillation, by tide-producing forces, but the periodic rises and falls of level will vary according to the size and shape of the different ocean bodies.

outcrop A part of layer or rock or rock formation which is exposed at the surface of the ground.

outer core The layer of the earth lying between depths of 2900 km and 5000 km below the surface.

outlet cave A cave-like hollow in a cliff or hillside marking the point at which an underground karst water course reemerges.

outlier A mass of newer rocks surrounded by older rocks often because it has been detached from a larger formation by erosion.

outport A subsidiary port developed, often at the mouth of an estuary.

outwash plain A gently sloping area, comprising sands and gravels deposited by melt water streams flowing from the stationary margins of a glacier or ice-sheet. The materials have literally been washed out by fluvio-glacial activity. There are outwash deposits in Norfolk (England) and on the 'corn belt' of the USA.

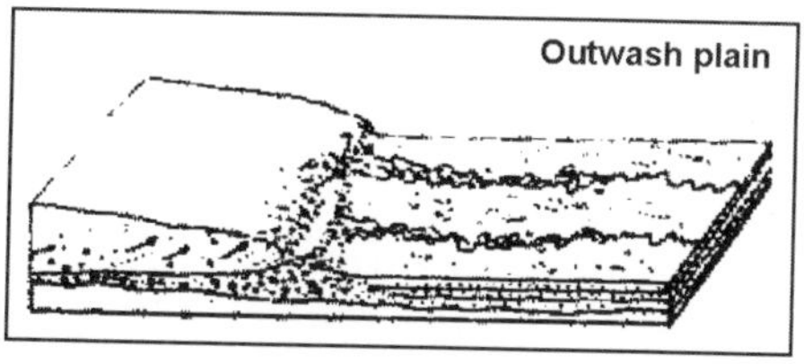

outwash terrace An outwash deposit that has been incised by melt water to form a terrace. A stepped sequence of outwash terraces may relate to successive periods of ice advance and retreat.

overbank deposit Flood sediments laid down by a river beyond its normal flow channel, generally fine materials deposited from suspension in floodwaters.

overflow channel A channel cut by water draining from a glacial lake. It can only be recognised with certainty where it is associated with deltas, lake shoreline and lake bottom deposits formed in the formerly impounded pro-glacial lake.

overland flow The proportion of total rainfall which is not intercepted by vegetation and does not infiltrate the soil but runs over the ground surface as a sheet or rill wash. It is also termed as **infiltration-excess, overland flow, saturation overland flow, sheet erosion, sheet flow surface wash, through flow.**

overlap When the beds above unconformity were deposited by transgressing sea. Thus, each successively younger bed extends further on to the older underlying rock series than does its predecessor.

overstep The disposition of rock strata at an unconformity when the younger series deposited during a transgression rests upon progressively older strata of the underlying rock series.

overthrust fault In these faults, one block rides over the other block and the fault plane is horizontal. Such faults are produced by compression and are accompanied by folds.

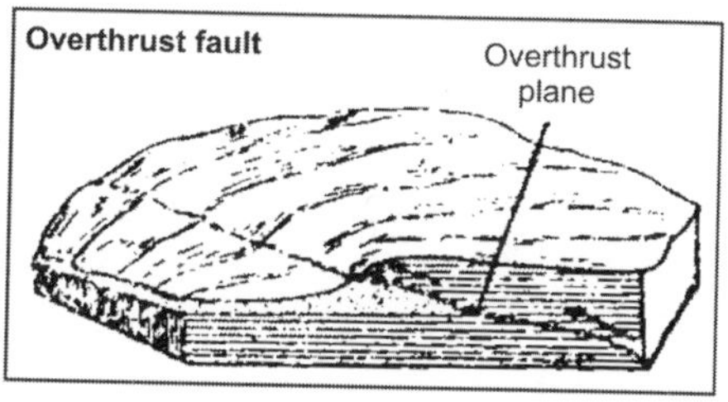

overturned fold A fold in which the beds in one limb have been rotated through more than 90° so that they are now inverted overturned.

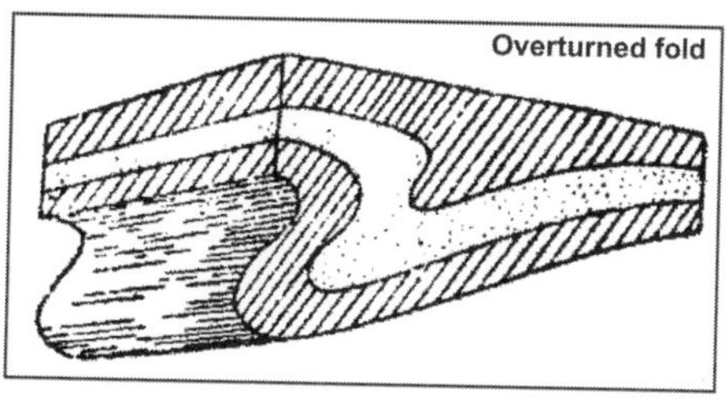

ox-bow lake A small lake which was originally a meander of a river. As sediment is deposited, the meander becomes cut off from the river to

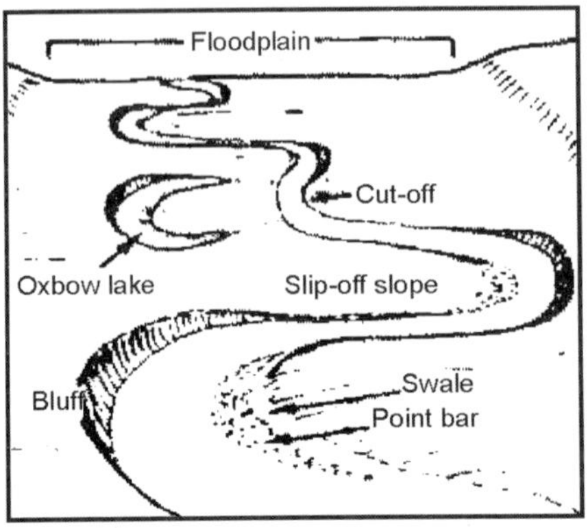

Oxbow lake

create a lake. Lakes of this type may become plugged with sediment where the adjoin the channel and there progressively fill in.

oxidation It is a form of chemical weathering, in which oxygen dissolved in water reacts with certain rock minerals, especially iron, to form oxides and hydroxide, which ultimately disintegrates. It is an important weathering mechanism.

ozone A gas, which is found in very small quantities in the atmosphere, mostly in the ozone layer, about 30 km to 60 km above the earth's surface. The ozone layer absorbs much of the ultraviolet wave energy, but allows longer ultraviolet waves to pass through. Ozone is constantly created and destroyed through chemical reaction. Its thinness is hazardous to life on earth. Ozone is believed to result from photochemical changes through the absorption of ultraviolet radiation by oxygen. Ozone layer acts as a filter, protecting the lower atmosphere and ground surface from harmful ultraviolet rays. Recent research has shown that concentrations of ozone in the ozone layer are being depleted, largely as a result of the build up in the atmosphere of chlorofluorocarbons (CFCs) from sources such as domestic refrigerators and aerosol spray, etc. So-called 'bodies' in the layer have been discovered, first over Antarctica and more recently, over the North Pole. Several world organizations and scientists are working over this problem because it is a serious threat to human life.

P

pacific type of coast also known as concordant coast, is a longitudinal coast where folded belts trend parallel to the coastline.

pack ice Any area of floating sea ice, which remains attached to the shore in contrast to fast ice.

pahoehoe The Hawaiian name for a fluid basaltic type of lava that solidifies into rope-like forms, also known as **ropy lava.**

paired and non-paired terrace A **paired terrace** may develop due to partial erosion cycle related to intermittent uplift or it may represent alternations between periods of aggradation and degradation, when lateral erosion is dominant. **Non-paired** terraces on either side of a valley were formed when there was continued downcutting during which the channel shifted laterally back and forth. C.A. Cotton (1940) has referred as cyclic and non-cyclic terraces respectively.

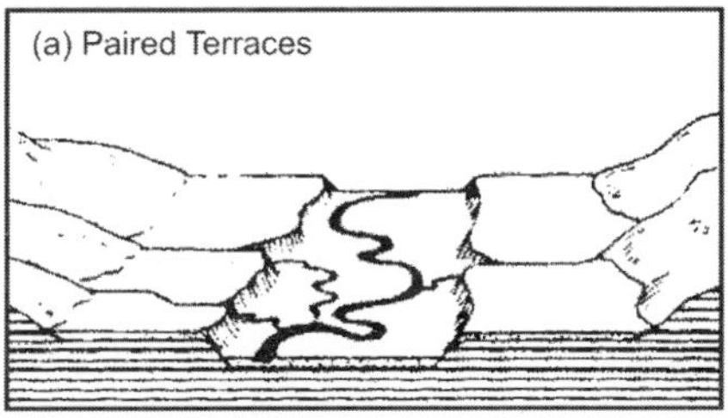

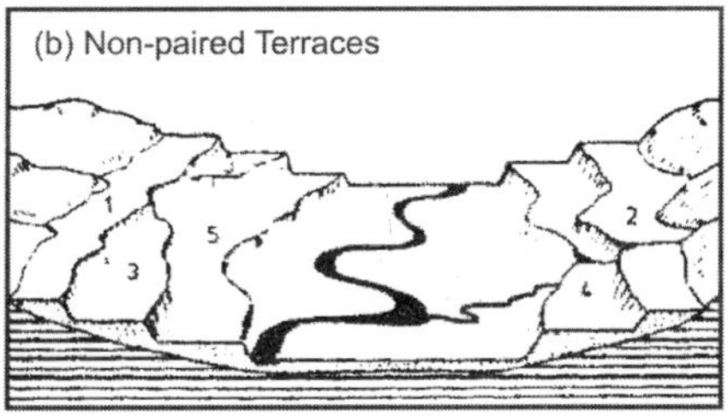

palaeochannels These river channels are those which no longer convey discharge as part of contemporary fluvial system. Rmanants of palaeochannels may be preserved on floodplains and they are evident only in sections in sediments because they have been infilled.

palaeoclimatology The study of the climate prior to the period of instrumental measurements. The palaeoclimatology is based on the use of climate dependent proxy data, such as ice cores, ocean cores, sediments, biological evidences, fossils, etc.

palaeoecology The science of the past-ecological conditions particularly during the Pleistocene and Post-Glacial periods.

palaeohydrology The study of the waters on the earth, their composition, distribution and movement on ancient landscapes from the occurrence of the first rainfall to the beginning of continuous hydrological records (Gregory, 1983).

palaeomagnetism The study of the old magnetic fields and the magnetism which can be found in ancient rocks. When igneous rocks cool and harden, they retain a certain magnetisation, resulting from the presence of particles of iron oxide which orientate themselves to the earth's magnetic field at that period. This study is also helpful to work out the location of these rocks at the time of their formation as well as in studies of drifting continents and plate tectonics.

palaeopedology The scientific study of fossil soil.

palaeosol A soil related with past environment. It was formed either by burial under later geological materials or because of a change in the climate or topographical conditions of soil formation.

palaontology The scientific study of fossil remains including fossils and trace fossils.

paleography The geography of a former time, especially a specific geological epoch.

Paleozoic The geological era which lasted from about 570 million years ago until about 225 million years ago. It has been divided into six periods: Cambrian, Ordovician, Silurian, Devonian, Carboniferous, and Permian.

pali ridge A sharply pointed ridge between two stream valleys on deeply dissected domes, especially in Hawaii.

pallid zone A widely identified weathering profile containing a cluricrust and lying between the motted zone. The pallid zone may reach a thickness of 60 m or more but usually less than 25 m. It comprises highly decomposed kaolin clay and quartz sand.

palsa A mound of peat or peaty earth containing perennial ice lenses and a core of permafrost. Generally, the feature is similar to a PINGO, but is restricted to peat boggy.

paludal It refers to a marshy environment and to the sediments deposited therein.

pampa A name given to the extensive temperate grasslands of Argentina and parts of Uruguay (South America). Originally more humid eastern part was covered in coarse grass (pampa grass) but much of this has now been replaced by farmland or used for cattle ranching.

pan A closed depression that can occur in arid and semi-arid areas such as the Kalahari, Patagonia, Western Australia, and the high plains of Texas. Pans may develop from such processes as solution and animal activity, but the prime cause of their development is deflation activity.

pan fan The surface produced when a hill or mountain is completely eroded so that the peripheral fans coalesce, as in the end stage of landscape evolution in an arid region.

Pangaea The name given to a supercontinent consisting of the whole land area of the globe. It consisted of **sial**, which gradually split to form the ancient continents of **Gondwanaland** and **Laurasia**. Pangaea began to break up about 200 million years ago, as described by Alfred Wagener in his theory of **continental drift**.

panplain It is a type of erosion surface. A flat or almost flat landscape that has been produced by lateral erosion by rivers and lowering of divides and interfluves. The process of panplanation by laterally eroding rivers should be contrasted with that

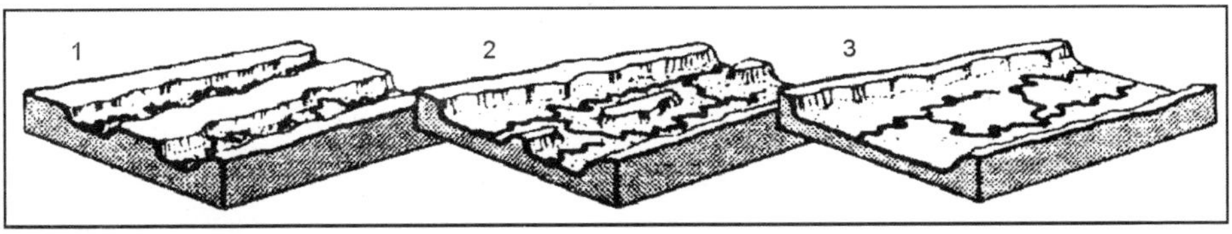

Formation of a panplain

of peneiplanation. This concept was proposed by C.H. Crickmay (1933).

parabolic dune A type of dune, characteristic of arid regions, which is curved in plan and with 'tails' that are pointing upward (opposite of **barchans**). It is usually formed by the process known as 'blow out' in which the centre of the dune is partly removed and carried downwind, leaving the horns behind and drawn out in an elongated form.

paralic A term referring to a littoral environment where shallow waters predominate, e.g., lagoonal environment.

parallel drainage Drainage pattern where the channels of streams and their tributaries run almost parallel to each other.

parallel retreat The phenomenon associated with denudation of a landscape by lateral erosion of scarp slopes and hills which maintain their slope angle as erosion progresses. It is one of three classic models of slope evolution as in the Davison cycle and slope replacement as in the model of W. Penck. L.C. King explained the great escarpments and inselberg of Africa through this process.

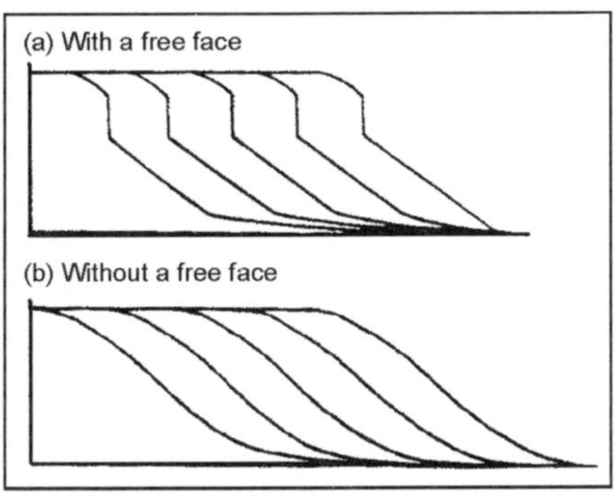

Parallel retreat of slope

parasitic cone A small cone occurring on the flanks of a large volcano.

parent material The bed rock from which a soil has been formed, it will be weathered, but little altered.

particle form The term used to describe several properties of the morphology of sedimentary particles. Thus, in a function expression:

$$F = f(Sh, A, R, T, F, Sp)$$

where Sh denotes the shape of the particle, A its angularity and R its roundness, T is the surface texture and Sp is the sphericity.

particle size analysis A geomorphological technique for analysing the sizes of particles: contained within a deposit, such as river alluvium, a coastal beach, or a glacial till. A sample of the deposit is taken in the field, the sediment sorted in laboratory and the weight of the sediment in each particle size class determined. The particle size distribution is then plotted on a frequency histogram. The main categories are: clay (0.00024 –0.004 mm particle diameter), silt (0.004–0.062 mm), sand (0.062–2 mm) and gravel (2–64 mm), etc.

passive glacier A glacier with low rates of accumulation and depletion, because it receives little snow to feed it in the winter and little melting in the summer.

paternoster lakes A series of lakes in a formerly glaciated valley, separated from each other by morainic deposits, but linked together by streams.

patterned ground A surface and sub-surface pattern of coarse and fine debris, most evident in areas with little vegetation cover, which is produced by lateral sorting processes, most frequently in a periglacial environment. Some forms often seen in patterned ground are—**circles nets polygons**, and **stone stripes**.

peat A dark brown or near black organic deposit, comprising partially decomposed plant materials which accumulated in waterlogged environment. Peat is widely used for fuel and as a conditioner for garden soils.

pebble A rounded or sub-rounded piece of rock intermediate in particle size between the smaller gravel and the larger cobble.

pedalfer A broad category of soils, associated with humid climates which favour leading and elevation and the accumulation of aluminium and iron oxides in the soil. In USA, pedalfers are found in the east, where annual rainfall exceeds 600 mm.

pedestal rock A rock which is shaped like a mushroom as a result of corrosion by sand. In arid landscapes, wind borne sand is moved along just above ground level. The sand erodes any rocks which are standing above the surface of the land and as the corrosion is most active, the eroded rocks become mushroom or pedestal-shaped. This is also known as **mushroom rock**.

pediment An area of eroded fragments of rock close to uplands in semi-arid or arid region. The accumulated fragments gradually build up a plain which slopes down gently from the mountains. Pediments, defined by

Pediment

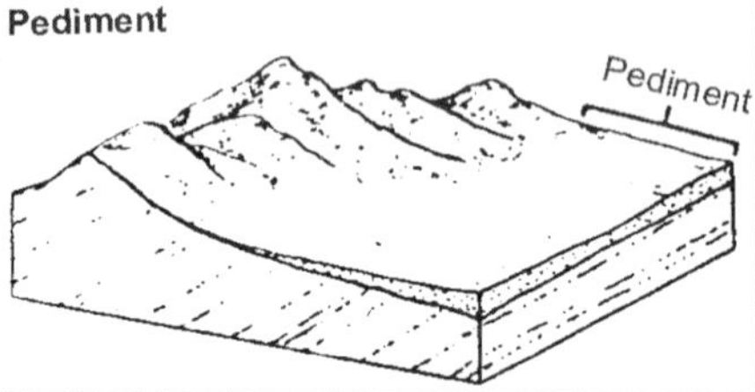

Adams (1975) as: a smooth plano-concave upward erosion surface, typically sloping down from the foot of a highland area and graded to either a local or more general base level.

Patterned ground

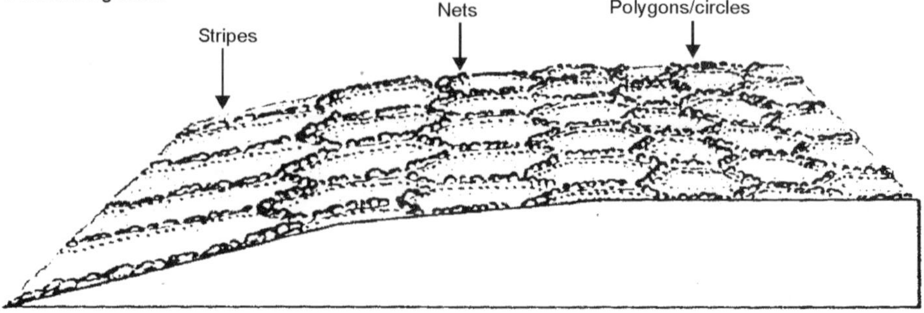

pediplain An extensive erosion surface, interrupted by the occasional inselberg and found in semi-arid and savanna landscapes. King (1949) stated that pediplains are the product of major cycles of pediplanation, involving the twin processes of scarp retreat and pedimentation. It is formed by merging of several areas of pediment.

pedocal A broad category of soils, associated with relatively dry climates where the processes of leading and eluviations are absent and calcification leads to the accumulation of calcium carbonate. It includes **chernozems** and **chestnut** soils.

pedogenesis The natural process of soil formation.

pedology The scientific study of soil and the processes of their formation.

peds The strength, size and shape, i.e., soil aggregates, that give the soil its structure.

pelagic A term applied to the water of the sea, and the organisms living there, independent of shore. Certain fish are described as 'pelagic fish' because they live in the moderately deep parts of the ocean. Pelagic deposits are fine muds which accumulate on the bed of the oceans.

pelean eruption Related to Mt. Pelee, an explosive and very acidic volcano on the island of Martinique in the Caribbean, which had a violent eruption in 1902. It sent cloud of gas down the mountain side after explosion. The gas killed everyone in the nearby town of St. Pierre, with the exception of one man, who was in a prison cell beneath ground level.

peneplain A gently undulating surface of erosion, resulting from a long period of mass wasting and fluvial erosion (see cycle of erosion). It is the end product of the normal cycle of erosion as defined by W.M. Davis (1889).

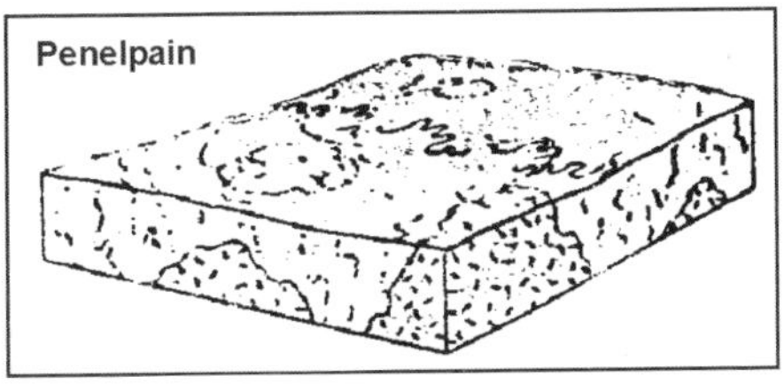

peninsula A headland or promontory surrounded by water but connected to the mainland by a neck or isthmus.

perceived environment perception See environmental perception

perched water table An isolated mass of groundwater, situated well above the level of the main water table and separated from the latter by an unsaturated zone.

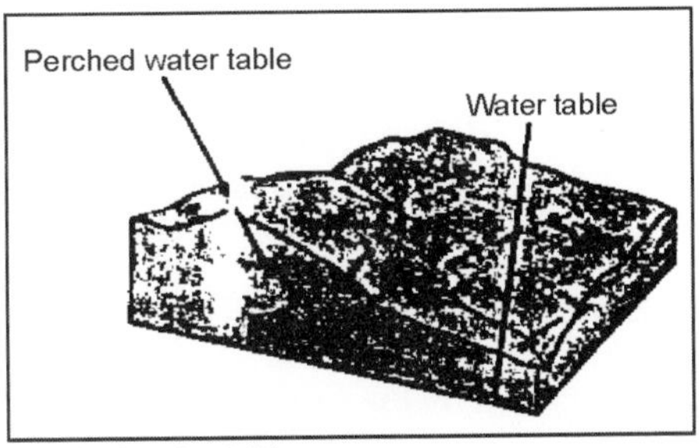

Perched water table

percolation The downward filtering of water through the joints, porous and bedding planets of a permeable rock.

percolation The downward movement of water under the influence of gravity, in other words, percolation water is the water that has passed through the soil or rock by this process. It can be measured by a Percolation Gauge.

percoline A path along which soil water moves laterally by seepage in the soil.

perennial stream A stream that flows throughout the year.

pericline A dome produced by folding. It is an anticline which pitches on both sides.

perigean tide A tide occurring when the moon is in perigee.

perigee The point in its orbit at which moon is closest to the earth.

periglacial A term first used by Walery Von Lozinski in 1909 to describe frost weathering conditions in the Carpathian Mts. of Central Europe. Subsequently, the concept of a 'periglacial zone' developed referring to the climate and geomorphic conditions of areas peripheral to Pleistocene ice sheet and glaciers. During the periglacial period there was permafrost, leading to the occurrence of frost bearing, freeze-thaw activity and solifluxion. Some of the dry valleys of Britain were probably formed at this time. Other features of periglacial include stone circles, stone strips, combe rock can be seen in the tundra regions.

periglacial cycle A cycle of erosion dominated by forst action and solifluction, proposed by L. Peltier (1950) as a counterpart to the cycles proposed by W.M. Davis.

perihelion The nearest point to the sun reached by a heavenly body during its orbit.

permafrost A perennially frozed ground in a periglacial environment. The top few centimetres will generally thaw out in the summer, but the meltwater will not be able to sink into the ground because of the frozen subsoil. Drainage takes place only if the water can flow down a gradient. The term was first introduced by S.W. Muller (1945) as shortened form of permanently frozed ground. Permafrost underlies approximately 20–25 per cent of the earth's land surface. It is widespread in Siberia, Northern Canada, Alaska, and China. Thickness ranges from a few metres at southern limits to approximately 500 m in parts of northern Canada and Siberia.

permeable A term describing rocks or superficial deposits that readily allow the passage of water by way of porus, joints and bedding planes.

permian The final geological period of the Palaeozoic era, extending from about 280 million years ago to 240 million years ago.

pH index (pH) The measure of the acidity or alkalinity of a substance measured as the number of hydrogen ions present in one litre of the substance and reported as a figure on a scale the centre of which is 7, representing neutrality. Acid substances have a pH of less than 7 and alkaline substances have higher pH.

phacolith A lens-shaped igneous intrusion usually situated beneath an anticlinal fold or in the base of a syncline.

phenocryst A crystal that is significantly larger than the crystals surrounding it. Phenocrysts form during an early phase in the cooling of a magma when the magma cools relatively slowly.

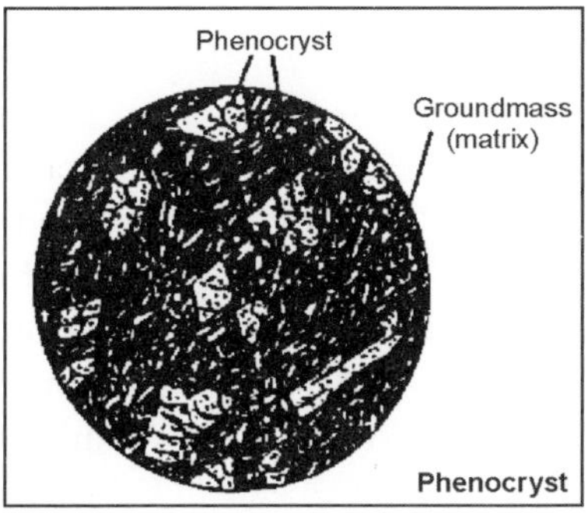

phenology The scientific study of the timing of recurring natural phenomena in the life cycle of plants and animals in nature.

phenomenal environment An enlarged concept of environment that embraces not only the physical and biotic environment, but also all manifestation of human activity. The concept focuses particularly the role of people as agents of change, acting and modifying the physical and biotic environments. It is also referred to as objective environment as well as the milieu.

phonolite A fine-grained lava with phenocrysts of feldspar and nepheline and the extrusive equivalent of nepheline.

photic zone The surface water of ocean sea and lake above the maximum depth to which sunlight penetrates.

photochemical fog It forms by the chemical reaction of sunlight on hydrocarbons in the atmosphere. Photochemical fog is really a haze produced when sunlight reacts with hydrocarbons and nitric oxides present in vehicle exhausts. Many world cities suffer from the problem of photochemic fog, but it is maximum in Los Angeles, where large quantities of nitrogen dioxide and ozone can be generated by road traffic.

photogrammetry The science of the study of aerial and ground photographs. It is very useful in topographical mapping as well as height and distance measurements.

photosynthesis The chemical process, where green plants build up complex organic compounds from atmospheric carbon dioxide and water using sunlight as energy. Oxygen is released as a result, which is source energy to all plants and animals.

physical geography A major branch of geography concerned with the study of the form of the land: surfaces (**geomorphology**), rivers, lakes and underground water (**hydrology**), seas and oceans (**oceanography**), atmospheric processes and phenomena (**climatology** and **meteorology**), soil types and formation (**pedology**) and the distribution of plants and animals (**biogeography**).

physiography Dana (1863) has defined: physiography, which begins where geology ends—that is, with the adult or finished earth and treats: (i) of the earth's final surface arrangements, and (ii) its systems of physical movements and changes.

phytogeography The scientific study of the distribution of plants on the earth. It refers to the study of geographical or spatial distribution of the surface of the earth. Now many aspects of plant ecology and plant biology have also been included in phytogeography.

phytogeomorphology The study of the relationship between landforms and vegetation. This is very useful in remote sensing as most satellite images of the land surface reflect terrain and vegetation types which have become the basis for the interpretation of many less clearly visible phenomena, including soils.

phytokarst The features produced by the weathering and erosive action of plants and animals in a limestone region.

piedmont It is a French word meaning, the foot of a mountain. The plateau at the eastern margin of the Appalachian Mountains in the USA is called the Piedmont Plateau and there is also a Piedmont (Piemonte) region in northern Italy. The term is also used as **desert piedmont**, where a sequence of landforms along the margins of uplands has been formed as in the arid southwest USA. There are also **piedmont glaciers**, the name given to the coalescence of a number of valley glaciers at the base of a mountain range. Good examples of piedmont glaciers are the Malaspina in Alaska and Skeidararjokull in S. Iceland.

pillow lava The form of lava which has solidified very rapidly in contact with water, to give a pillow structure, resembling a pile of sacks or pillows.

pingo A cone-shaped mound that stands alone and has a core of ice. Pingos are upto 60 m in height and 300 m in diameter, and comprise a sediment layer upto 10m in thickness, the latter may be breached at the summit of the pingo to expose the ice beneath. Two main types of pingo are open system pingos form a valley bottoms where the water seeps into the upper layers of the ground and freezes (these pingos are common in Greenland) and closed-system pingos develop on the sites of small lakes or lake beds, such pingos are believed to have been formed in Mackenzi Delta.

PINGOS

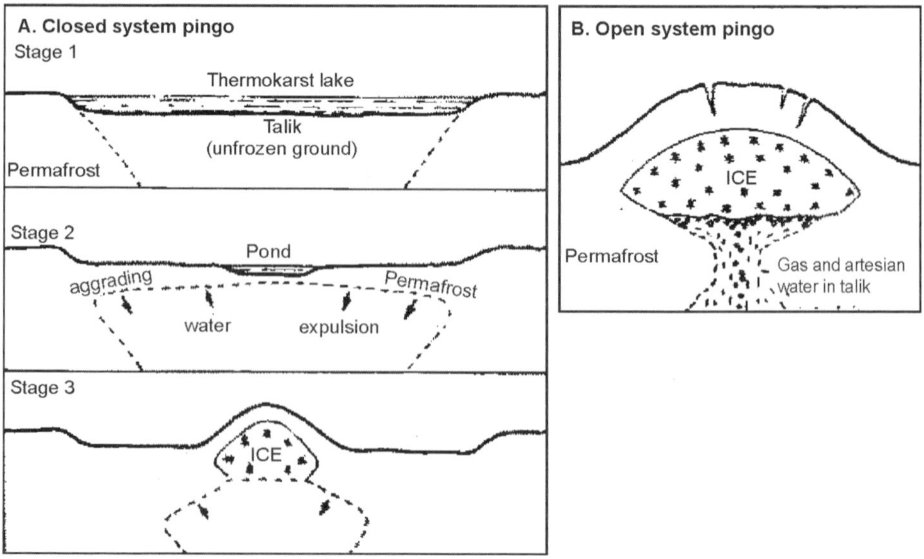

pinnate drainage Feather-like drainage pattern containing a large number of closely spaced tributaries.

pipe The word used variously as: (1) In geology, for the feeder or vent which allows the lava to rise into the crater of a volcano. (2) A cylindrical pit, usually filled with sandy or gravelly sediments in the chalk lands, formed by solution. (3) A deep circular void now filled with a mass of mineral ore. (4) A sloping tube-like feature in the soil created by the process of seepage, usually along the boundary of two horizons, and (5) In general a tubular opening in a solid body.

pitted-outwash It refers to the terrain created by the burial of a stagnant ice sheet by glaciofluvial deposits, in which a large number of kettle holes have been formed.

pivot fault Fault which has allowed one block to rotate around a point on the fault plane. It is also known as **scissor fault.**

pixel The basic unit from which a satellite image is build up in remote sensing, employing the technique of multi-spectral scanning.

plagioserve A plant community from the deflection of the normal plant succession by the direct or indirect activities of man and his animals.

plain These are extensive tracts of flat land or a gently undulating terrain without prominent hills or depressions. The various types of plain are **flood plain, coastal plain, belted outcrop plain, pan plain, pediplain, periplain, sandur**. Some of the plain areas have been reclaimed from the sea, as in the **polders** of The Netherlands and in the Fens and the Somerset levels in England. Flood plains or alluvial plains are the river valleys and now very fertile areas of the world. Glacial plains occur where large expanses of till or glacial drift have been deposited. There are several large plains which are covered by grasslands, such as prairies, steppes and pampas.

planeze A triangular wedge of resistant lava, tapering towards a point near the cone, which protects the underlying materials from erosion.

plankton These are small freshwater and marine organisms; the movement of plankton mainly depends upon the motion of the water in which they are suspended. Barnes (1980) described three major planktonic categories: phytoplankton, zooplankton and bacterioplankton. Phytoplankton (algae) account for the bulk of primary production in aquatic ecosystems.

plate tectonics An important theory, which proposes that the earth's crust consists of a number of mobile rigid plates. Over a long period of geological time, these plates move relative to each other, causing (i) ocean basins to change in size and form, (ii) the continental drift, and (iii) major landforms to be formed. According to plate tectonics theory the earth's surface is divided into eight major and several minor internally rigid plates, which are in motion with respect to each other and the underlying asthenosphere. The plates, which are thought to be moved by large-scale thermal conviction currents, are composed of

PLATES

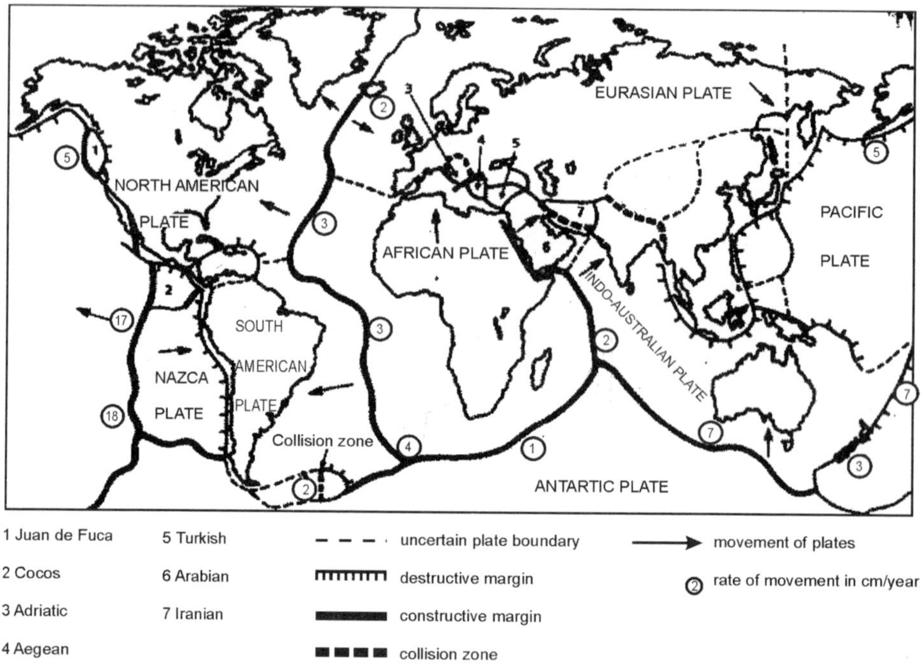

continental or oceanic crust or a combination of both, i.e., sial and sima. They are bounded by plate margins, which are of three types: constructive, destructive and neutral. These margins are marked by transform faults. It has been claimed that many earth's structures can be explained by plate tectonics, especially, mountain belts, island arcs, ocean deeps, while global patterns of earthquakes and volcanoes can also be explained. This theory also include the hypotheses of sea-floor spreading and now considered as replacement of the old theories of continental drift.

plateau An elevated area of land which is fairly flat. Where its relief is broken by deep river valleys, it makes a 'dissected plateau' such as the Mesata of Spain. Some plateaux are very high, over 3000 m above sea level, as Tibet (China). These plateaux are surrounded by mountains and are called intermountain plateaux. Plateaux of Peru, Ecuador and Bolivia in the Andes are also of these categories. Some plateaux may be the result of uplift between faults to form block mountains, such as the Massif Central on the Vosges in France.

playa A closed depression in an arid or semi-arid region that is periodically inundated by surface runoff. However, the lake that forms is rapidly depleted by evaporation, and its site may be occupied for most of the time by a saline crust or mud-flat.

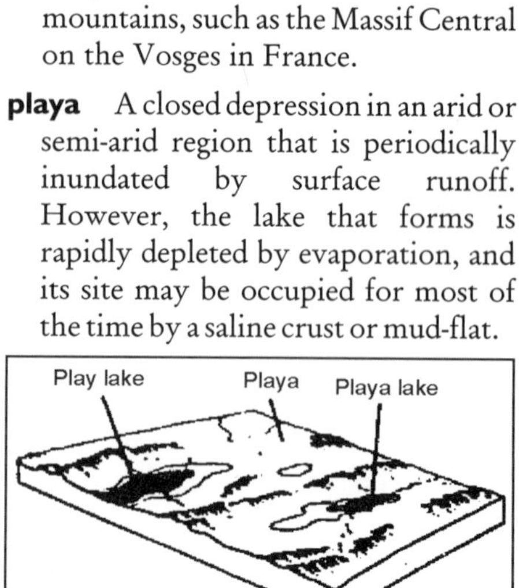

Playa

Playfair's Law In 1802, John Playfair stated that every river will flow in a valley proportional to the size of the river and that where rivers join their levels will be accordant. The implication of Playfair's law is that slope processes are in equilibrium with fluvial processes, and that the individual components of the stream network are nicely adjusted to each other.

Pleistocene The first epoch of the Quarternary, preceded by the Pliocene and succeeded by the Holocene. The Pleistocene was composed of alternations of cold with stages of relatively greater warmth, during which worldwide sea levels fluctuated in response to the formation and melting of ice sheets. In glaciated regions, these eustatic changes were accompanied by isostic depression under the weight of ice cover during glaciations and recovery during the inter-glacial period.

plinian eruption A highly explosive volcanic eruption with dense clouds of gas and tephra being propelled upwards for many kilometres.

plinthite A hard capping or crust at the surface of an unconsolidated soil. The term used to denote surface layer of laterite.

plume An upwelling of molten rock through the asthenosphere to the lower lithosphere. The hot spots thus formed show up in volcanic activity at the surface.

pluton A mass of rock which has solidified underground from intrusions of magma. **Batholiths, dykes, laccoliths, lopoliths, sills** and **stocks** are the main forms.

plutonic rock This is an igneous rock which has formed beneath the surface. Molten magma will have cooled slowly, and there will have been time for large crystals to form. The most common plutonic rock is granite.

pluvial A lengthy period of time when rainfall was considerably heavier than in preceding and succeeding stages. It has been most commonly used about Africa, where it was once suggested that the tropics experienced Pleistocene pluvial stages which coincided with glacial maxima at higher latitudes. Many geomorphologists have assumed that desert water courses and valleys (wadis) were eroded during the Pleistocene pluvials. Pluvials used to be equated in a simple temporal manner with glacials, but this point of view is no longer acceptable.

podzol The commonest type of acidic soil, often occurring in areas of coniferous forest. Podzol soils are developed in cool temperate climates where rainfall is adequate for forest cover (coniferous trees). In this environment, soil bacterial activity is limited, clay-humus production weak and leaching highly effective. A typical podsol has clearly demarked soil horizons—perhaps more so than any other type of soil.

point bar An accumulation of silt, sand and gravel, often beach-like appearance formed by fluvial deposi-

Point bar

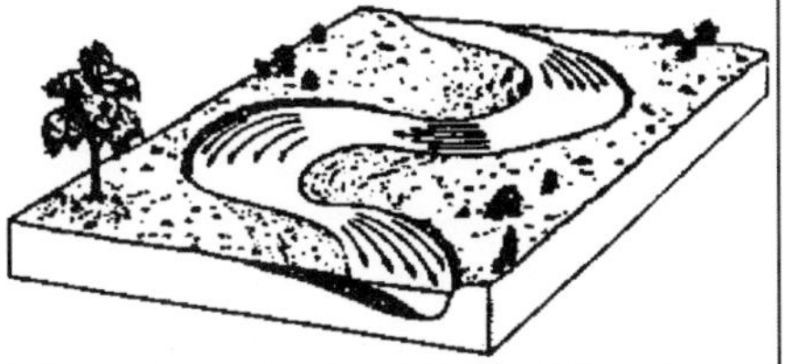

tion on the inside of a river meander. It is a crescent-shaped accumulation of sand and gravel deposited on the inside of a meander bend.

polar front The zone or front separating air of polar origin from that originating within the subtropics. In winter, it can often be traced as a band of cloud over thousands of kilometre. Between 40° and 50° latitude especially over the oceans. Extratropical cyclones may be initiated along the strong thermal gradient of the front.

polder A low-lying area of land that has been reclaimed from the sea or a lake by artificial means. The largest scheme ever attempted in Europe is the partial reclamation of the former Zuider Zee in the Netherlands, about 2227 km^2 or 60 per cent of the total area has been reclaimed in five polders.

pole Geographically, the extremities at the northern and southern ends of the earth's axis, known as North and South Pole. There are also two magnetic poles.

polje In karst topography, a flat-floored, steep-sided, enclosed basin/depression upto 65 km long and 10 km wide. If it is drained by surface water courses it is termed as open polje, but if it drain by means of shallow holes it is a closed polje.

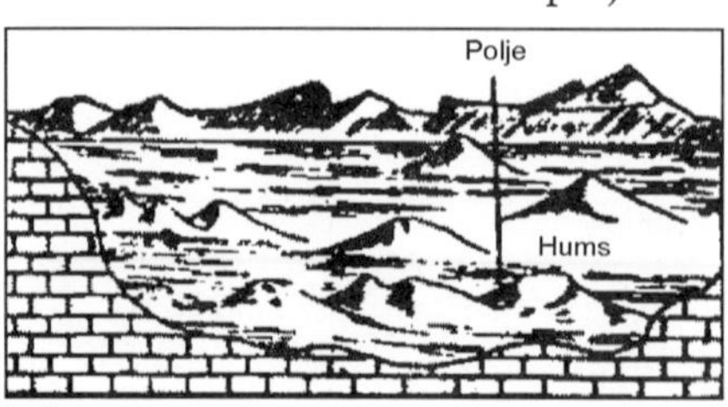

Polje

pollution A substance which causes an undesirable change in the physical, chemical or biological characteristics of the natural environment. Pollution generally occurs because of human activity, although there are some natural pollutants such as volcanoes. The various types of pollution are: air, water, soil and radioactive pollution. Pollutants can be divided as biodegradable pollutants and non-biodegradable pollutants which are more harmful. Pollution now has become a worldwide problem and several steps have been taken to control or at least to minimise its adverse effect.

pollution dome A dome-shaped layer of polluted air occurring within the warm air layer trapped beneath an inversion of temperature. It is usually generated by a large city in association with the heat island effect, when winds are very light or calm. When winds increase in strength the pollutants are carried downwind to produce a pollution plume.

polycyclic landscape When a landscape developed under several cycles or part cycles of erosion, which may be initiated by either changes by base-level or by climatic change. Different terrains exhibit elements from several cycles, not only of the current cycle.

pool and riffle A sequence of gravel bars separated by sections of deeper water, formed in stream channels where bed-load of a heterogeneous nature is being transported. Stream velocity is higher over the bars, producing a riffled surface to the water but is reduced through the pools. Riffles tend to migrate to the inner side of a bend in a sinuous stream where they become linked with point bars. Conversely, pools become deepest on the outer side of the bend.

pore A void or space in a soil or rock not occupied by solid mineral material.

porosity A property of a rock or soil concerned with the extent to which it contains voids. It is defined as a ratio of the aggregate volume of voids to the total volume of the rock or soil. Different rocks and soils vary considerably in their porosity: alluvial gravel 25–35 per cent, till 20–40 per cent, conglomerate 2–25 per cent, slate 0.001–1 per cent.

post-glacial That part of geological time which commenced at the termination of the Pleistocene ice age. Some scientists believe that post-glacial time merely represents an interglacial within the Pleistocene.

potential energy The energy stored by a body in relation to a fixed datum. The energy possessed by a body due to its position, as defined by

$$E_p = m\,g\,h$$

where m= mass, g = gravity and h = height. For example, in streams the potential energy of the water is stated by

$$E_p = W_z$$

where W is the weight of the water and z the height above base level. Potential energy is converted into kinetic energy when the water actually flows downhill.

pot-hole (1) A circular bowl, cut into bed rock, in the channel of a high-velocity stream, carried by the scouring and grinding effect of pebbles rotated in an eddy in a stretch of rapids. (2) In karst region, a steep-sided shaft, down which a surface stream disappears.

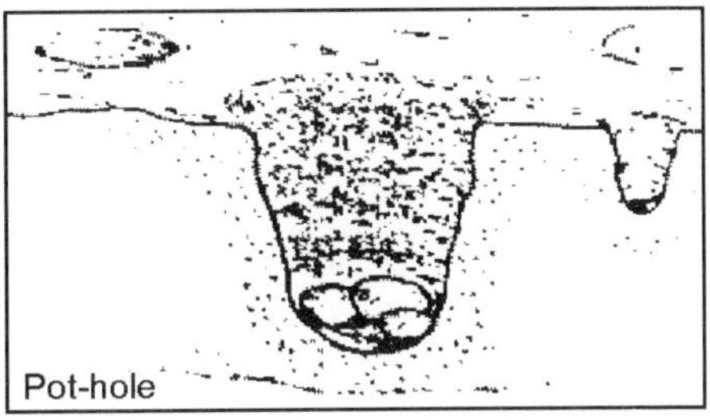

prairie The extensive grassland of North America to the east of Great Plains. This region is having rainfall under 250–500 mm with high temperature in summer and low winter temperatures. Thus, suitable only for grass as well as for the production of wheat.

Pre-Cambrian The geological period prior to the Cambrian and the rocks which were formed before 600 million years BP.

precipitation The various forms of condensation in the atmosphere which fall on surface of the earth including rain, snow, hail, sleet, dew and hoar frost.

pressure The term is used mostly as air pressure that is force per unit horizontal area exerted at any given level in the atmosphere by the weight of the air above that level. The air pressure decreases most rapidly with height near sea level where the air is most dense. It decreases by about 50 per cent for every 5 km of ascent.

pressure belts and permanent winds On the earth surface, the pressure belts have been created due to solar radiation, axial rotation of the earth, land-sea distribution etc. These belts

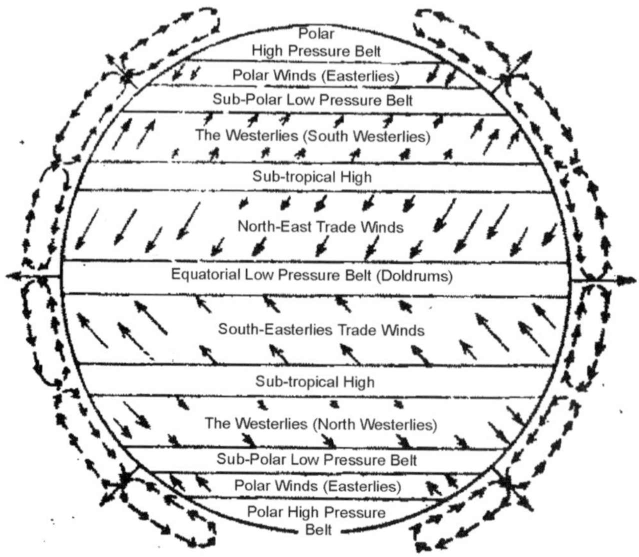

Permanent Winds and Air Belts

are: equatorial low pressure (doldrums) between 5°N and 5°S latitudes, subtropical high pressure between 30°–35°N and S of equator, sub-polar low pressure between 60°–65°N and S of equator and polar high pressure belts around North and South Poles. These pressure belts control the permanent wind system of the world, i.e., trade winds, westerlies and polar winds.

prevailing wind The wind which blows most frequently, though not necessarily most strongly, from a specified direction. Some parts of the world have very persistent winds, for example, the trade winds. In temperate latitudes, the winds are more variable, especially where depressions occur. In South and South East Asia, monsoon winds change their direction with seasons.

primarrumpf W. Penck has introduced this term to describe the initial progressively expanding dome in which the tectonic uplift is so slow that denudation is able to keep pace with it. He stated that there would be no actual net rise of the surface nor increase in its relief. Accelerated uplift would elevate the primarrumpf above sea-level, even though it undergoes degradation.

principle of superposition William Smith (1769–1839) has introduced this stratigraphic principle upon which the whole geological chronology is based. It states that underlying rock strata must be older than overlying strata where there has been neither inversion nor overthrust.

process lapse-rate The rate of temperature decrease of a parcel of air as it moves upwards in the atmosphere, following by either dry adiabatic lapse-rate or the saturated adiabatic lapse-rate.

profile (1) A vertical section of soil, which shows the changes in the soil, including the different horizons from the surface down. (2) The long profile of a river. (3) A topographical profile which depicts a vertical section of the ground along a given surface line. Deposits of glaciofluvial origin or any sediment which is laid down in a water body having been carried by melt pro-glacial deposit water streams beyond the limits of a glacier.

progressive wave theory of tides It was an early concept which attempted to explain tidal behaviour but which has now been replaced by the oscillation theory of tides. The earlier theory postulated that in the southern ocean two large tidal waves of low amplitude and great breadth travelled westwards around globe every 24 hrs and 50 minutes. It was further believed that the progressive waves further generated northwards into the Atlantic, Indian and Pacific oceans respectively, their speeds being governed by the shape of the oceans and depths of a water rather than by the lunar period, thereby reaching particular coastlines at progressively later times.

protalus rampart A narrow ridge, usually a metre or so high and tens of metres long in front of a mountain rock face, composed of rock fragments. These are also known as winter talus ridges and 'naval ridges' but Bryan (1934) has called them 'protalus rampart' which is now the accepted name.

protectionist theory E.J. Garwood (1910) has introduced this concept in which he stated that glaciers are not as effective erosive agents as streams and that a glacier lying within a valley actually protects it from the more aggressive sculpturing by stream erosion.

proximal trough It is a depression around a steep rock body caused by the increased velocity that can occur when a moving element, e.g., ice, water or wind flows round an obstruction.

pseudokarst landforms Produced in non-carbonate rocks which are morphologically similar to those normally associated with karst rocks.

pumice A highly porous fine-grained volcanic rock produced when numerous gas bubbles are trapped within the lava when it solidifies.

puna A cold desert, especially one at high altitude as in the Andes.

puy A volcanic plug, taking its name from Le Puy, a town in the Massif Central in France, where there are several examples, including one which has a monastery built on its narrow summit.

pyramid of numbers, ecological pyramid A diagram showing a food chain which depicts each tropic level as a horizontal bar. The bars are centred about a vertical axis and the levels are drawn in proportion to the biomass at each level.

pyramidical peak A steep-sided isolated mountain summit produced by convergence of the back walls of adjoining cirques. The precipitous cliffs of this horned peak (horn) are separated by steep arêtes on the intervening ridges. The Matterhorn of the Swiss Alps is the best example.

pyroclastic Refers to fragmental rock products, such as ash, volcanic bombs, ignimbrite, etc. ejected by volcanic explosions.

Pyroclastic flow

Q

qaid This is a dune massif in which the summits rise high above the general level of the dunes to form irregularly shaped conical hills, with hollows and terraces on their steep sides.

qanat An underground irrigation channel which brings water from the foothills to the nearby plains.

quartz A very common mineral. It consists of silica and grains of sand consist largely of quartz. It is very common in granites and other acidic igneous rocks. Pure quartz is colourless and glassy and it is the major source of glass.

Quaternary Second period of the Cainozic Era. It followed the Tertiary and comprises the Pleistocene and Holocene series, Bowen (1978) stated that Quarternary was characterised by widespread glaciation.

Quaternary ecology The study of the ecological events and processes that have occurred during the most recent geological interval, the Quaternary, which started some 1.6 to 1.7 million years ago.

quick clay Clay-rich deposit, comprising glacially-abraded particles laid down in water which when subjected to stress becomes thixo-tropic or changes from a solid to a liquefied state and then undergoes rapid flow. In other words, it is a water-saturated clay which has insufficient cohesion to prevent heavy objects from sinking into its surface.

quick flow The past of the stream hydrograph which lies above an arbitrary cut-off line drawn on the hydrograph. It represents the most rapidly responding hydrological processes and parts of the catchments.

quicksand Area of sand and mud which contains a high proportion of water. It generally occurs near a river or the sea, where the water table is high, but drainage is poor. The quicksand is almost liquid that animals may sink into it.

R

radial drainage A pattern of drainage in which the streams radiate outwards from the central highpoint of a structural dome. Mounts Elgon, Kenya and Kilimanjaro are good examples of radial drainage.

radial dyke A dyke of igneous rock which radiates from a volcanic vent, following fracture lines due to the stress imposed by crustal swelling prior to the eruption.

radiation Emission of energy in the form of heat. Radiant energy is given off by the sun, and some reaches the earth is known as solar radiation. The earth gains heat from radiation from the sun, but also loses heat by radiation from the ground. During the day, more heat is received from the sun, by insolation, than is lost by ground. At night, the loss is much greater. The atmospheric gases absorb much of the terrestrial radiation (absorption) except any radiation with wave-lengths between 8.5 μm and 14 μm, which passes through the atmosphere and is lost into space. Owing to absorption, the atmosphere is warmed from below and in turn reradiates energy back to the earth's surface.

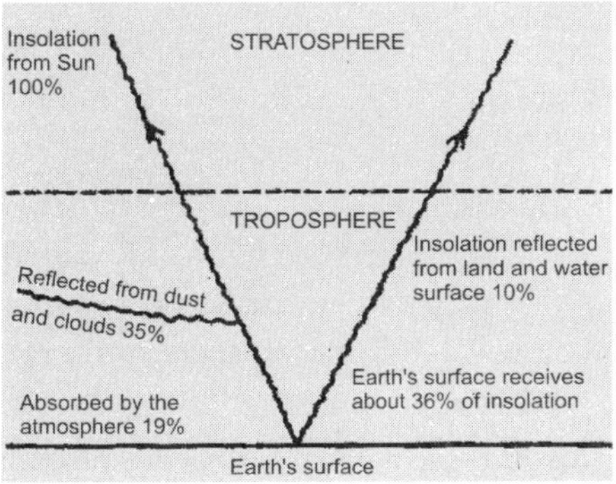

Solar Radiation

radiation fog The fog which occurs as a result of heat loss from the ground by radiation during the night. Clear skies and calm weather conditions are ideal for the formation of radiation fog.

radioactive decay The gradual breakdown of an element by the emission of the charged particles from the nuclei of its atoms. The rate of radioactive decay helps in determining age of the certain rocks.

radiolarian ooze A deep-sea siliceous ooze, containing the minute skeletal remains of radiolarian.

radiosonde An instrument for measuring the pressure, temperature and humidity of the air at heights greater than 10 m above the ground. The instrument is attached to a balloon which rises through the troposphere, usually bursting near the tropopause. The data are transmitted to the ground as the balloon ascends.

rain Drops of water falling from the sky, due to condensation of water vapour in the atmosphere. Drop sizes vary upto a maximum of about 0.5 cm in diameter to the smallest diameter 0.02 cm.

rain day A period of 24 hours in which a measurable amount of rain has fallen. The minimum measurable amount is generally taken to be 0.25 mm.

rain shadow An area of relatively low rainfall to the leeward side of the uplands. The incoming air has been forced to rise over the highland, causing precipitation on the windward side, and thus decreasing the water content of the air which descends on the lee side. The descending air is subject to adiabatic warming, and this increases its capacity to hold much of the remaining water vapour, thus further reducing rain on the lee side

rain splash It is a process of raindrop impact which contributes to raindrop erosion. On level ground, it merely rearranges the soil particles but on a slope there is a net transport of material owing to a longer down slope for rebounded droplets and to the down slope component of that impact force.

raindrop erosion It is a type of soil erosion created by the impact of large raindrops falling at high velocities.

rainfall The amount of precipitation received by a given area over a given period. All rain is formed by rising air, which is then cooled to enable condensation to take place. The four types of rainfall are: **convectional**, **cyclonic**, **frontal** and **orographic** rainfall.

rainfall intensity The amount of rain that falls in a given time period, usually expressed in millimetres per hour. Rainfall intensity is thus a measure of how 'heavy' the rain is.

rainfall reliability The mean annual (or mean monthly) rainfall at a place actually being received in any particular year (month). Mean values of rainfall are calculated over an extended period of time (30 years or more), if individual annual and/or monthly values for this period are plotted on a Dispension Diagram, and median and quartile values calculated, a measure of rainfall reliability can be derived. Thus, the greater the inter-quartile range, and the greater the differences between either quartile values and the median, the less will be the reliability of the rainfall.

rainforest A dense forest, which has developed where rainfall exceeds 1000 mm per annum. Rainforests are found in tropical and equatorial regions and also in some temperate regions, such as the Pacific Northwest of the USA. Tropical rainforest is developed in hot-humid conditions, where rainfall is abundant. These forests are dominated by broad-leaved evergreen trees.

rain-gauge An instrument used to measure rainfall at a given point on the ground. It comprises a colleting funnel, the rim of which is a standard 30 cm from the ground surface, and an interchangeable aperture leading into a collecting container. The accumulated water is emptied and measured in calibrated cylindrical flask.

rainwash This is a type of surface wash characterised by the movement of soil particles and loose materials down a hillside as a result of rain splash.

raised beach An emerged shoreline represented by stranded beach deposits, marine shell beds, and wave cut platforms backed by former sea cliffs.

raised bog Variety of bog, which occurs in wetter, cloudier climatic zones. The raised bog starts its growth in an oligotrophic environment of shallow lakes in which a fen or carr vegetation may creates a basin of fen peat.

randkuft The gap between the back wall of a cirque and the glacial ice that fills the cirque.

range (1) A line of mountains. (2) An open area used for grazing as on the High Plains of the USA. (3) The difference between the maximum and minimum of a series of numerical values, especially climatic elements, such as, seasonal and daily temperatures.

ranker Name given to a soil when it reaches a stage in its formation just beyond the original primary soil stage. It is mostly found in recently glaciated areas, on recently deposited alluvium and on sand dunes.

rapids Waterfalls and turbulent water along a stretch of river. They are generally the result of a different and harder type of rock outcropping on the bed of a river. Steep rapids, such as, those on the upper reaches of River Nile are called 'cataracts'.

recession col It is used to describe a wind-gap on the crest of a cuesta, created when scarp recession breaks into the valley of a dip-slope stream.

rectangular drainage pattern A drainage network having right angled bends and right angled junctions between tributaries and the trunk stream. It is mainly the result of structural formation.

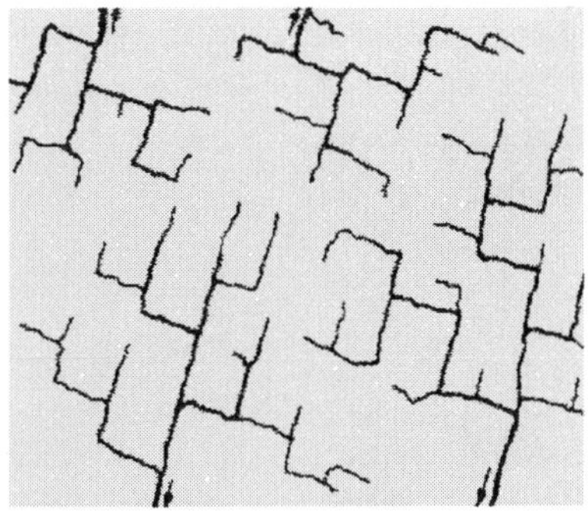

Rectangular drainage pattern

rectilinear slope A slope which has a constant angle. It is straight and has uniform conditions on all its parts.

recumbent fold This is an overturned fold in which the axil plane is virtually horizontal.

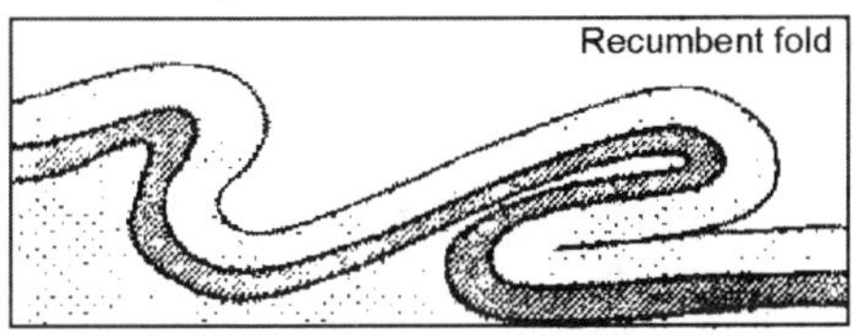

red clay An ocean floor deposit of the abyssal zone, comprising a fine-grained hydrated silicate of alumina, coloured brownish-red by ferric iron and manganese compounds. It is the most widespread of the deep-sea deposits, covering about 28 per cent of the ocean floor.

red mud Deposits of clay/silt on the continental slope. It is different from red clay, as it derived directly from the continents.

red rain Rainfall coloured by red sand or dust transported on high level winds from arid to more humid areas. In Southern and Western Europe; this has happened due to incursion of dust from North Africa.

reef (1) A line of rocks in the tidal zone of a coast, submerged at high water but partly uncovered at low water. (2) Coral reef, which may be in the form of atoll, a barrier reef or a fringing reef. (3) A gold bearing conglomerate of Pre-Cambrian age in S. Africa. (4) A quartz vein containing gold or other precious metal.

reef-knoll A dome-like mass of limestone that has grown upwards from a reef in order to keep pace with the deposition of surrounding sediments.

refolded folds A complex fold, which have undergone more than one period of folding, giving rise to cross-folding in which the axes will intersect each other at an angle.

reforestation The replanting of trees in an area where forest has been denuded either by natural causes or by human activities.

refraction The process by which ways undergo a change of direction as they approach headlands and beaches and pass the distal ends of spits and bars.

reg An Arabic term referring to a stony desert, comprising a layer of tightly packed gravel overlying sand and clayey sand.

regelation The refreezing of melt water. This commonly occurs at a glacier surface to form superimposed ice and at a base of sliding glacier to form regelation ice. The melting point of ice is lowered by the pressure caused by the weight of the ice. If the pressure is released, then the melting point will rise and the water will freeze again. Regelation is very important in explaining the movement of a glacier.

regime The seasonal changes in climatic conditions. The way in which temperature and rainfall change from month to month. There is also a river regime.

region Geographical area having distinctive characteristics. It may be very small or very large, i.e., micro, meso and macro region. The regions based on physical characteristics are: geomorphic regions, climatic regions, vegetation regions, soil regions, etc. On the contrary there are economic regions including agricultural, industrial, mining, transport regions. There are also demographic, cultural and historic regions. Regions can be delimited on the basis of single element or a number of elements also. The study of regions has for long been regarded as a principal aim of geography.

regolith A layer of decomposed rock material overlying unweathered bedrock. Regolith is best developed in areas of low relief and in tropical humid climates, where rates of chemical weathering are high.

rejuvenation The process through which a river, a result of a fall in base level or a climatic change, regains its powers of downcutting, thus resulting in the formation of valley in valley forms, in trenched meanders, knick points and river terraces.

relative humidity The amount of water vapour actually present in the air, expressed as a percentage of the maximum that could be held by air at a particular temperature relief. The features of the landscape, i.e., mountains, plateau, plains, river valleys as well as all the associated features including altitude slopes, etc. The term includes only physical features. The study of relief features is called geomorphology.

remote sensing The scientific collection of mass data information relating to objects without being in physical contact with them. The method used for recording is the electromagnetic radiation which is reflected or radiated from the objects. The four most important sensors are the aerial camera, multispectral scanner, sideways looking airborne radar and thermal infrared line scanner. The two most important forms are the aircrafts and satellite. Today, remote sensing is regarded as a most powerful and useful technique in physical geography.

rendzina It is a dark organic-rich soil horizon developed upon unconsolidated calcareous materials in areas of chalk and limestone bedrock.

renewal of exposure The continual re-exposure of rock face to weathering processes. The weathered products are transported away by gravity fall, soil creep, rain wash, etc. Renewal of exposure is most effective on steep slopes and helps to promote rapid slope recession, and possibly parallel retreat of slopes.

repose slope A term proposed by Strahler to describe a slope whose steepness is controlled by the angle of repose of the superficial debris layer.

resequent drainage This is a type of drainage in which, after a lengthy period of development and change. Streams 'reseek' their initial consequent courses. This might apply to streams on the back slope of a cuesta of resistant rock which did not outcrop on the originally exposed land surface which guided initial stream development.

residence time It is the mean length of time that a water molecule remains in a particular section of the hydrological cycle, e.g., in the river in the ocean, in the atmosphere or in the groundwater. It is calculated by dividing the volume of water in a particular section by the rate at which the water passes through it.

residual In geomorphology, a remnant of a formerly extensive mass of rock or land surface that has been partly destroyed by denudation, such as, inselberg, mesa, monadnock.

residual soil A soil formed *in situ* by the decomposition of the underlying solid rock by weathering.

resistant rock Any rock withstands the forces of denudation more successfully than a less resistant rock.

resultant wind Vectorial average of all wind directions and speeds for a given level at a given place for a specified period. It is obtained by resolving each wind observation into components from north and east, summing over the given period.

resurgence The emergence of a stream from underground usually after flowing through permeable rock strata and having come to an impermeable rock layer.

reversed drainage Type of drainage in which, as a result of earth movements, rivers are forced to flow in the opposite direction to their original courses. Examples are rivers Kafu and Katonga in Uganda.

reversed fault A fault whose fault plane is inclined towards the up thrown side; a right-angle thrust fault.

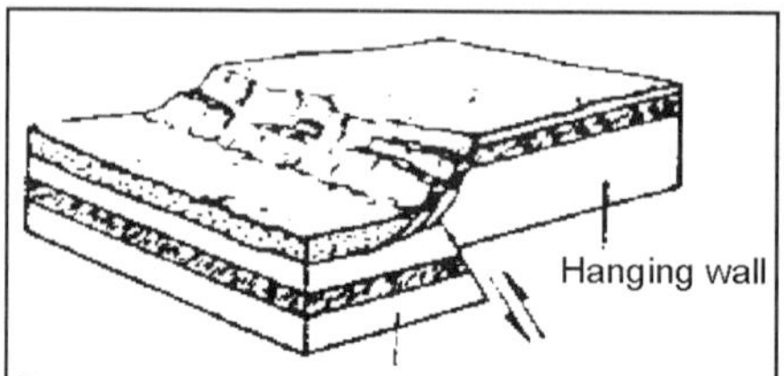

Reversed fault

ria A drowned river valley in a hilly landscape. Where a river valley has been flooded, either by a rise in sea level or because the land has sunk, quite small rivers can develop large estuaries. Ria coasts provide sheltered harbours, and have been used for fishing but in many places they are important for sailing. Devon and Cornwall in Britain are good examples. Similarly in SW Ireland and NW Spain ria coasts have been formed.

ridge A narrow elongated area of high ground with low ground on both sides. The sides are generally quite steep, and a ridge is likely to consist of harder rock than the neighbouring rock.

ridge of high pressure An enclosed area of high pressure which is an extension from an anticyclone. It gives weather similar to that of anticyclone but is unlikely to last as long.

ridge-and-valley terrain Relief that develops in an area of parallel folding, where parallel ridges and river valleys develop on the contrasting lithologies as exist in province of Appalancohious, USA, which varies in width between 40 and 120 km.

riegal A step in the rock floor of a glacial valley. It is a German term for the transverse rock bar or rock step which causes irregularity in the long profile.

rift valley A valley resulting from earth movements along fault lines, where two or more roughly parallel faults have caused land to sink by down faulting. They can be quite small or very extensive; for example, the Great Rift Valley of East Africa extends for nearly 5000 km. Some

Rhine Rift Valley

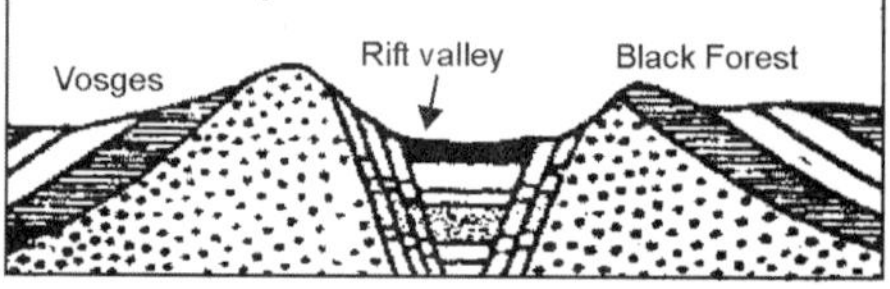

other rift valleys are the Rhine Valley near Strasbourg, the Central Valley of Scotland, the Great Glen in Scotland, etc. Plate tectonic theory suggests that rift valleys are the result of large-scale doming above a mantle plume, followed by fracturing along crest of the dome as plates diverge.

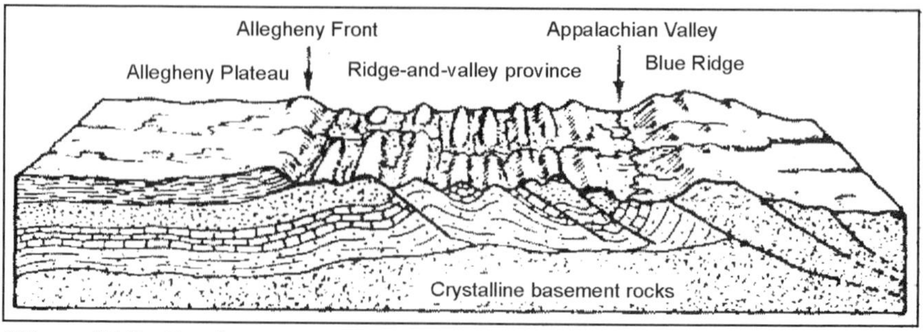

Ridge and Valley Terrain

rill These are small channels between 5 and 2000 mm in width and very closely spaced. They develop mostly in areas with heavy rainfall, especially upon weaker rocks. Rills are often seasonal features and few contain enough load to be regarded as miniature rivers.

rill erosion The removal of surface material by the action of running water. The process creates a number of tiny channels (rills), most of which carry water only during rain storms. R.E. Horton stated that initial parallel rills are enlarged by a process of coalescence know as cross-grading. Further enlargement leads on to the formation of a gully and ultimately to soil erosion.

rip It is a strong current moving offshore in the sea. A rip may occur when two tidal streams meet, when a tidal stream flows over an irregular floor or when a tidal streams flows into shallow water.

rip current A strong seaward moving current which develops as a compensation mechanism, whereby sea water that is piled up along coastlines by wave action can be returned seaward. In plan the current is seen to run initially parallel to the coast before flowing out through the breaker zone.

ripple marks The ripple marks are produced by an oscillatory current, but where the current direction is well-defined asymmetric ripple marks will be formed. Ripple marks are in more or less perpendicular alignment with the flow direction and are found in the intertidal zone.

rise In oceanography, a broad, gently sloping, elevated portion of the sea-floor, similar to a mid-oceanic ridge but without the median rift valley.

river capture The process by which a river by a combination of down cutting and headward erosion, enlarges its catchment and is thereby able to takeover part of the drainage of a neighbouring river. Figure depicts how the tributary x to stream A extends to capture B, giving rise to (i) an elbow of capture at the point of diversion (ii) a col or abandoned valley and (iii) the misfit stream resulting from the beheading of stream B. A famous example of capture is that of the R Wey in Surrey. River capture is also known as **river piracy**.

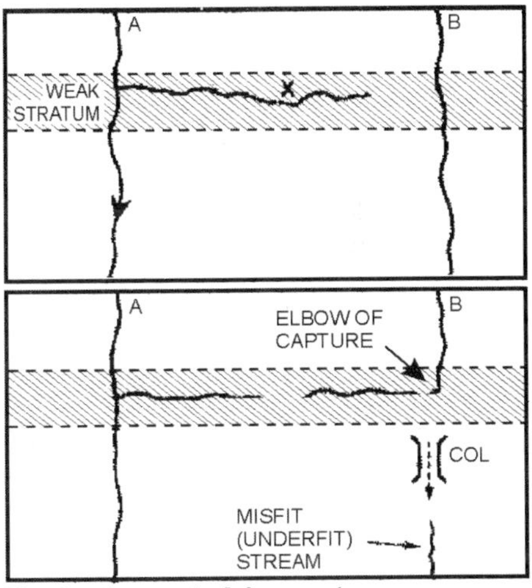

The development of river capture

river metamorphosis It refers to the change of river channel morphology that can occur when changes of discharge and sediment exceed a threshold condition. S.A. Schumm (1969) introduced the term river metamorphosis to signify the range of changes that may arise. It may involve changes in size, shape or composition of aspects of river channel morphology.

river profile A drawing of a section of a river, whether a long profile or a transverse profile which is cross-sectional and can be very variable.

river terraces These are formed in old and abandoned flood plains. When rejuvenation gives a river renewed powers of erosion, it is able to cut its valley down to a lower level. When this happens, the former flood plain is left at a higher level, and a new plain is created. The old plain is called terrace, and in many valleys flattish areas can be seen above the level of the river.

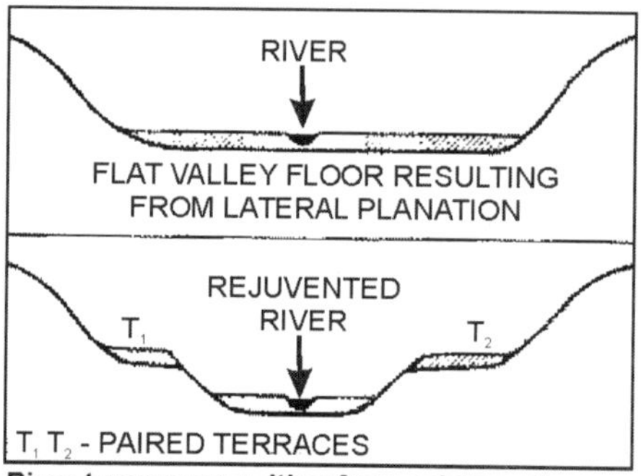

River terraces resulting from rejuvenation.

riverscape The landscape character of the river and the adjacent area.

Roaring Forties In the southern hemisphere, the area between latitudes 40°S and 50°S, where the westerly winds blow uninterrupted past any land mass. These westerly winds are persistent and very strong. The frequency of depression brings storminess, cloudness and roughseas responsible for a notoriously hazardous shipping zone. Since these winds below in forty to fifty latitudes these are known as 'roaring forties'.

roche moutonnee A French term for a glacially moulded rock mound exhibiting an asymmetrical form, with the gently sloping and smoothly abraded, up valley face (stoss), ice-plucked, down-valley face (lee). Roches moutonnees can be only a few metres in height, but some are much bigger reaching 30 m to 40 m. It is said that roche moutonnees were so named because they look rather like sheep lying down or perhaps like old-fashioned wigs called moutonnees.

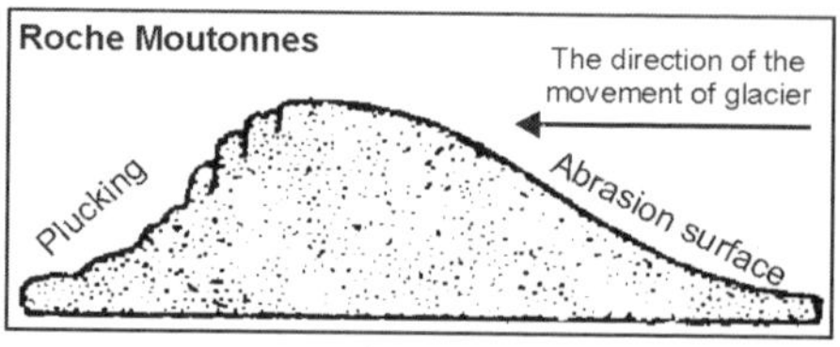

rock A coherent consolidated and compact mass of mineral matter. Rocks make up the surface of the earth, and they are very varied in appearance, hardness and mineral composition. All rocks can be classified into one of the three groups: sedimentary, igneous and metamorphic.

rock creep The movement of rock fragments across an inclined surface by means of intermittent and the ground surface. Such movement may be initiated by heating and cooling or by the growth of ice crystals beneath the rock fragment. It is also referred as superficial rock creep.

rock dome Type of inselberg, characterised by its vertical sides and 'sugar-loaf' shape.

rock fall The most rapid type of mass movement in which rocks ranging from large masses to sample fragments are loosened from the face of a cliff.

rock flour The fine debris created by abrasion beneath a glacier. The material, generally less than 0.2 mm in diameter.

rock glacier Term which has been used to describe a feature, common in many alpine areas, which looks like a glacier, but is composed of rock debris. Rock glacier usually exhibits slow movement, often less than a metre per year.

rock mass strength It is a quantitative measure of the resistance of a rock mass to erosion as given by Selby (1980) and others. It involves giving a rank of importance to a range of different rock parameters and then summing them to come up with a total rating of strength.

rock pavement L.C. King has introduced this term to describe a low, whaleback rock exposure in the flat alluvial plains of tropical areas. It represents the initial exposure of a **bornhardt**, during an early stage of stripping away of the regolith to the reveal of the basal surface of weathering.

rock slide The sliding of a detached block or rock along a plane of failure. The process is usually aided by the presence of water, which reduces friction along the sliding surface.

rock step A pronounced break in the long profile of a glaciated mountain valley. Frequently the down valley face of the step is precipitous, and bears the mark of powerful glacial plucking whilst the summit of the step has been smoothed by glacial abrasion. Rock steps are also related to the outcrops of hard masses or reflect subtle structural variations.

roller These are large ocean waves that breaks on exposed ocean coastlines where fetch is considerable. Rollers occur regularly on the coasts of oceanic islands even during calm weather.

ropy lava An extrusion of lava which has solidified in extended linear fashion, like pieces of thick rope. It is also called pahoehoe.

Rossby waves These waves were discovered by C.J. Rossby (1939). These waves are ridges and troughs in the westerly movements of the upper air, with a wavelength of around 2,000 km. Rossby waves are the result of reaction to the unequal heating of the earth's surface and also in response to relief barriers, like those east of Rockies and east of the Hima-

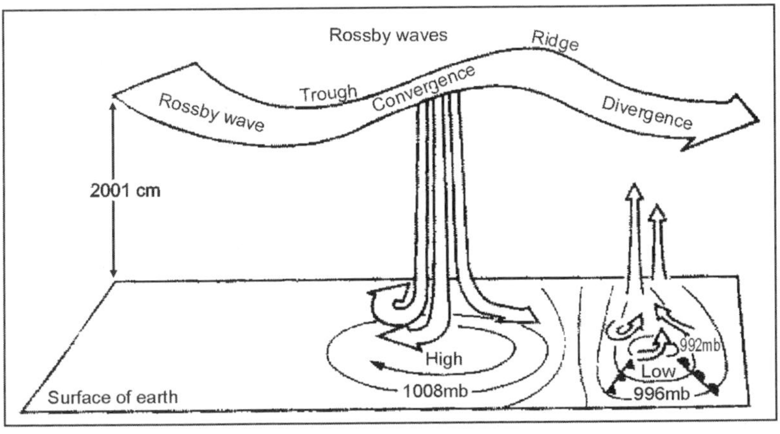

layas. These waves are associated with the formation of depressions and anti-cyclones in the middle latitudes of the N. hemisphere.

Rossi-Forel Scale A scale for measuring the intensity of earth-quakes, devised in 1878 by M.S. de Rossi and F.A. Forel. It was commonly used until 1931, when the modified Merclli Scale become more popular. Both have now been super-seded by the Richter Scale.

rotation of the earth The turning of the earth about its own axis. One complete rotation takes 24 hours, during which day and night are experienced, as different parts of the earth turn to face the sun. Because of the tilt, of the earth's axis at 2.5°, the length of day and night varies. At the equator, day and night are approximately of equal length throughout the year. At the North Pole there is 4.5 month's daylight and 4.5 month's darkness, while for the rest of year there is twilight. The same is true of the South Pole, although the periods of day and night are reversed. All places north of 66.5°N and south of 66.5°S receive at least one 24 hour period of continuous day light. On equinoxes—21st March and 22 September every where has 12 hours' darkness and 12 hours' day light.

rotational slip The downward movement, in a semi-circular motion, of a mass of rock, soil or ice along a concave face.

rotor streaming A form of turbulent air motion in a vertical plane, characteristic of a lee wave. It sometimes produces a lenticular cloud.

roughness It is an expression for the degree to which a stream channel is marked by irregularities, which by increasing friction slow downstream flow. The factors affecting roughness include the size and angularities of bed load particles, rock obstructions on the channel floor, the presence of ripples and sand bars, channel sinuosity, bank vegetation and man-made obstructions.

roundness A term used to describe the extent to which clasts have had their points smoothed by abrasion.

roundness index Roundness can be determined by various methods. Cailleux roundness is calculated from formula $\frac{2R}{a} \times 1000$ where R is the minimum radius of curvature in the principal plane, and a is the long axis of the fragment.

rudaceous Coarse-grained sedimentary rock, either consolidated as in conglomerate or unconsolidated as in till.

run-off The surface discharge of water, derived mainly from rainfall but also at a times from melting snow and ice, in the form of sheet wash and rivulets on slopes. Run-off is generated when rainfall intensity exceeds the infiltration capacity of the soil. It is affected by vegetation cover, surface roughness and the gradient and length of the slope.

ruware A low, dome-shaped exposure of bedrock projecting from a cover of alluvium, or weathered bedrock. Ruwares are common in savanna regions, as in Central and N. Nigeria and E. Kenya. They appear to represent 'domical rises' in the basal surface of weathering, which are in the process of being exposed by removal of overling chemically weathered rock.

S

saddle A slightly lower area between two summits. It is often flattish, and provides a gap for a route between two valleys. It is similar to a col, but is slightly broader.

saeter In Scandinavia, an upland pasture, usually used in summer only.

sahel An east-west zone along the southern margins of the Sahara desert of Africa. Sahel is subject to severe drought bringing crop failures, death of live-stock and famine. In this region droughts are partly natural but more due to human activities such as over-stocking, destruction of natural vegetation, etc.

salars Inland drainage basins located in desert, which are only occasionally inundated with saline water. They are also known as **playas**, **sabkhas** or **chotts.**

salina A Spanish term used to describe saline deposits in salt-marshes, playas and salars. These are extensively found in S. America, Central America, Mexico and SW states of the USA.

salinization The term is used to describe the salinity problems caused by the presence of salts in the water and soil. Salinization is the built up of salts at or near surface of a soil. In hot-dry climates,evapotranspiration exceeds precipitation so that the surface water evaporates rapidly. This caused the soil moisture, together with its dissolved salts, to come to the surface by capillary action. This water then evaporates, leaving behind a crust of salts on the surface. In coastal districts, salinization can occur when sea water percolates into the soil as a result of the over-pumping of groundwater.

salt dome A mass of salt which is injected thereby piercing and deforming them. It is in the form of a circular plug (salt play) upto 1.5 km in diameter but of considerable depth. It is of great economic importance because it assists in formation of a 'trap' structure for oil accumulation and also associated with deposits of gypsum, sulphur and anhydrite.

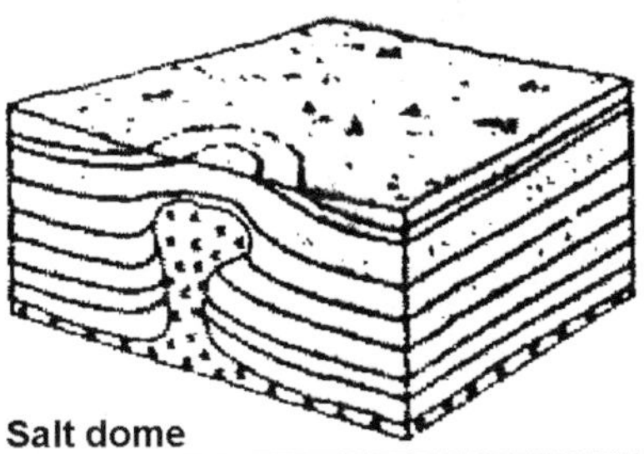
Salt dome

salt lake A lake containing water of high salinity, which occurs in areas where evaporation is high, mostly in arid and semi-arid regions. The salinity of salt lake is considerable higher than that of sea water, e.g., 238 per cent in the Dead Sea, 220 per cent in the Great Salt Lake, Utah (USA). In India, 'Sambhar Lake' is a salt lake, located in Rajasthan state.

salt marsh These are vegetated mud-flats in the high inter-tidal zone found commonly on many low-lying coasts. Salt marshes are also found around salt lakes in deserts and near large lakes and estuaries in warm regions. Many of the coastal marshes can be reclaimed once the land has become well-vegetated.

salt pan A shallow salt-lake occurring in a small enclosed basin.

salt weathering A type of weathering, results from the crystallization of supersaturated solutions of salts occupying pore spaces and cracks. As the crystals form and grow, expansive stresses are exerted on the rock, followed by granular disintegration or surface flaking. This process has been recognised in hot deserts and also in tropical humid climates.]

saltation A mechanism by which sediment is transported by bouncing along the surface. It is the most common form of wind transport in arid environments but is also significant in the movement of solid particles along the bed of a stream.

Saltation

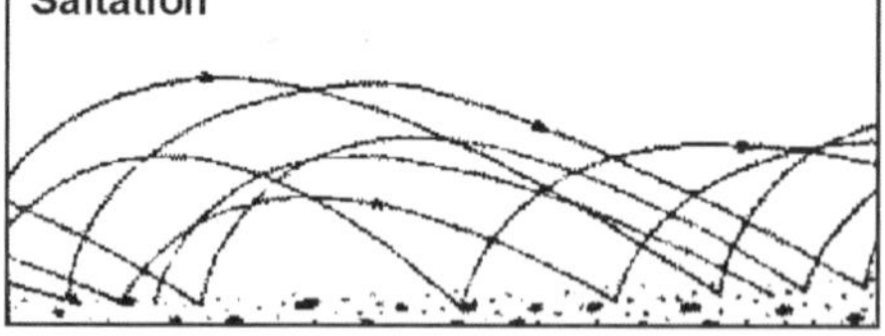

samoon (samun) An Iranian term for a hot dry wind of **fohn**-like character, descending from the highlands of Kurdistan.

San Andreas fault A 1,125 km long fault, stretching through California in the USA. It marks the boundary where two crustal plates move alongside each other, with the friction causing earthquakes on the surface. The cities of San Francisco and Los Angeles lie near the fault and have suffered many earthquakes and subsequent damage.

sand Unconsolidated material composed of small grains particles ranging in diameter between 0.06 and 2 mm. Sand is divided into fine (0.06–0.2 mm), medium (0.2.06 mm), coarse (0.6–1 mm) and very coarse (1–2 mm).

sand bank (1) An accumulation of sand in an estuary or in a coastal environment, created by the tidal flow working on an abundant supply of land-derived sediments. (2) A bank of sand built by currents in a review channel.

sand dune The deposition of sand by wind action in arid and semi-arid regions. Dunes formed in the lee of some obstacle are **topographic dunes, lunettes** are formed in the lee of a deflation hollow, **nebkhas** form in the lee of bushes, **wind shadow dunes** form in the lee of hills and plateaux. **Parabolic** dunes are hairpin-shaped with the bend pointing downwind and originate around patches of vegetation. When the supply of sand is limited, **barchans** form with the horns pointing downwind. There are also **star dunes** and linear **seif dunes**.

sand storm A mass movement of sand driven by a strong wind in desertic climates. It differs from dust storm because in this case sand particles are bigger and transported to greater distances.

sandstone It is a sedimentary rock composed of compacted and cemented sand. The sand grains are chiefly quartz and feldspar. It is also characterised by a wide variety of colours ranging from dark brown through red, yellow and grey to white.

sandur A large glacial outwash plain in Iceland formed from glaciofluvial material carried out from the front of an ice sheet by melt streams. Sandurs appear as extensive accumulations of gravel, sand and silt, crossed by braided streams.

sapping The undermining of the base of a cliff, with subsequent failure of the cliff face. It can be accomplished by a variety of processes, including lateral erosion by streams, wave action and groundwater outflow. The last is often called 'spring sapping.'

sapropel An aquatic ooze, highly charged with carbonaceous matter, that forms in swamps and shallow marine basins as a result of rotting. It has a higher percentage of hydrocarbon than peat and is considered to be a source of natural gas and petroleum.

saprophyte A plant that flourishes in an environment of putrefaction and draws its nourishment from rotting vegetable matter.

sastrugi These are the ridges of ice particles, transported by wind and lying across ice sheet. They are oriented at right angles to the prevailing winds.

satellite, artificial A space vehicle having various scientific instruments, that is placed into orbit around the earth or other planets in order to collect and transmit data. Since their inception, satellites have been greatly assisting in weather forecasting as well as in monitoring cloud patterns. There use now has become multipurpose with the introduction of remote sensing techniques.

saturated adiabatic lapse-rate The reduction in temperature of a pocket of saturated air which rises spontaneously. When a parcel of saturated air rises through the atmosphere it cools by expansion, but the amount of latent heat released with condensation depends on the amount of water vapour and therefore on the temperature of the air. If the air is very moist, the rate will be at the low end of this range, since the adiabatic cooling will be partially affected by the latent heat released by condensation.

savanna An area of tropical grassland, found between the major desert regions and the tropical forests. Savannas occur mainly between 5° and 20° north and south of the equator, although, in East Africa, savanna regions are situated on the equator as well. This is because the land is elevated, and the East African plateau is at the appropriate height for grassland vegetation to grow. Savanna grasslands are widely distributed in seasonally arid parts of Africa, S. America (**Llanos** of the Orinoco Basin and **Campos** of Brazil) and Australia, Pastoral farming is the main activity on the grassland and this may be large-scale commercialised farming in northern territory of Australia.

scar A short steep; slope, which is generally bare rock. The term is used in northern England, and it is generally associated with areas of carboniferous limestone.

scarp A steep edge to an upland such as cuesta or plateau. Owing to the effective evacuation of rock debris released by weathering, scarps are relatively rapid recession (scrap retreat) without significant loss of angle.

scarp-and-vale terrain The alternating pattern of **cuestas** and intervening lowland corridors which is created by differential erosion of gently tilted sedimentary rock structures. The vales are carved out in the less-resistant rocks, such as clays, while the scarps are formed on more resistant limestones and sandstones.

scattering The break-up of solar radiation by dust and molecules of waste vapour in the atmosphere.

schist A metamorphic rock characterised by thin plates of minerals, overlapping rather like the tiles on a roof. Green schists like **slates** and **phyllites** are formed from clay sediments altered by low-grade metamorphism. Crystalline schists are formed at higher metamorphic grade, producing quartzitic schists as **mica schists**, **hornblende schists** and **biotite schists**.

scree These are loose fragments of rock which have accumulated on a hillside. They are likely to slide down the hill when lubricated. The rock fragments may be very varied in size and when the move downhill large particles tending to move to the bottom.

scroll It is a low, curving ridge on a flood plain running parallel with the lops of a meander. It is sometimes referred to as a scroll bar and is part of the floodplain micro relief.

scrub A vegetation type associated with poor soil and low rainfall. Scrubs are mostly found in semi-arid regions in which species are stunted, gnarled to seasonal drought.

sea arch An arch cut by wave erosion through a headland.

sea breeze A wind which blows from the sea on to the land. In tropical areas, it is usually a cooling breeze, as the sea is cooler than the land. In temperate latitudes, it is not only cool and raw but bitterly cold. During the day, as the land is warmed by solar

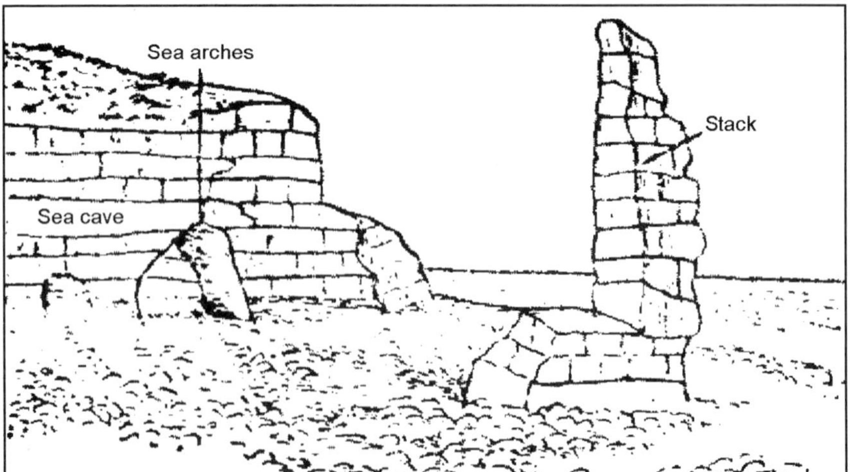

Sea arches stacks, Sea cave etc.

radiation, atmospheric pressure is slightly lowered by compensation with the relatively high pressure maintained over the cooler sea; this leads to compensation by way of a movement of air landwards.

sea cliffs These are the steep slopes that border ocean coasts. They are ubiquitous and occur along approximately 80 per cent of the ocean coasts of the earth. According to their stage of development, sea cliffs can be classified into three main types: **active cliffs**, where bedrock is exposed by continuous retreat under the influence of marine and subaerial forces, **inactive cliffs**, where their bases are mantled by **talus** and **former cliffs**, where the influences of marine erosion have disappeared. Strees (1962) stated that many cliff profiles display complex forms brought about by long continued changes in sea level, climate, and the balances between subaerial and marine processes.

sea ice Ice which forms on the sea surface when water temperature falls to around –1.9° C. Such conditions apply to considerable areas of the Arctic and Antarctic.

sea level The mean surface elevation of the sea, determined by averaging all tidal levels recorded at hourly intervals.

sea stack A small pillar-shaped, rocky island formed by wave erosion through a headland near a sea cliff.

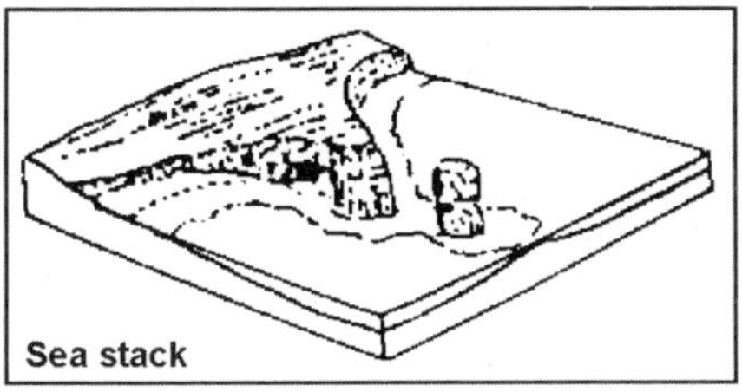

sea surge A large movement of oceanic water, which is likely to cause large waves or high tides when it reaches land. Britain is affected occasionally by a North Sea surge, where a large volume of water is moved southwards through the North Sea.

sea-floor spreading The movement in plate tectonics whereby some ocean floors are extended laterally as plates move apart from a zone of separation, which corresponds to a mid-oceanic ridge, as a Mid-Atlantic ridge and other mid-ocean divergences.

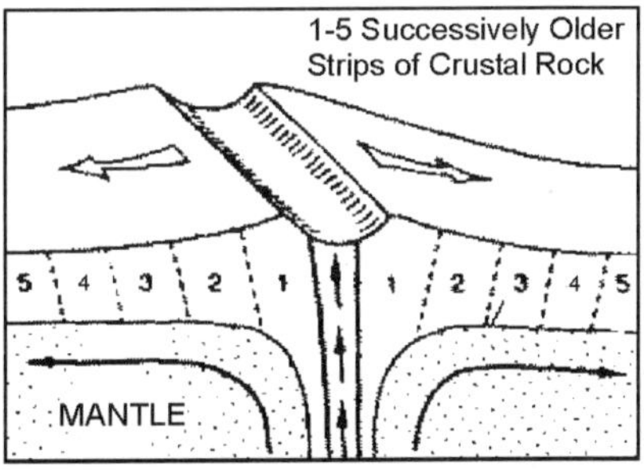

Sea-floor spreading

seamount A mountain on the ocean floor which does not break the water surface. Its origin is mostly volcanic and its conical summit contrasts with that of a guyot. Seamounts can be isolated, clustered into seamount groups or in a linear pattern as seamount range. They occur most frequently on the floor of the Pacific Ocean.

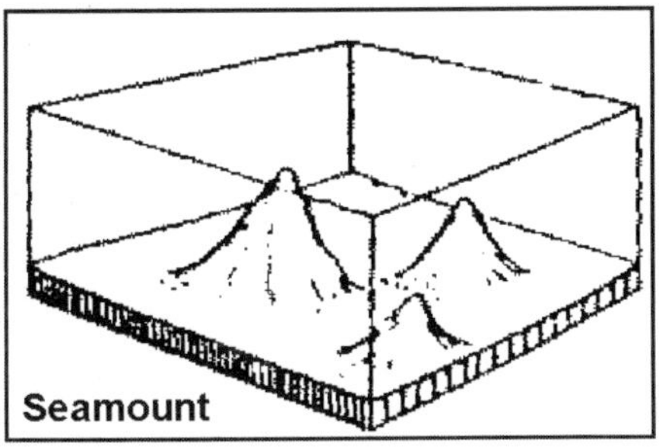

seasons A time period generally based on differences in duration and intensity of solar radiation, resulting in different climatic conditions. The four seasons of the year are: spring, summer, autumn and winter. The resulting contrasts of daylight hours and temperature differences in middle and high latitudes are used to divide the year into seasons, often linked with the farming calendar. In equatorial regions, seasonal changes are minimum. Similarly, in very high latitudes the spring and autumn seasons are barely noticeable, so that a two-season year (winter and summer) prevails.

secondary A geological era, following the **Primary** and preceding the **Tertiary**. Now generally known as the **Mesozoic**.

secondary depression A low pressure small area which is found on the edge of a major depression. Secondary depressions often become deeper and more intense than the primary depression, and in Western Europe they often move around the primary in anti-clockwise direction. A secondary depression may start life, often at a **front**, simply as a bulge in the isobars, but it may develop into a distinct pressure system in its own right, with closed isobars and with a central pressure sometimes lower than that of associated primary.

sedentary soil A soil formed *in situ* form a regolith derived from the underlying rock.

sediment Particles of rock which have been transported and laid down by agencies such as rivers, glaciers, waves, wind.

sedimentary rock Rock formed by deposition of sediment derived from preexisting rocks. Most of the sediments accumulate on the land of the sea, having dumped thereby rivers. The accumulated sediment will be consolidated and compressed. Earth movements uplift the sediments and they may be tilted, folded or faulted and the resulting rocks are sedimentary. The layers of sedimentary rocks are called **strata**. Sedimentary rocks are usually classified, according to their mode of formation into: (i) **clastic sediments**, (ii) **organic deposits calcareous rocks** and (iii) rocks formed by chemical processes dolomite, evaporite, gypsum, limonite, rock salt, etc. These rocks are also classified on the basis of their geochemical nature: **resistates**, **hydrolystates**, **oxidates**, **reduzates**, **precipitates** and **evaporates.**

sedimentology The scientific study of sediments, sedimentary rocks and of the processes by which they were formed.

seepage (1) The gradual soaking away of surface water into the soil. (2) It is the slow emission of groundwater at the surface when the rock structures permit, but in lesser volume for it constitutes a spring.

seiche The oscillation of a body of water at its natural period. The period of oscillation varies from a few minutes to several hours and is set up by rapid changes in wind direction, in atmospheric pressure or occasionally by an earthquake.

seif It is an elongated type of sand dune, found in parts of Libya and other areas of sand desert. These are often occurring in groups, seifs may be extended for several kilometres.

seismic waves These waves are produced from the focus of earthquake. Three types of waves are: (1) **Primary** (P) **waves** of short wavelength and high frequency that are longitudinal waves, which not only travel through solid crust but also through liquid part of the core of the earth. (2) **Secondary** (S) **waves** of short wavelength and high frequency, that are transverse waves which travel through all solid parts of the earth but not the liquid part of the core. (3) **L waves** of long wavelength, low frequency and confined to the skin of the earth's crust.

seismograph An instrument designed to record seismic waves.

seismology The scientific study and interpretation of earthquakes.

selva The equatorial rain forest of the Amazon Basin, now also applied to similar types of vegetation elsewhere in the world. They are characterised by a rich variety of plants and insects, many of which are not found elsewhere.

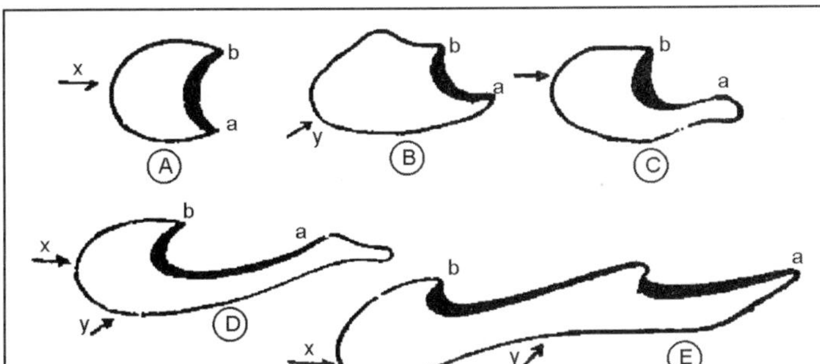

Development of a seif

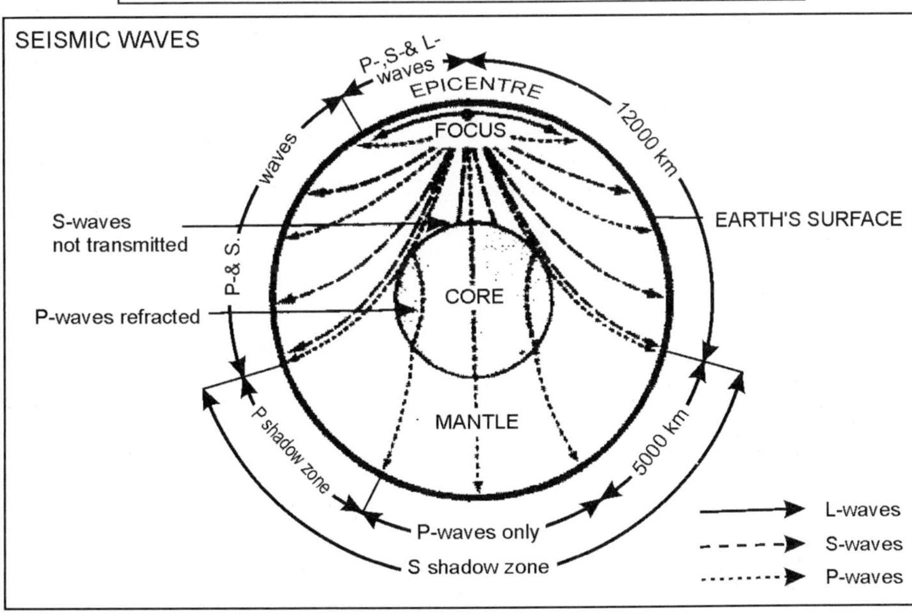

semi-desert It is a transition zone between true desert and more thickly vegetated zones. The climatic conditions of the semi-desert areas are similar to that of arid regions but with some wildness. These areas also received some rainfall, therefore, having patches of vegetation, mostly thorny in nature.

sensible temperature The temperature not recorded by the dry bulb thermometer, but as 'felt' by human beings. This is influenced by atmospheric humidity and wind speed. If humidity is high (in the order of 90%), very warm temperatures (of 25°C and above) feel extremely uncomfortable.

sensor A term used in remote sensing, referring to any device that is sensitive to electromagnetic radiation and can convert its measurements into a form suitable for input an information gathering system.

seracs An irregular surface of ice, generally found at the foot of an icefall on a glacier. It is the result of a crevassed section of a glacier becoming compressed when the gradient decreases or the seep of ice movement is reduced.

sere A sequence of communities, usually plant communities, at a particular site. It is a series of stages that follows from the process of succession.

shake hole A depression in which water drains into underground limestone cave system. Warwick (1976) stated that the term is used synonymously with Doline and swallow hole, but should be restricted in usage to a depression formed by callapse of underlying limestone strata.

sheet erosion A very slow-acting form of erosion whereby a thin film of water-sheet wash transports soil particles by rolling them along the ground sheet erosion is effected by (i) the impact of raindrops, (ii) a surface layer of rain wash, and (iii) the subsurface removal of fine particles by through flow. Generally, sheet erosion is most effective on the upper and lower parts of convexo-concave slopes, at points where gradients are relatively gentle.

sheet flood A form of surface run-off developed on gently sloping, unchannelled areas in deserts.

sheet flow Unchannelled flow of water over the soil surface. This is an overlain flow on the smoothest surfaces, and on all natural hillsides. In the sheet flow areas between sediment detached or dislodged by rain-splash will be transported further than in the absence of flow, so that sheet flow has a small influence on sediment transport even though it cannot initiate erosion.

sheet-joint A joint running parallel to the surface of a rock, usually granite or gneiss, resulting from the process of dilatation. Sheet-joints result in a 'layered' structure.

shield These are large rigid blocks of ancient rocks, which belong to Pre-Cambrian in age. Shields have been affected in the distant geological past by intense folding movements, powerful metamorphism and volcanicity. Examples include the Canadian Shield of North America and the Baltic Shield in northern Europe.

shield volcano A volcano which has been built up by eruptions of fluid basic lava. Because the lava has been free flowing the slopes of the volcano are quite gentle, allowing the lava to flow out over a large area. Eventually, the massive mountain can be created. The largest example is Mauna Loa in Hawaii, which has risen over 9000 m from the seabed. The name 'shield' is given because basic volcanoes are shaped like an upturned shield or a saucer.

shingle Pebbles which accumulate on a shore. The stones have been smoothed and rounded by water action and by attraction.

sial The uppermost lighter, granitic layer of the earth's crust consisting largely of silica and aluminium. The layer of sial rocks forms the continents, which floats over sima layer, because its density is less than sima layer.

sideways looking airborne radar (SLAR) This is a microwave remote sensor used to derive images of the earth's surface. The SLAR senses the terrain to the side of an aircraft track. The SLAR possesses collectors, detectors and recorders and, in addition, a transmitter and antenna. The imagery received from SLAR is used in geomorphology, the mapping of soil moisture and vegetation.

silicon An element which comprises about 28 per cent of the earth's crust. It is the second most common element in the crust, after oxygen with which it combines to form silicon dioxide (SiO_2), generally known as silica or quartz. It is also found in feldspars, mica and many other minerals.

sill A tabular sheet of igneous rock injected along the bedding planes of sedimentary or volcanic formations. Sills can be a few centimetres, or many metres in thickness.

silt A sedimentary deposit laid down in a river or lake. It consists of particles which are finer than sand but coarser than clay, with a diameter from 0.002 mm to 0.02 mm.

siltation The accumulation of fine sediment in a body of water. In fact, it is a deposition of fine particles in lakes, reservoirs or water courses.

sima A term introduced by Adward Suess to describe that past of the earth's crust which is composed of material rich in silica (Si) and magnesium (Mg). It is denser (2.9 to 3.3) than **sial**. It forms the crust of the ocean basins and the lower portion of the crust of the continents.

simoon An Arabic term for a warm and excessively dust-laden wind in the N. Sahara, accompanying an occasional cyclonic storm which invades the desert.

simulation A means of representing a real world situation or real system in an abstract form. Hypothesis-testing in physical geography often makes use of the vicarious experiment procedure of simulation in which genuine phenomena are represented by scale, analogue or mathematical models. Humans can use and manipulate the parts of a model in a fairly subjective manner. Computers can be used in simulation and they will be objective. Simulations can look into the future and make predictions about what is likely to happen.

sinkhole This is a hollow or hole in an area of limestone where solution has enlarged a joint. There may have been collapse of surface material. Water may seep down a sinkhole after periods of heavy rain there will not be a permanent flow of water down the hole. Some sinkholes are large enough for potholers to descend.

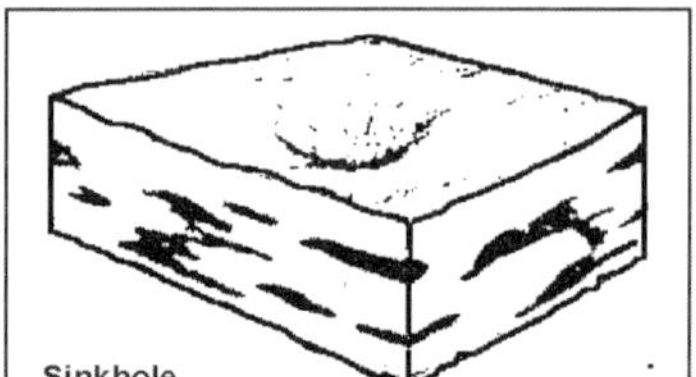

Sinkhole

sinter A deposit of minerals, notably of silica and sulphates, precipitated in layer it deposits from the gases released in an area of volcanic activity. **Sinter terraces** can result from this deposition.

sinuosity The amount that a river meanders within its valley, calculated by dividing stream length by valley length. It may be defined as the ratio of actual channel distance between identified points compared to the straight or down-valley distance.

siphon A vertical U-shaped portion of a subterranean stream channel in which the water is in hydrostatic equilibrium.

sirocco (scirocco) A warm, drywind which blows northwards from Sahara Desert towards the Mediterranean. Sirocco is also known as **Khamsin**.

slide Generally referred as landslide. The landform produced by mass movement under the influence of gravity. The sliding mass generally breaks into many blocks as it moves.

slip face The steeply sloping portion of some sand dunes on their down-wind side where overstepping causes sand to cascade downwards and rest as its maximum angle of repose.

slip-off-slope The relatively gentle slope at the inner edge of a meander.

slope In physical geography, the word slope is used with two senses: (1) It refers to the angle which any part of the earth's surface makes with a horizontal datum. (2) In geomorphology, it refers to any geometric element of the earth's solid surface whether that element is above or below sea level. Slope elements thus form entire landscapes. A slope may be **concave, convex, steep, gentle, constant, denudation slope, foot slope, free face, non-relict slope, parallel retreat of slopes, talus slope, wash slope**, etc.

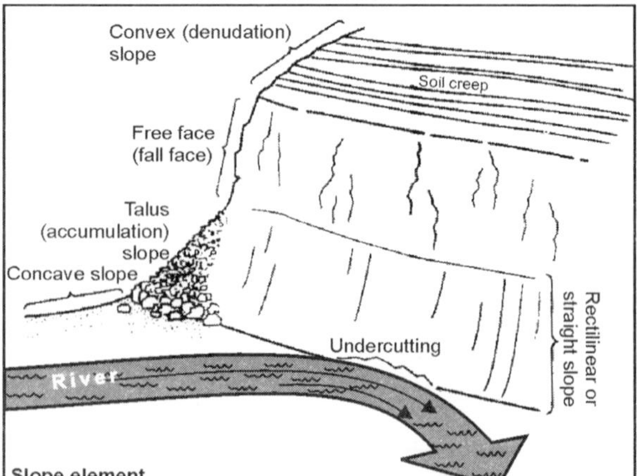

Slope element

slope decline This is an hypothesis referring to a type of slope evolution in which the steepest part of the slope progressively decreases in angle. The

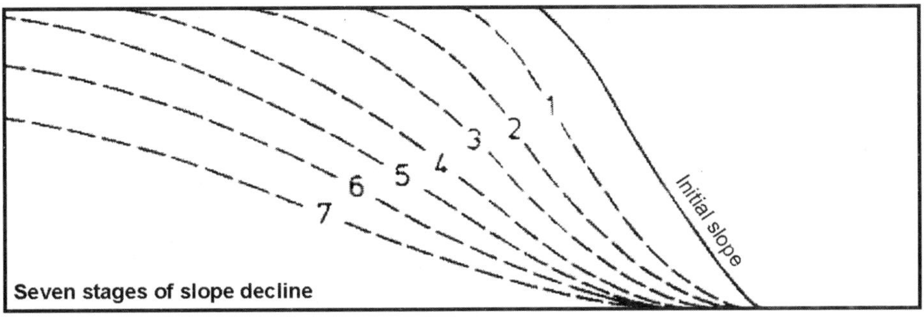

Seven stages of slope decline

concept to slope decline was originally formulated by W.M. Davis as part of his 'cycle of erosion.' Slope decline was earlier considered the characteristic only of the normal cycle of erosion in a humid temperate climate, but now it has been proved that this process is operative under different climates.

slope genesis It refers to the evolution or formation of a slope. The character of slope will depend on the following variables: lithology, geomorphological process, climate, vegetation cover, tectonic movement, and base-level changes. The three important hypothesis relating to slope evolution are slope decline, parallel retreat of slopes and slope replacement.

slope replacement W. Penck has suggested a hypothesis in which the maximum slope angle decreases through time as a result of replacement from below by gentler slopes, causing majority of the slope profile to become occupied by a concavity

slope unit The individual element of a concave or convex slope or the segment of a rectilinear slope. When examined in detail most slopes are found to be composed of a number of slope units with different angles, rather than of a single unit of uniform gradient.

slump A form of mass movement where rock and soil move downwards along a concave face. Slumps are most common in thick regoliths and large mud stone rock units, but can also occur in hard rock which has been shattered.

smog A mixture of smoke and fog. It is a photochemical smog which forms when hydrocarbons, originating from vaporised gasoline and other petroleum products, combine with nitrogen oxide molecules and water vapour in presence of ultraviolet sunlight. Smogs were common in most industrial cities until the smoke-abatement laws of the mid-20th century led to cleaner air, but not before there had been several disasters, the greatest of which was the London smog of 1952 when some 4000 people died from bronchial disorder exacerbated by the smog.

snout The downstream end of a glacier. It will be an area of melting for much of the year, and a river is likely to be flowing from underneath the ice, often through a cave.

snow Type of precipitation formed when atmospheric water vapour condenses at temperatures below

0°C. The resultant minute spicules of ice join together to form larger crystals, which themselves aggregate into snow flakes.

snow field Accumulation of snow in a mountainous area. If there is enough snow it compress the snowfield, it will gradually turn into ice, and eventually glaciers will be able to form.

snow line Line on a mountain side above which there is perpetual snow. The altitude of the snow line varies with latitude. At equator it is about 5000 m. In the Alps, the snow line is about 3000 m, but it varies from north to south, and there is further variation depending on the aspect of the slopes.

soil The thin surface layer of the earth, comprising closely intermixed mineral and organic substances. It consists of tiny particles derived from the broken down fragments of rock, together with accumulations of plant remains. The organic remains provide the humus and the inorganic particles provide vital minerals. There are many types of soil, and various methods of classification. A common classification is into **zonal soils** (for example, **podzols**), **azonal soils** (alluvium) and **intrazonal soils** (as peat).

soil conservation The protection and preservation of the soil, so that it will not be ruined or eroded. In order to conserve the soil, it should not be left bare for too long, otherwise, wind and rain may start to erode it. Soil can be conserved by planting grass or trees around an exposed area. In many areas, hillsides are terraced in order to prevent the soil being washed away down the hillside.

soil creep It is a slow downhill movement of soil particles. The movement can take place if the gradient is as slight as 2°. It is very slow, but over long periods large quantities of soil slide to the bottom of hillsides.

soil erosion Removal of soil by processes such as rain wash, deflation and gulling by wind more rapidly than it is formed by natural pedogenic activity. Soil erosion is thus a form of accelerated erosion, resulting from such activities as deforestation, overgrazing and ploughing which leave the soil surface bare and unprotected from rainfall. Accelerated soil erosion has been occurred by human activities. Erosion may be sheet or gully, and may be caused by wind or water. Wind and water erosion are likely to occur at the same time in the same place. In addition to the soil erosion caused by deforestation and agriculture, urbanisation, fire, mining are often significant in accelerating erosion of the soil.

soil horizon Visible layers within a mature soil, identifiable in the first instance by its colour but on closer examination seem to reflect differences in mineral composition and humus content. Soil horizons are sometimes clearly demarcated, but on other occasions grade into each other.

soil landscape systems The system in which soils are organised within the landscape. The interaction between soils and topography can be treated at several levels of scale. Dalrymple (1968) has developed a hypothetical nine-unit land surface model which shows the relationship between slope position and dominant soil and land forming processes.

soil moisture The moisture which is contained within the pore spaces of a soil, thus providing one of the vital needs for plant growth.

soil profile This is a vertical section through a soil, revealing its crudely layered structure. The formation of distinctive soil profiles is due to the operation of pedogenic processes, particularly leaching, eluviation and illuviation.

soil structure The manner in which soil grains are aggregated together into larger pieces known as **peds**. **Blockly** structures are commonly found in loams, **prismatic**, (or columnal) structures develop in clayey soils, **platy** structures, consisting of 'flat' peds in a horizontal position are associated with compacted horizons and **crumb** structure are characteristic of humus rich soils.

soil taxonomy A hierarchical classification of soils ranging from the most general grouping (soil order) to the most detailed (soil series). There are 10 orders and 47 sub-orders of soil. Similarly, the number of great soil groups is 180 and their sub-orders are 960 with 4,700 families and 11,000 series.

soil texture The make up of the soil according to the proportions of sand, silt and clay present. Twelve different textural classes are recognised, and the structure of soil can be determined when the percentage of three soil constituents are plotted on a taxonomy diagram.

solar day This is the time interval between two successive transits of the sun across a given meridian. The mean solar day is 24 hours, the time varies slightly with latitude, since the earth's orbit is ellipsoid and inclines towards the equator. This is also known as **lunar day, mere solar time, sidereal day**.

solar energy Any energy source based directly on the sun's radiation. Solar heat is trapped by an absorbent material, usually a black metal panel. The heat is then transferred to pipes which carry warmed air or water. Direct conversion systems of solar energy include: (i) **solar cells**, which convert directly to electricity by thermoelectric and photovoltaic means; (ii) **flat-plate collectors**, in which a fluid, such as water is heated; (iii) **focusing collectors**, which concentrate direct radiation on to a small area, (iv) **chemical and biochemical systems**, which are largely at the experimental stages. The chief advantage of solar energy is that, to all intents and purposes, it is inexhaustible.

solar system The system in which nine planets plus the asteroids revolve around the Sun following almost circular orbits.

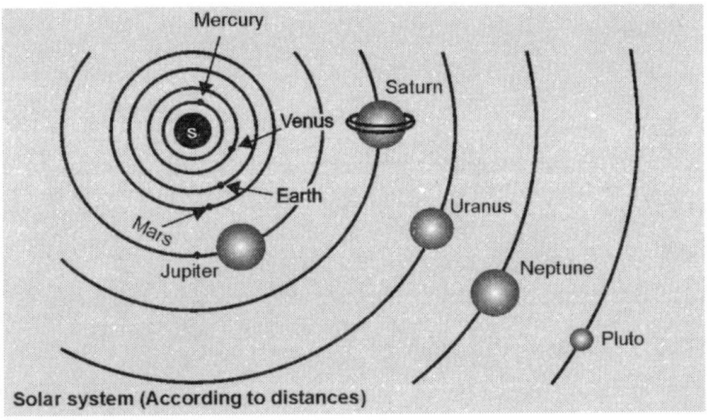

Solar system (According to distances)

solar year The time period during which the earth completes one orbit around the sun. It comprises 365.2422 mean solar days or 365 days 5 h 48 min and 45.51 sec.

solfatara A vent through which steam and volcanic gases are emitted. The phase is characterised by a gentle emission of sulphurous gases and steam but without violent explosions.

solifluction Its actual meaning is the flowage of soil which has become saturated or super-saturated with water. Solifuction is a major transportational process on slopes in periglacial regions. As water freezes at night and then thaws during the day, particles of soil are moved and then slip down hill because of the effects of gravity. Both water and ice make these movements easier. Solifluction is most likely to occur in Tundra regions.

solstice The day of maximum or minimum declination of the sun. It is either the longest or shortest day of the year. When the sun reaches the tropic of Cancer on 21 June, this is summer solstice. Similarly, when it reaches the tropic of Capricorn on 22 December, this is winter solstice. On the summer solstices, the sun is directly overhead at the tropic of Cancer, and it is the longest day of the year for the northern hemisphere. At the winter solstice, it will be the shortest day of the year in the northern hemisphere.

solution The process by which matter is changed from a solid or gaseous state into a liquid state by combination with solvent. It is a form of chemical weathering, in which solid rocks are dissolved by water. This process is more effective in limestone topography and responsible for the development of karst landforms.

solution collapse The collapse of the roofs of underground caverns and passages, particularly in limestone, resulting in the formation of surface depressions.

spate A sudden increase in the velocity and volume of a river due to heavy rains or sudden melting of snow in the upper reaches.

spatter cone This is a small but steep-sided mound of lava (below 15 m) built by small eruptions around a volcanic vent or along a fissure.

species A collection of similar organisms that are capable of interbreeding to produce fertile off-springs. It is the basis of any taxonomic classification of organisms.

spheroidal weathering A process in which spherical boulders develop from original joint bounded rock masses, frequently composed of granite, gneiss or basalt. The rounding is due primarily to chemical weathering, operating beneath the ground surface with a partially decomposed layer.

Spheroidal weathering

spit A bank of sand and/or shingle, projecting from the shoreline into the sea or partially across the mouth of a river estuary or deep coastal inlet.

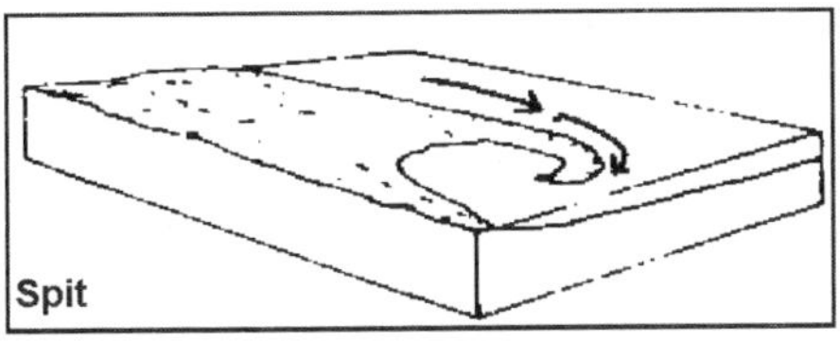

spongwork A pattern of shallow depressions and intervening rim-like partitions occurring on the walls and ceilings of caves in a karst topography.

spot height It is an exact point on a map with elevation printed along side. The dot represents height above mean sea level.

spring line A line of springs indicating the level at which the water table intersects the ground surface. A spring line, comprising numerous individual springs spaced at intervals of 1–2 km.

spring tide A tide with a range higher from that of the mean tidal levels. It occurs twice each month around the time of the new moon and the full moon. It is due to the complementary gravitational effects caused when the earth, moon and sun are in a straight line.

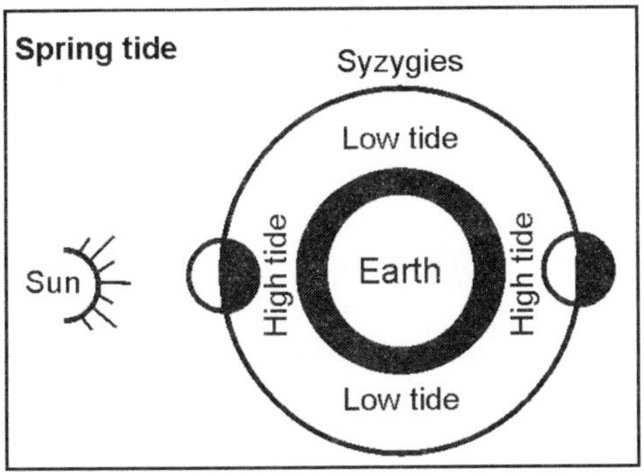

springs The emergence of groundwater at the earth's surface, usually at a clearly defined point marking a fault or major joint which has guided the underground flow of the water. Springs are classified in many different ways. According to temperature one may distinguish—**thermal** or **geothermal** and **coldwater springs**; while on the basis of water quality springs are **mineral** and **saline**. There are **permanent springs**, where the climate is wet throughout the year or where underground source of water is enough to support it throughout the year. **Intermittent springs** are found in climate characterised by long droughts, springs occur in a variety of situations, for examples at the base of scarp faces (scarp-foot springs, dip-slope springs, etc.)

spur An area of high ground which protrudes into an area of lower ground, such as river valley. Spurs alongside valleys are gradually eroded by the river, but in the upper and middle courses especially, they often form interlocking spurs.

squall A sudden increase of wind-speed, often accompanied by turbulence and showers of rain, but for a short duration. A series of squalls sometimes accompanies the passage of a cold front.

stability (1) The condition of the atmosphere in which, if a parcel of air is given an upward impulse, it will return to its original position because it remains cooler and heavier than the surrounding air. (2) The ability of an ecosystem to maintain or return to its original condition following a natural or human-induced disturbance.

stabilised dune A dune that has become fixed or anchored by vegetation and is therefore protected from further aeolian action.

stack A free standing pinnacle of rock usually in the sea which represents an outlier of a coastal cliff. It was formerly part of the adjoining land but has become isolated from it by wave erosion, probably after having formed part of a marine arch.

stage (1) The point to which a landform has evolved during a cycle of erosion (youth, maturity, old). (2) In geology, a stratigraphic division of rocks formed at the same age. (3) The level of water in a channel. (4) The position reached in a sequence such as Rostow's stages of growth.

stalactite The stalactite is formed by the precipitation over a long period of time, of calcium carbonate contained within water which has percolated from above by way of joints and fissures. The process of deposition involves both evaporation of water and the escape of carbon dioxide, with the result that dissolved calcium bicarbonate is transformed into soluble calcium carbonate and because a mass of calcite, often in

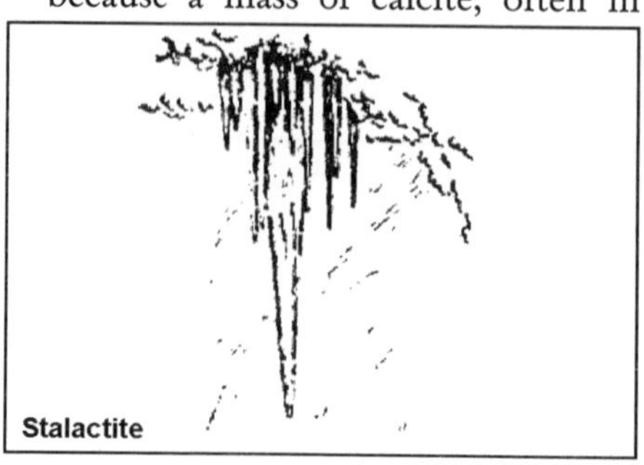

columnar form, suspended from the roof of a limestone cavern.

stalagmite A column of pure limestone, formed on the floor of a cave when the dissolved calcium carbonate in the underground water is deposited and the water evaporates as it splashed onto the cave floor. The process of formation are similar to those producing stalactites. However, the stalagmite is often thicker and less high the finer, more cylindrical formations of stalactites. Both these features are common in limestone topography.

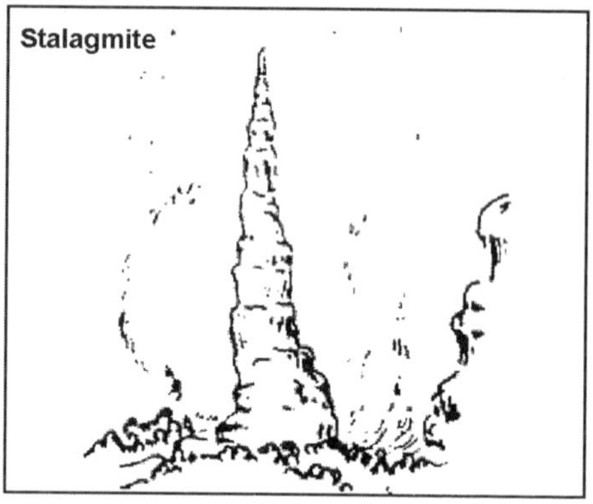

stand An area occupied by a collection of plants or trees which are structurally and floristically homogeneous.

standard hillslope A hillslope which has four components: an upper convexity or waning element, a free face, a debris slope, and a basal concavity or waxing element. This term was proposed by L.C. King (1962) to describe the 'fully developed' hillslopes found particularly in areas of massive bedrock under semi-arid conditions, where the production of debris by weathering and its removal by various slope processes is at an optimum. Other writers have proposed slope models in which these four components are further subdivided.

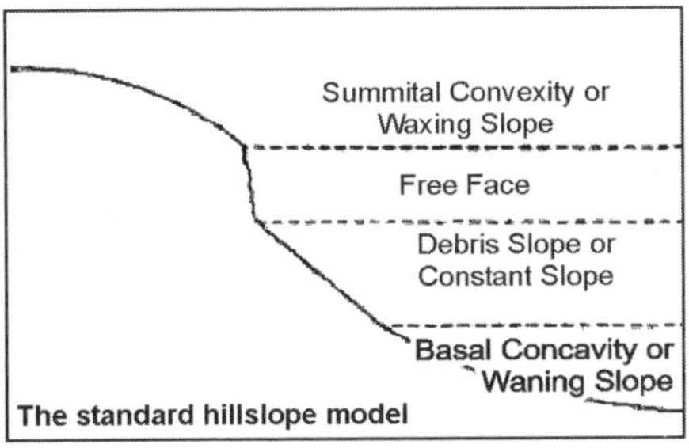

The standard hillslope model

star dune A pyramidal dune of star shaped, with three or more radial arms extending in various directions from a high central cone.

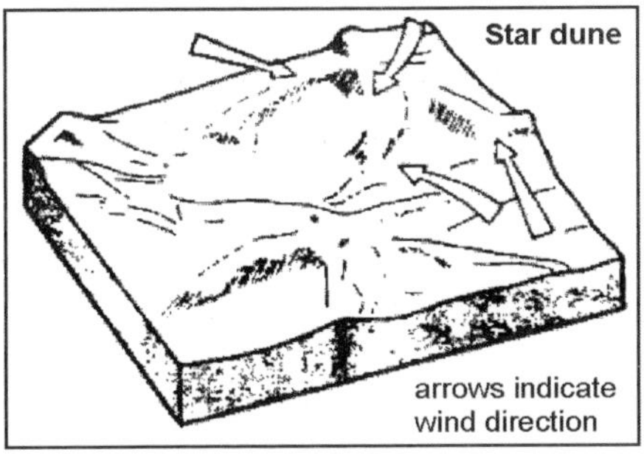

Star dune

static rejuvenation The rejuvenation of a river resulting from an increase in its discharge rather than a fall in base-level. Such an increase in a river's erosive capability may be caused by climatic change or by river capture.

static slope A slope on which there is no addition and no removal of material, because of resistance of the materials forming slope is greater than the strength of the forces which tend to carry them away. Although in practice, this condition can never be attained in a natural environment.

steady time S.A. Schumm and L.W. Lichty (1965) proposed one of the categories of the 'geomorphological time-scale'. It is the time when minor landform features are created, i.e., a period of short duration.

steam fog This is a type of mist or fog, usually developed on a small scale, which is associated with the passage of cold air over a warm water body. Evaporation from the surface of water results in instant condensation in the cold overlying air to give innumerable tiny water droplets. Thus, the water surface gives the appearance of 'steaming'.

step fault It is a series of parallel faults with a downthrow of the rocks in the same direction. Step faults are characteristic of the margins of rift valleys and result in a morphological staircase, as in the E. African Rift Valley at Lake Naivasha, Kenya.

Step fault

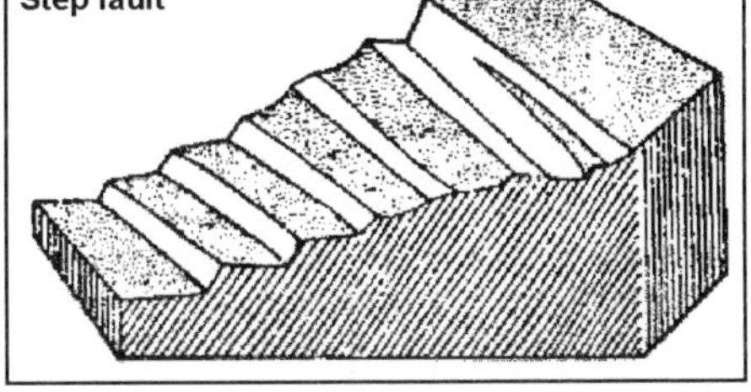

steppe The temperate grassland found in Kazakhstan, Ukraine and in neighbouring parts of southern Europe. The natural vegetation has now been removed or much altered by cultivation and grazing.

steps A type of micro-relief created by frost-action on slopes. They consist of either stone-banked terraces or turf-banked terraces, which have been derived from circles or polygons.

steptoe A term referring to an isolated hill or tract of land that has been surrounded by a lava flow.

Stevenson screen It is a white box on legs, designed to house weather instruments. It stands about 1.3 m above ground level and has louvred sides to allow free circulation of air. It is painted white in order to reflect heat. It will contain wet and dry bulb thermometers and maximum and minimum thermometers. It may also contain a thermograph and a barograph.

stock A granite intrusion which is like a small version of a **batholiths.**

stone pavement The word used variously such as: a hard mineral matter, a small piece of rock, a gem or precious stone. Also used as suffix, e.g., sandstone, limestone, soapstone, etc.

storm beach A beach ridge situated well above the normal limit of high tides, formed as unusually large waves fling ashore shingle, cobbles and bounders.

storm surge A rapid rise of sea-level caused by storm winds pushing sea water on to a windward coast. If the storm occurs during a period of high (spring) tide the sea-level may surpass the highest predicted tides and cause coastal flooding.

stoss A German term for the slope facing the direction from which a glacier moves. Also referred for the less steep side of a ripple.

strain A measure of the deformation of a body where a load or stress is applied. It is usually expressed as a dimensionless value (ratio or percentage) as, for linear strain, it is the change in length divided by the original length.

strait A relatively narrow waterway linking two larger bodies of water, e.g., the strait of Dover links the North Sea with the Channel.

stratigraphy The study of the order and arrangement of geological strata. Its sub-branches are **lithostratigraphy** (rock stratigraphy), **biostratigraphy** (related with fossil content), and **chronostraitgraphy** (time stratigraphy).

stratosphere The layer of the atmosphere, occurring between the **troposphere** and the **mesosphere** at a height of about 50 km. The base of the stratosphere is at 16 km above the equator and at 9 km above poles. At this level, the temperatures range from –80°C to –90°C over equator, but from –40°C (in summer) to –80°C above poles. The lower part of the stratosphere is virtually isothermal, but in the upper layers temperature increases with height to reach its secondary maximum of 0°C at **stratopause**. A great part of the atmospheric ozone is found in the stratosphere but very little water vapour.

stratus A cloud formation, developed mainly in the lower levels of the atmosphere (at less than 2500 m). These clouds often forms under stable atmospheric conditions, from the turbulent mixing by wind of warm and cold air in the surface layer. The result is a low grey-looking cloud.

stream order This is a morphometric classification of a drainage system according to a hierarchy of orders of magnitude of the channel segments. The scheme of stream ordering was devised by R.E. Horton and later modified by A.N. Strahler. The very smallest headwater streams within a basin are identified and designated as **1st order streams**. Where two such 1st order streams join, a **2nd order stream** segment results; extending down to meet another second order channel to create a **third order channel** and so forth. Thus, the highest order stream within the drainage basin is necessarily the largest in terms of discharge. When the task of order designation is complete, order analysis can begin.

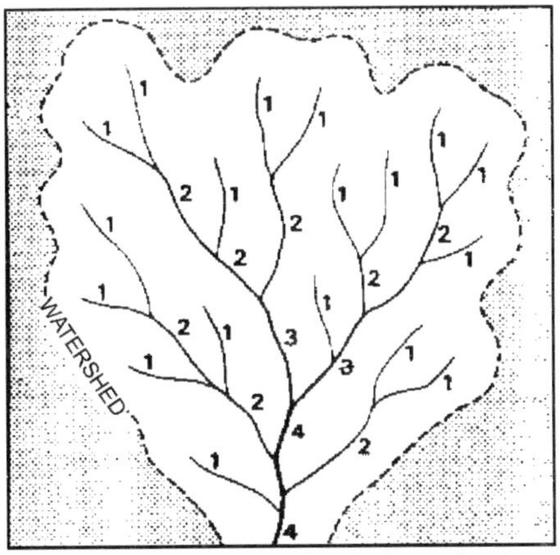

Stream order in a 4th order drainage basin

stream piracy A diversion of the head-waters of one stream into another stream. The process occurs by headward erosion of a stream having greater erosive power than the stream it captures.

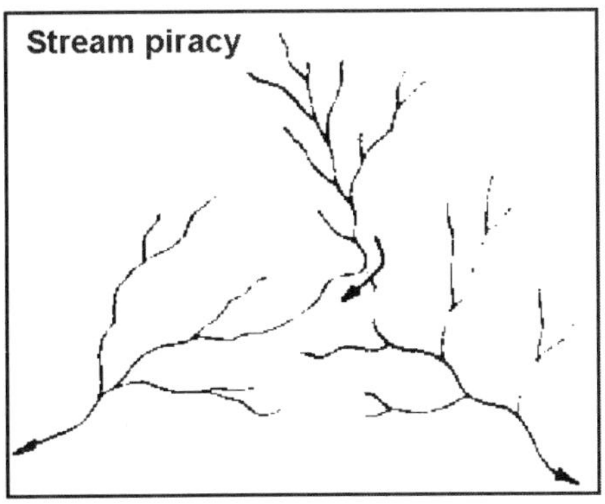

stream-flood A form of surface run-off develops in dissected upland areas in deserts, following brief but heavy rain-storms, as occurred in Libyan desert.

stress The force applied to a unit area of a substance measured in newtons per square metre. Stress can be produced in tension or compression, by hydrostatic pressure or by shear stress. **Compressive stress** crushes the rock which may collapse as the air pockets within it are compressed. **Tensile stress** is a force which tends to pull a rock or soil apart and which may cause features. **Shear stress** is a force which deforms a mass of material by one part sliding over another along one or more failure plains.

strike The direction taken by a horizontal line on the plane of an inclined rock stratum, fault or clearage. It is at right angles to the true dip of the rock structure. It also indicates the direction of a morphological feature developed along a structure.

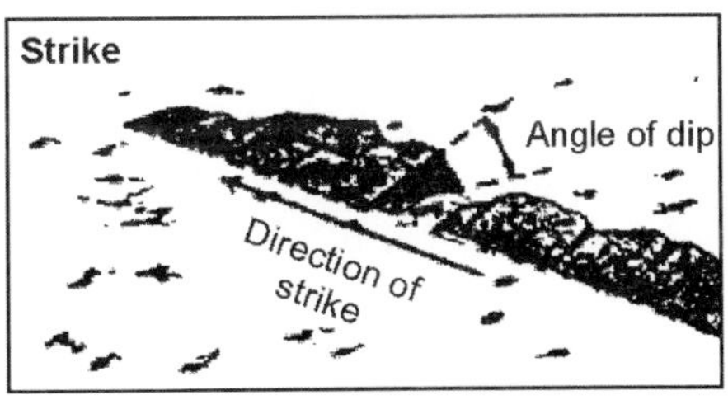

strike line A line joining points of equal elevation on a rock stratum above or below a selected stratigraphic datum. It is also known as structure contour.

strike slip fault A fault whose strike is parallel to the strike of the rock strata through which it passes.

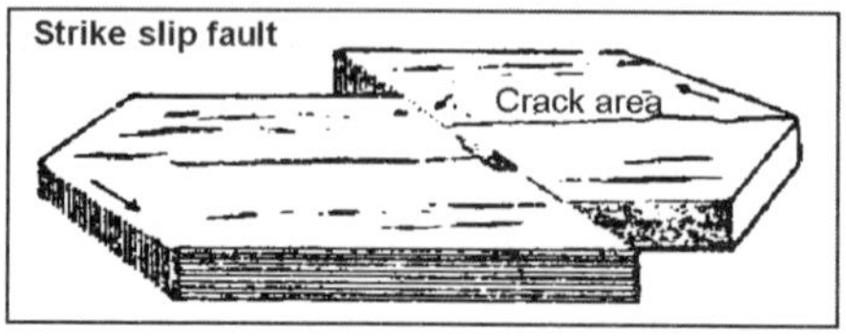

strike valley A valley which has been eroded along structural lines that lie parallel with the regional strike of an area.

string bog An area of waterlogged land characterised by ridges of peat separated by water filled troughs.

stripes This is a type of patterned ground oriented down the steepest slope gradient of a terrain influenced by marked thermal activity in the soil. Their two types are: **sorted strips** (stone strips) consist of parallel lines of stones and **intervening strips** of fines and non-sorted stripes consist of parallel lines of vegetation covered ground and intervening strips of bare ground.

stromatolite or stromatolith This term was first used by E. Kalkowsky (1908) to describe sedimentary structures in the Bunter of North Germany. Walter (1976) has given its definition as 'organosedimentary' structures produced by sediment trapping, binding and/or precipitation as a result of the growth and metabolic activity of micro-organism principally cyanophytes. They can develop in marine, marsh and lacustrine environments.

strombolian eruption A type of volcanic eruption in which molten lava and gases escape at frequent intervals but without violent explosions. It is named after the volcano on the island of Stromboli of N. Sicily.

structural surface A plane surface formed by the erosional striping of a weak stratum from an underlying resistant stratum, in such a way that the latter is either undissectual or only slightly dissected.

structure The word 'structure' is used variously, such as: (1) a combination of connected elements or an organised body. (2) In geology, the overall relationship between rock together with their large-scale arrangements and dispositions. (3) In petrology, the more detailed internal relationships between the different parts of rocks. (4) In biography, the spatial distribution of the vegetation in an association. (5) A soil structure and (6) also used in urban geography, regional planning and demography.

subaerial The term is used to describe the process of atmospheric weathering and surface erosion by rain wash, mass transport and rivers.

suballuvial bench The lower portion of a rock pediment where it is overlain by alluvial sediments.

subclimax Any plant community related to and closely preceding the true climax community for an area. The term is also used simply for a long persisting serial community that appears to be climax.

subduction zone A zone where rocks of an oceanic plate are forced to plunge below much thicker continental crust. As the plate descends, it melts and is released into the magma below the earth's crust. Such a zone is marked by volcanoes and earthquakes.

subglacial channel This is a melt water channel cut by glacial melt waters beneath a glacier or ice sheet.

sub-humid A climatic type referring to any area lying outside the humid zones and the desert zones where aridity prevails. It is a zone too dry for tree growth but humid enough for grass growth.

sublimation The process of direct deposition of atmospheric water vapour on the ice surface or evaporation from an ice surface.

sublittoral The area of the seas between the intertidal zone and the edge of the continental shelf.

submarine canyon A canyon-like valley form cut into the continental shelf, continental slope or continental rise. Most submarine canyons contain sediments deposited by high density flows initiated by submarine slides known as turbidity currents.

submerged coast A coastline formed when the sea level rose and submerged the former area of land. Features like raise and fjords may be found in these areas.

submerged forest An area of forest that has become submerged. Such features are commonly associated with coastlines that have experienced a recent submergence.

subsequent stream A stream that has developed its valley along an outcrop of less-resistant rocks. Because they are developing along relatively weak structures subsequent streams develops mainly by headward erosion.

subsidence The sinking of land surface due to such processes as the withdrawal of groundwater, geothermal fluids, oil and gas, the extraction of coal, salt, sulphur and other solids through mining, the hydrocompaction of sediments, shrinkage of organic deposits. Subsidence also occurred due to large-scale structural readjustment of the earth's surface, as in basin down warping or rift valley formation. In climatological sense, the term refers to the widespread downward movement of air associated with dry weather.

subsoil The weathered parent material which lies above the bedrock and below the true soil.

succession The term is used to describe the complex processes producing a gradual directional change in the structure and species composition of ecosystems at a particular site. Clements (1928) has pointed seven basic successional processes: nudation (initiation), migration of propagules ecesis (establishment), competition between species, action of environmental factors on organism, reaction of organism on their environment, and stabilisation.

suffosion An erosional process occurring in areas where limestone bedrock is overlain by unconsolidated superficial materials.

summer solstice It occurred on 21st June in northern hemisphere and 22nd December in the southern hemisphere.

sunspots A dark area on the sun's surface, which gives out a higher level of solar radiation. The number of sunspots varies, but seems to reach a maximum every eleven years or so.

superimposed drainage A term used for a river system which does not follow the trend of the rocks over which it is flowing, this is because it is continuing to follow the trend of other rocks which have since been eroded.

superimposed profile A diagrammatic representation made up of a series of topographical profiles taken at regular intervals across an area.

supersaturation When relative humidity in the air exceeds 100 per cent, i.e., when it contains more water vapour than is required to saturate it.

surf A mass of broken water in the breaker zone of a shoreline. Surf is formed on coasts as waves initially break on the shore.

surface wash When transport of surface soil has been occurred across the ground surface through the agency of running water. It consists of two processes, rain splash and overland flow and results in gullying, rill or sheet erosion of soil.

surge A large or massive movement of water, generally in the sea.

surging glacier When a glacier flow at a velocity of an order magnitude higher than normal. Whereas normal glaciers flow at 3–300 m $year^{-1}$, surging glaciers may flow at velocity of 4–12 km $year^{-1}$.

suspension The state in which small particles of sediment are carried by rivers. The particles are held up in the water by turbulent upward eddies, which prevent the tiny particles from following to the bed of the river.

swale A depression in regions of undulating moraine, a trough between beach ridges produced by erosion, or an area of low ground between dune ridges.

swallet This is a hole in limestone bedrock which has been produced by solution and through which stream water disappears. Also known as sinkhole or swallow hole.

swamp A waterlogged area, often associated with tree growth such as mangrove. When it is associated with decayed waterlogged vegetation it is termed as bog or carr.

symbiosis It refers to the organism living together in close association. It occurs when one or both members of two species come to depend on the presence of the other for survival and reproduction.

syncline The downfold infolded strata. The upfold is called anticline. The rocks in the syncline are often compressed, which makes them harder and more resistant to erosion and they often erode more slowly than the rocks in the anticline.

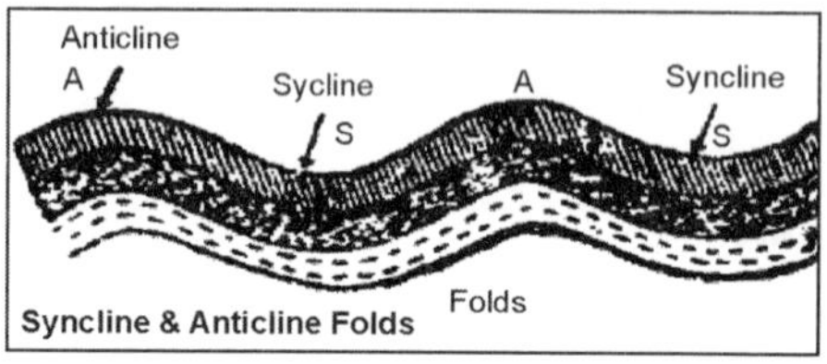

Syncline & Anticline Folds

synecology The ecology of whole communities, i.e., plant, animal or biotic communities as opposed to individual or single species.

synoptic climatology The climatology which deals with the description of local and regional climates in terms of properties and motions of the atmosphere rather than with reference to arbitrary time intervals, such as months.

system Any set of related elements. The term is very commonly used in geography. This can consist of only abstract parts, but is usually used as a concept by geographers to look at features in different areas, such as ecology and geomorphology.

T

taiga The coniferous forest which extends over thousands of square kilometres in Siberia, Russia, Finland, Sweden, Norway and Canada. The climatic conditions in these areas are not suitable for deciduous woodland, because the growing season is seven months or less but coniferous trees can survive. The main types of trees are spruce, larch, fir and pine. These trees are a source of softwood for pulp, which is used to make paper.

tail-dune An accumulation of wind-blown sand leeward of an obstacle such as an outcrop of rock in a desert region.

talus This is an accumulation of angular fragments on a slope. Talus may comprise a relatively thin veneer of debris, resulting largely from weathering of the underlying rock or a thick deposit of fragments which has collected at a base of free face.

tear faults A fault characterised by lateral movement, transverse to the strike of the rocks.

tectonics The movements which affect the features of the earth's crust. The processes which bend or crack the crustal rocks to cause folding or faulting are the result of tectonics, resulting from the movement of earth's plates.

temperate Those areas which are located between Tropic of Cancer to the Arctic Circle and from Tropic of Capricorn to the Antarctic Circle. The summers in temperate climate tend to be warm, but winters are cool. These areas are having a moderate climate.

temperature It is measured by a thermometer and generally recorded in degrees Celsius or Fahrenheit; 0°C (32°F) is the freezing point of water and 100°C (212°F) is the boiling point of water. Temperatures generally decrease from the equator towards the poles because of the corresponding decreases in the angle of the sun away from the equator. It also decrease with the height above mean sea level because the air becomes thinner.

temperature inversion Normally, the temperature of the air decreases from ground level to upwards, but in certain conditions, the normal situation is reversed, i.e., increase in temperature with height. On a clear, calm anticyclonic night, the cool air may roll downhill and accumulate in valleys, and the air temperature will be lower near the valley bottom than it is 100 to 200 m higher.

tephra The air, dust and cinders thrown out by a volcanic eruption. This does not include lava. When tephra is thrown up into the air by an explosion, it will soon fall again, quite close to the creator. In 1973 eruption of Heimacy (Iceland) large deposits of tephra were left. So much tephra fell on to some houses that roofs collapsed under the weight.

terminal moraine The collection of boulders, gravel, sand and silt formed at the terminus of a glacier or ice-sheet. It is like a linear ridge of glacial debris. Its inner slope is usually steeper than its outer slope. Also known as end moraine.

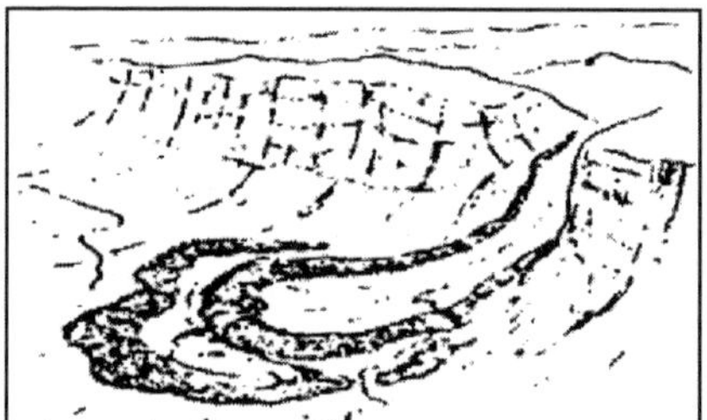
Terminal moraine

terra rossa A red intra-zonal soil developed in Mediterranean regions by weathering of limestone. The soil has a clay-loam texture and its red colour comes from the dissociation of clay from iron hydroxide. It is thus a residual soil, though there has been some lateral movement of the soil particles into valley bottoms and enclosed hollows such as dolines.

terra roxa A reddish soil, very rich in iron which formed on the basaltic rocks of the Sao Paulo state of the Brazil. This soil is suitable for growing coffee.

terrace A bench-like feature in the landscape, developed either by erosional or depositional processes. Terraces are flat or gently inclined land surface bounded by a steeper ascending slope on its inner margin

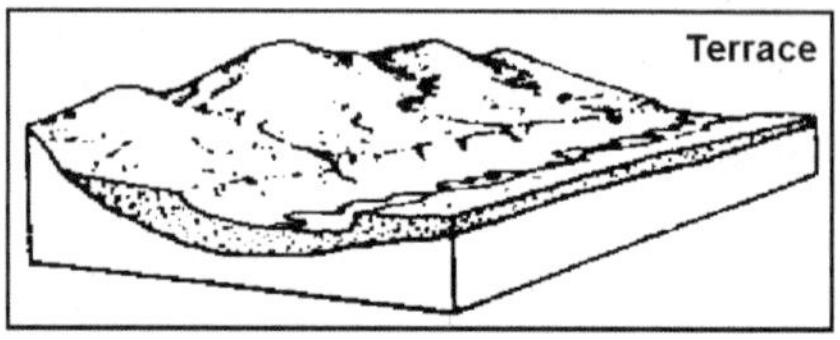
Terrace

and a steeper descending slope on its outer margin. Various types of terrace include: **alluvial terrace, kame terrace, marine-built terrace, meander terrace** and **river terrace**.

terrain The physical characteristics of the natural features of an area. It includes landforms, vegetation and soils.

terrain analysis The scientific interpretation of the landforms, vegetation and soils of a given area in relation to the uses to which it may be put.

terrestrial environment The environment having continental characteristics distinct from the marine and atmospheric environments. It is the environment in which terrestrial organisms live and in which terrigenous sediments are formed.

territorial seas/waters The coastal waters, over which a state has its sovereignty. Under international law, this was originally defined as a distance of 5 km. In 1958, the Law of the Sea Convention extended this to 22.2 km or 12 nautical miles and also clarifying the rights of states to the resources of the continental shelf to a distance of 370 km now known as the Exclusive Economic Zone.

Tertiary The first geological period of the Cainozoic era, preceding the

Quaternary period. It commenced at the end of the Cretaceous, some 65 million years ago and lasted for about 63 million years.

tethys The Tethys sea was a geosynclines of Mesozoic Era located between Gondwanaland and Laurasia. Within it sediments were deposited for a long period and these were later to be compressed and uplifted to form mountain ranges of Southern Europe and South Central Asia during alpine orogeny. It is a shallow, rounded depression containing a circular or semi-circular pond, in a lowland periglacial region. These lakes are very common features of thermokarst, and result from the surface thawing of frozen ground. This produces localised collapse and the formation of an irregular depression which becomes occupied by melt water.

thermal depression A region of low pressure generated by strong solar heating over land in the summer. Such depressions may be large as that of South Asian monsoon or short-lived that sometimes form over England on hot summer days.

thermal efficiency index
C.W. Thornthwaite has devised this index as a measure of the amount of heat received by a particular area in terms of the potential evapotranspiration that would be produced. It is expressed as: [T-32]/4, where T is the mean monthly temperature in degrees F.

thermal equator Also known as **heat equator**, is a line encircling the low latitudes of the global joining up the point on each meridian which exhibits the highest mean surface air temperature. Its average position lies to the north of the true equator because there are large land masses in the N hemisphere and land absorbs more heat than do the oceans.

thermal fracture When rocks cracked due to alternate heating and cooling is known as thermal fracture. It is also referred as insolation weathering and may lead to exfoliation.

thermal pollution The pollution caused due to release of heat from thermal power plant and industrial and domestic heat into the atmosphere. In industrial towns, it creates heat islands and also responsible for discharge of hot water into rivers, where it caused de-oxygenation, adversely affecting fish and other aquatic fauna.

thermal stratification of water
This is the division of water temperatures in the oceans and deep lakes. Three layers are: the uppermost layer (epilimnion), affected by wind and convection currents; the middle layer (thermocline), exhibiting a rapid decrease of temperature with depth, the lowest layer (hypoliminion), with the lowest temperatures.

thermal wind (Vector wind)
In general, the shear or change in direction of the geostrophic flow with increasing elevation, determined by distribution of mean temperature in a specified atmospheric layer. The thermal wind blows parallel to the mean isotherms with the effect that low mean temperatures remain on the left in the northern hemisphere and on the right in the southern hemisphere. Therefore, the poleward

decrease causes a large westerly component in the upper winds. The result is the polar front jet. The force of a thermal wind may be strengthened by any pressure gradient at ground level.

thermocirque A large hollow on a hillside formed from the coalescence of nivation hollows. The sides usually experience parallel retreat. Thermocirques are shallow because the centre of the hollow is protected from further erosion by covering of snow.

thermograph A meteorological instrument which is housed in a weather screen and provides a continuous record of air temperature.

thermokarst Topographical depressions resulting from the thawing of ground ice (Washburn, 1979). There are many kinds of thermokarst, including collapsed Pingos, ground ice mud slumps, linear and polygonal troughs, thaw lakes and Alases. The term 'karst' is used to indicate the numerous features formed by subsidence, and does not imply the presence or development of a limestone and landscape.

thermoline A layer of water within a lake or ocean through which the rate of decrease of temperature with depth is much greater than in adjacent layers. It is particularly well-developed in tropical oceans where permanent thermocline upto a few hundred metres in thickness, lies with its upper boundary some 25–150 m below the ocean surface. A seasonal thermocline also develops in both lakes and oceans where surface temperature undergo a significant annual variation.

thermometer An instrument used for measuring temperature. The most common model consists of a column of mercury enclosed in a sealed glass capillary tube rising from a small bulb which acts as a reservoir from the mercury. The glass tube is graduated so that temperature can be measured either in Celsius (centigrade) or Fahrenheit. The various types of thermometer include Maximum and Minimum Thermometer, Dry and Wet-bulb Thermometer, Black Bulb Thermometer, Resistance Thermometer, and Gas Thermometer.

thermosphere That part of the atmosphere, starting at about 85 km above the earth. Here, temperature increases with height. The warming of the thermosphere comes from the photo-dissociation of oxygen molecules.

threshold In geomorphology, threshold is used to describe the submerged rock bar at the mouth of a fiord. It is also used for a low neck of sand dividing for a lake into two basins. A third use of the term is a condition marking the transition from one state of operation of a system to another.

through flow or subsurface flow The downslope flow within the soil. It may follow natural percolines in the soil. Through flow is a major factor in the hydrology of a drainage basin where the rocks underlying the soil are impermeable.

thrust fault It is a type of reverse fault that is characterised by a low angle of inclination with reference to the horizontal plane.

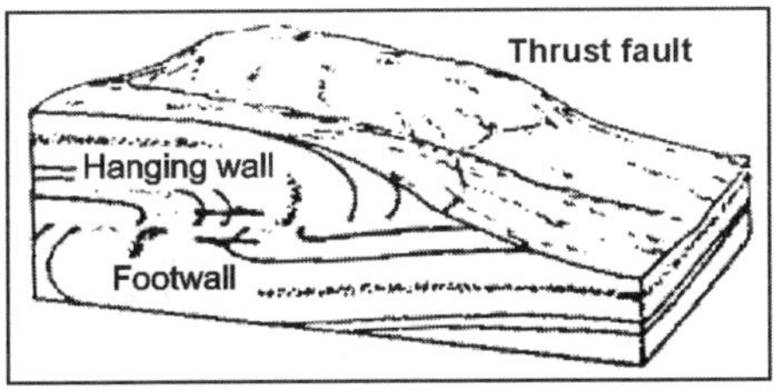

thufur A soil **hummock** found in periglacial environments.

thunder When a stroke of lightning passes through the atmosphere, the air becomes intensely hot. The violent expansion thus cause a shock wave heard as thunder. As the speed of sound is considerably less than the speed of light the noise of thunder always follows the flash. A storm associated with thunder is called thunderstorm, often followed by very heavy rain or hail. The greatest thunderstorm activity occurs in low latitude where there is both intense heating and moisture availability.

tidal limit The highest point in a river inlet which is ever reached by the sea.

tidal range Difference between the water level at high tide and water level at low tide.

tide The periodic rise and fall of the sea water caused by the gravitation pull exerted on earth by the moon, and, to a lesser extent by the sun. In many parts of the world, there are two high tides and two low tides every day. The time of each high tide is 12 hours 20–25 minutes later than the preceding tides because the position of the moon relative to the earth will have changed by a small amount after 12 hours have elapsed. See neap tide.

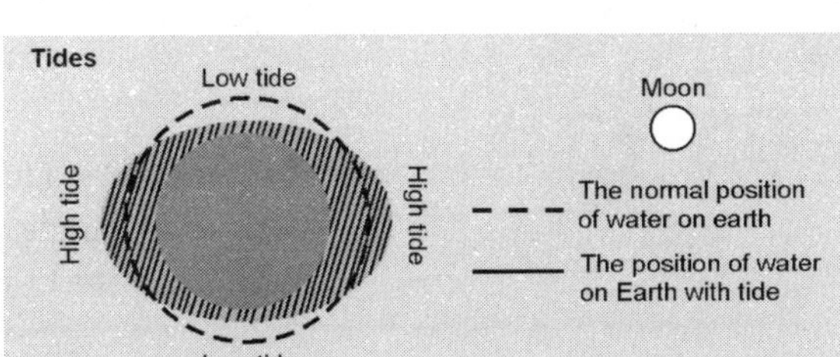

tierra A Spanish term adopted to define climatic regions based on altitude in South America, Central America and Mexico. These are: **Tierra helada,** the mountain zone of snow covered peaks; **Tierra fria,** the upland zone above 1800 m; **Tierra templada,** the intermediate zone between 900 and 1800 m; and **Tierra caliente,** the lowest, coastal zone below 900 m.

till A Scottish word, used by A. Geikie (1863) to describe a coarse, bouldery soil, which was adopted to describe unstratified glacier deposits. It also refers to the sedimentary material deposited directly by the action of glacier. Sometimes till and moraine are regarded as synonymous, however, till refers to the deposit itself and moraine to the surface form of the till.

till-plain A wide area of low relief created by till sheet.

timberline or tree line The upper (altitude) or polar (latitudinal) margins of tree growth. In the northern hemisphere, the timberline is approximately coincident with the Arctic Circle. While on mountains, the timberlines generally decrease in altitude with distance from the equator, where they are found around 3300–4000 m, but the weight varies greatly with local conditions.

time-dependent form A landform in which the form of a particular slope depends on the time. W.M. Davis stated in the cycle of erosion that slope form is time-dependent.

time-independent form A landform, in which the relative form (slope) remains unchanged with time even though the altitude of the entire slope may be lowered.

tolerance The ability of organism to withstand environmental conditions. Plants and animals within a particular environment have limits of tolerance beyond which they cannot exist.

tombolo A bar or spit resulting from longshore drift which joins an offshore island to the mainland. In Tuscany, Italy, Monte Argentario is linked by three tomboli to the land.

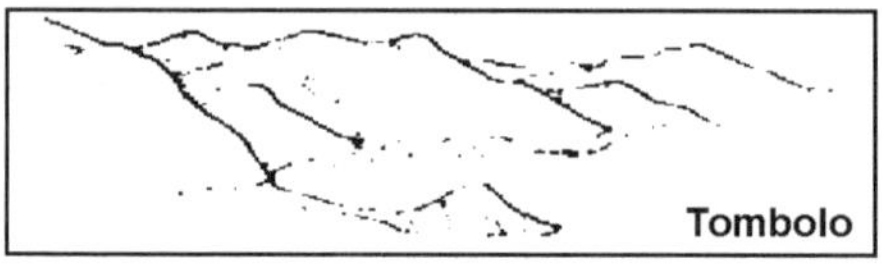
Tombolo

topography The relief and configuration of a landscape, including its natural and man-made features. The hills, valleys drainage, forests, transportation, settlements, etc. are part of topography. The maps which depict all the physical and cultural features are known as topographical maps. In India, Survey of India published topographical maps (Topo-sheets) in different scales.

topsoil It is the top layer of the soil or 'A' horizon of the soil profile. It contains fine particles, humus, nutrients and the roots of the plants.

tor An outcrop of granite which stands above the general level of the landscape. It is an isolated mass of rock, showing vertical as well as horizontal lines of weakness where weathering has attacked the rock. It is a residual rock, the remnant of a layer of granite that has been largely worn away.

toreva blocks These are large masses of relatively unfactured rock that have slipped down a cliff or mountain side, rotating backwards towards the cliff in doing so.

tornado A violent rotating storm under a funnel-shaped cloud which advances over the land along a narrow path. These are created due to intense heating in continental areas in late summer, when the land masses are at their hottest. The centre of a tornado is an area of low pressure which sucks up dust to give a blackish funnel rising

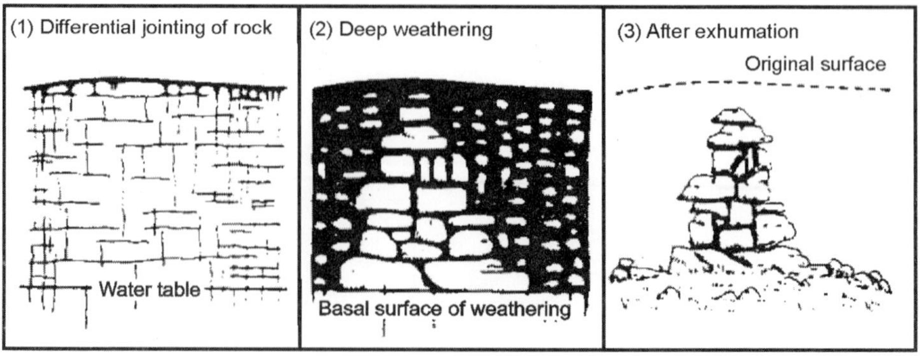

Formation of a tor

to the sky. Around this funnel of rising air are very strong winds which destroy crops and sometimes buildings. Tornadoes are common in the interior of the USA, but also occur in India and many other countries.

torrent It is a swift, turbulent flow of water or lava.

tower karst These are residual limestone hills rising from a flat plain. Limestone towers, from 30 to 200 m in height with nearly vertical walls and gently domed or serrated summits. Tower karst is well-developed in south-east China.

trade winds Winds which blow from the subtropical high pressure areas around 30° latitude towards the equatorial low pressure zones (doldrums). In the northern hemisphere, the air moving from the Tropic of Cancer towards the equator is deflected to the right, making it a north-easterly wind. In southern hemisphere, the air moving from the Tropic of Capricorn to the equator is turned to the left to become a south-easterly wind. These winds are persisted; they are all easterlies and tropical.

transcurrent fault It is a tear-fault of massive proportions in which the net slip is in the direction of the fault strike. It is a fault which frequently offsets mid-oceanic ridges.

transect This is a transverse section across a region, mainly used by geographers to show the spatial relationships between vegetation, soils and relief.

transform fault A type of massive tear-fault on a continental scale in which the fault on a continental scale terminates sharply at a place where the movement is transformed into a

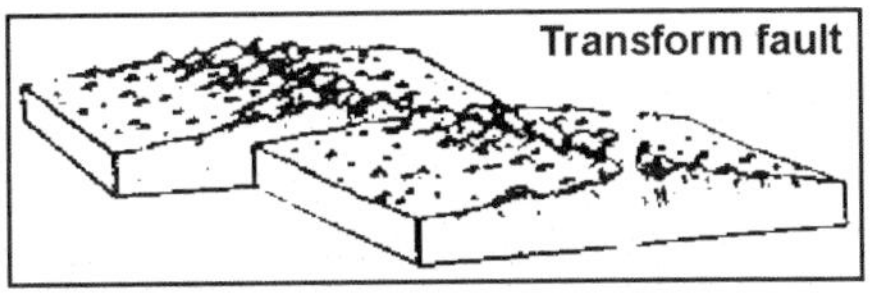
Transform fault

structure of another type, i.e., mid-oceanic ridges, island arcs or fold-mountain belts.

transpiration The process by which plants give off and lose water vapour. Plant perspiration mainly occurred from the cells of the leaves through pores (stomata) but also from the leaf cuticle and through lenticels of the stem, and goes out into the atmosphere. From the atmosphere, the moisture may be recycled as rain, and then taken up into the plants again, via the roots. Transpiration contributes to the formation of rain.

transportation slope A type of slope on which at each point the amount of material brought in from upslope is more or less balanced by the amount of material carried away downslope.

transverse dune In desert topography an asymmetric sand ridge has been formed, in which the long dimension is normal to the dominant wind direction. It has one slip face.

trellised drainage This is a drainage pattern associated with scarp-and-vale landscapes and characterised by

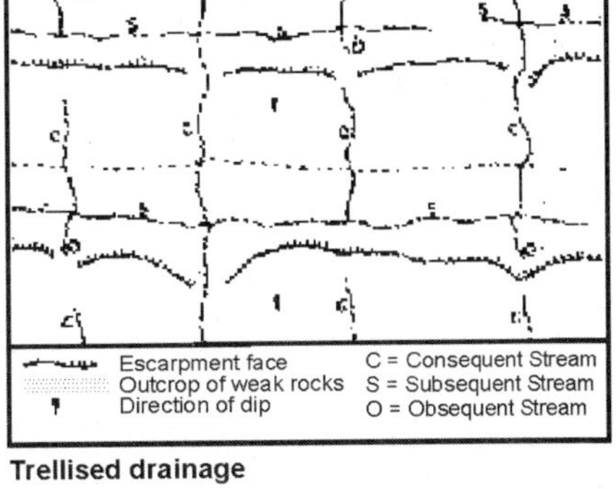

Trellised drainage

right-angled stream junctions. The main consequent streams flow across bands of resistant and unresistant rocks, following the direction of dip. The weaker strata are eroded into strike-vales by subsequent streams. Trellised drainage is thus well-adjusted to geological structure.

Triassic The first geological period of the Mesozoic era, years ago.

tributary A subsidiary stream or river which flows into a larger extending from about 240 million years ago to about 195 million years ago. Most of rivers have a number of tributaries, all of which collect the rain and ground water from a part of the river basin, but rivers which flow through deserts may have few tributaries.

trim line The former limit of a glacier, indicated in the present landscape by a sharp change of vegetation, often most clearly visible on the valley sides. It separates vegetated lateral moraine on a valley side from bare moraine lower down the slope.

trophic level In a **food chain** of each ecosystem there is a level of good production and consumption, which is termed as trophic level. Most food chains comprise three main trophic levels, i.e., **primary producers (autotrophs)**, **primary consumers (herbivores)** and **secondary consumers (carnivores)**. A fourth trophic level is the top carnivores. For example, in the grazing food chain, all producer organism (green plants, blue-green and other algae, phytoplankton, etc.) are placed in the first trophic level, which contain maximum store of energy, which is available for the use in any given system; the second is comprised of grazing animals (herbivore); the third of humans (carnivore).

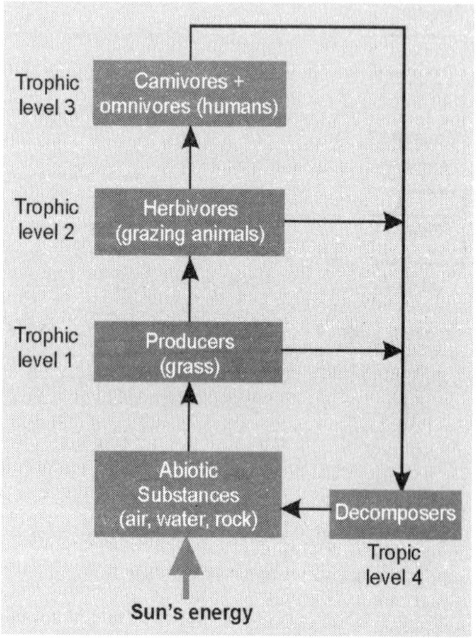

Trophic levels in an ecosystem

Tropic of Cancer An imaginary line of latitude at 23.5° north of the equator. It is the most northerly latitude reached by the overhead sun, on 21st June, which is the Summer Solstice.

Tropic of Capricorn An imaginary line of latitude at 23.5° south of the equator. It is the most southerly latitude reached by overhead sun, on 22nd December, the Winter Solstice.

tropical cyclone It is a type of violent cyclone mostly occurred in tropical areas in which wind speeds exceed 150 km/h and several millimetres of rain may fall. They are known as **hurricanes** in the Caribbean, **typhoons** in the northwest Pacific and **cyclones** in the Indian ocean. They are formed from pre-existing disturbances most frequently in the 10°–15° latitude band. These cyclones are character-

Tropical cyclone

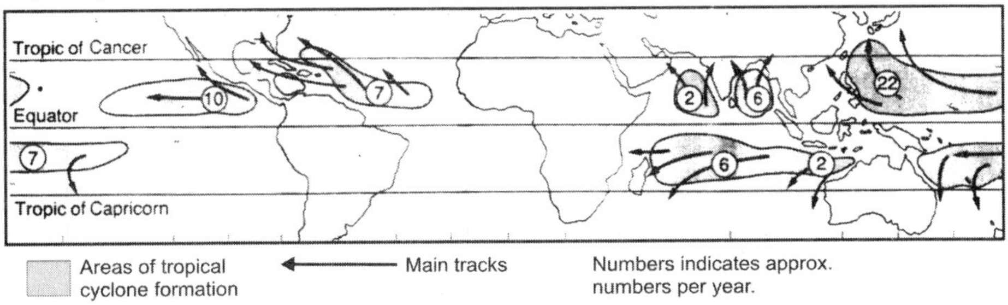

ised by cyclonic inflow, with inward spiralling cloud band hundreds of kilometres long and ten kilometres or so wide, followed by heavy precipitation

tropical grassland The grassland which is found in tropical area. These areas are those which do not have rainfall throughout the year, but only in the summer months. Tropical grassland is found in Kenya, Northern Nigeria, Northern Australia, the Brazilian plateau and the Orinoco basin in South American. These grasslands are located mostly between 5° and 20° north and south of Equator, but in East Africa they are also found on the Equator.

tropical rainforest These forests are distributed in equatorial regions of the world, mostly in areas between 5° north and 5° south latitudes. Tropical rainforest are most luxuriant and evergreen forest. These forests have been developed in these regions because of year-round rainfall and constant heat. Amazon basin and Zaire are having largest areas of tropical forest. The other areas are located on the coasts of Colombia, northern equator, north-east Brazil, the lower lands of Panama and Costa Rica, northern coastal region of Mozambique, equatorial regions of Africa and South-East Asia. Tropical forest trees are characterised by buttress roots, long straight lower trunks and leathery leaves.

tropopause Located between troposphere and stratosphere at a height of about 18 km above the equator, 9 km at latitude 50°N and S and 6 km at poles.

tropophyte A plant which is adopted to stand seasonal periods of cold or draught, i.e., it acts as a xerophytes during dry season but becomes a hygrophyte during the rainy season.

troposphere The lowest part of the atmosphere, below the tropopause. It is the zone where atmospheric turbulence is at its greatest and where the bulk of earth's weather is generated. In the troposphere, there is generally a steady fall of temperature with an increase in height, this is called lapse rate.

Tropical Rainforest

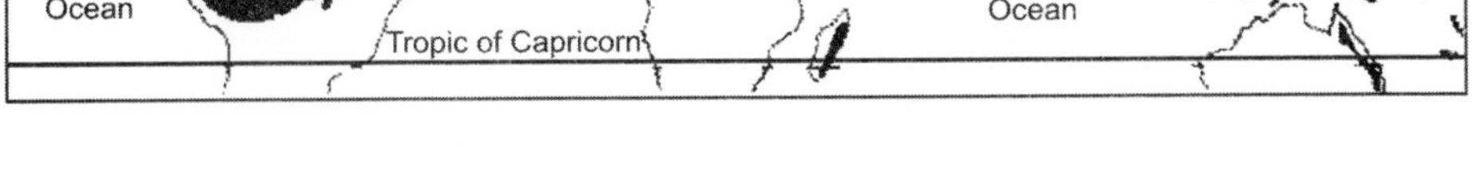

trough This term used variously as: (1) A low pressure area which is an elongation or extension of a depression. (2) A deep trench on the ocean floor. (3) A valley that has been over-deepened by glacial erosion, such as a glacial U-shaped valley. (4) A term referring to cross-bedding (trough-bedding) in sedimentary structures.

trough end The precipitous rock wall which closes off the upper end of an over-deepened glacial valley and around which a group of cirques occurs.

truncated spur A spur which has been cut off by glacial erosion. Truncated spurs are found in glacial U-shaped valleys and truncation leaves very steep valley walls. In some locations, there are waterfalls while at the foot, there is likely to be accumulation of scree.

tsunami Japanese term which has been universally adopted to describe ocean waves caused beneath the sea or a volcanic eruption. Tsunamis travel for considerable distances across the sea as long, low waves having wave-length may be over 100 km and forward velocity as much as 600 km hr^{-1}. As tsunami approaches the shore, the wave height increases upto 15 m or so, thus capable of causing immense destruction, as happened in 2005 tsunami in Indonesia, India and Sri Lanka.

tufa It is a deposit, resulting from the precipitation of calcium carbonate from calcium bicarbonate dissolved in water. Tufa is formed at points where springs emerge from underground passages in limestone or chalk region.

tuff A rock which consists of consolidated ash and dust thrown out by a volcanic eruption.

tundra The barren plains of northern Canada, Alaska and Eurasia. The tundra lies between the polar region of perpetual snow and ice and the northern limit of tree growth. Its surface is stony or marshy and is characterised by mosses, lichens and a low vegetation of dwarf shrubs. The lower soil is permanently frozen so that drainage is poor.

turbidity It is a sediment deposited in water by a turbidity current. Its particle size decreases upwards to create a type of graded bedding and it is most commonly formed in a geosyncline.

turbidity current A submarine flow of a dense mixture of water and sediment, capable of moving at a very high speeds (50–80 km/h) for distances of tens of kilometres. It transfers large volumes of sedimentary materials into the deeper zones of the ocean. The sediment which finally settles out on the sea-bed is termed a turbidite.

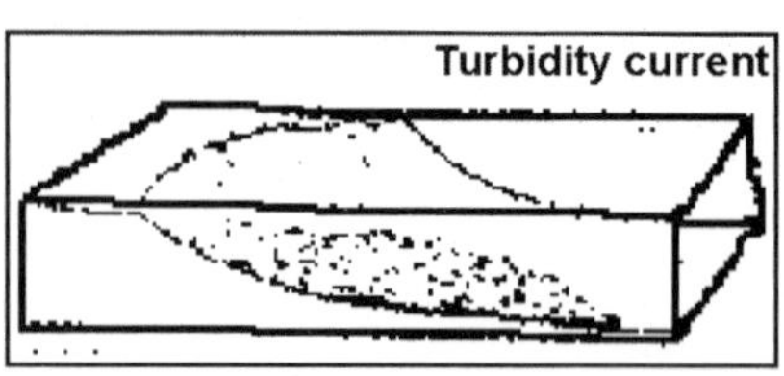

turbulence (1) In meteorology, it is an irregular movements of the air in the lowest layers of the atmosphere. Turbulent activity causes mixing of the air, and often causes condensation as the rising air-cools. Turbulence can also cause showers of rain, as well as bumpy conditions for air craft. (2) In geomorphology, turbulence is a part of the stream flow mechanism and is important in the transport of sediment by suspension. (3) In oceanography, it is a feature of an ocean drift.

turlough It is a small, enclosed surface depression in a limestone terrain in which intermittent water bodies occur after heavy rainfall has raised the water table and increased surface runoff.

turmkarst It is one of the major form of tropical **karst** terrain, which is also referred as **tower karst** or **pinnacle karst**. It consists of steep-sided hills or pinnacles, termed **mogotes**, rising abruptly from an alluvial plain to heights of 200–300 m. Tower karst is well-developed in Puerto Rico, Central China, Vietnam, Sarawak and Indonesia.

twilight The period of faint light before the sun has risen above the horizon in the morning and after the sun has descended below the horizon in the evening. It lasts for few minutes near equator, but can last for an hour or more in the Arctic and Antarctic.

typhoon A tropical storm, characterised by winds of very high velocity and torrential rainfall, which affects the coast of the Philippines, Hong Kong and China. Its wind-speed exceeds force 12 on the Beaufort scale causing widespread destruction. Typhoons are a local type of tropical cyclone.

U

ubac The shady side of the valley, normally the north facing side. It receives the maximum amount of sunshine and warmth.

ultrabasic An igneous rock containing very little quartz and are usually plutonic in origin.

ultraviolet radiation That part of solar radiation, which has wavelength between 10^{-2} μm and 10^{-1} μm. Only small percentage of ultraviolent rays reach the earth surface, the majority of it is absorbed by **ozone** in the atmosphere.

unconformity A break in the continuity of a sequence of rocks. The older rocks beneath the unconformity are often folded or tilted, whereas the younger rocks above are usually of simpler structure. Also a discontinuity between sedimentary strata which testifies to a temporary interruption in the process of accumulation.

undercliff A mass of slumped material at the foot of a sea-cliff, which is being affected by large-scale mass movements. The presence of an undercliff indicates that sub-aerial cliff recession is at least as effective as marine erosion.

undercutting Erosive action of a river or a ocean wave at the base of a cliff. It also refers to the mechanical wearing effect of wind-carried sand on the base of a rock pillar.

underground stream A stream which runs below the surface of the ground. Such streams occur in limestone areas, where streams travel many kilometres underground.

unequal slopes The law of unequal slope states that unequal angles of hill slope are the result of variations in the rate of slope erosion, which in turn result from differences in geological resistance and from variations in the power of erosional processes.

uniclinal Related to a formation of rock strata which dip uniformly in one direction.

uniclinal shifting The process whereby a river flowing in an asymmetric valley in an area of gently dipping rocks migrates down the dip slope of the valley, cutting back the steeper scarp slope.

uniformitarianism James Hutton (1785) first suggested and further elaborated by Charles Lyell (1830) a geological principle, which states that the earth processes observable at the present time are essentially the same as those that operated in the past, producing the rocks and structures stretching back to Pre-Cambrian times. In brief, rocks are being formed today in the same way as they were hundreds of millions of years ago.

unit hydrograph A characteristic or generalised hydrograph for a particular drainage basin.

unloading The removal of overlying materials by erosion, resulting in release of pressure on the land surface.

unsable equilibrium Is a condition where a small displacement leads to an even greater displacement. For example, the behaviour of an air parcel under certain weather conditions when the environmental lapse-rate becomes greater than the dry adiabatic lapse-rate. This is found in the formation of cumulonimbus cloud.

unstable channels A river or tidal channel that is shifting through erosion and deposition.

updraught A powerful, ascending current of air, often associated with cumulonimbus clouds and thunderstorm conditions.

upslope fog A type of **advection** fog common in upland areas, where moist air is forced upwards by the terrain and its temperature is cooled below the saturation point.

upthrow The side of a fault that has moved upwards by the relative displacement of the down throw on the other side of the fault plane.

upwelling The vertical movement of deeper water towards the sea surface. It occurs where a divergence of surface currents must be compensated for by vertical flow. Such transfers are common in the zones of divergence of ocean currents, thereby replacing the warmer surface layers by upwelling colder water below and giving rise to cold currents, rich in nutrients.

Ural type glacier A small glacier developed in the lee of prevailing winds of a mountain or plateau. It is named after Ural mountains but also found in other mountains.

urstromtaler A German term for melt water channels in northern Germany. These channels may be hundreds of kilometres long and often more than 100 m deep with irregular long profile.

U-shaped valley Valley created by glacial erosion having 'U' shaped. These valleys are having flat floor and steep sides of the pre-existing river valley, for example, in the Lake District in England or the Alps. The slopes of the former fluvial valley are considerably over-steepened by the glacier, which occupies the entire valley rather than merely the channel on the floor of the former river valley. The U-shaped valley may terminate headwards in a trough end or it may form part of a breached divide. Its flanks may be marked by debris fans and often tributary streams formed **hanging valleys**.

uvala It is a large surface depression in limestone (**karst**) topography, formed by the coalescence of adjoining dolines. It has an irregular floor reflecting the former elevations and character of degraded slopes of the dolines.

V

valley It is an elongated depression created by either rivers or glaciers. It may be deep for a few metres or as much as 5000 m, as in the case of some Himalayan valleys. Their width may be just a few metres or several kilometres. The valleys created by river erosion are 'V' shaped while the valleys created by glacial erosion are of 'U' shaped. The shape will also depend on rock type and the amount of river erosion. Some valleys have lost their river because it has been captured, or has gone underground, and they are called 'dry valleys'.

valley glacier A glacier situated in an upland valley or basin.

valley train A plain within a valley sloping down and away from the site of a glacier snout, composed of sands and gravels and containing pebbles and boulders.

valley winds The air movement at night down the gradient of valleys and the enclosing mountain sides; alternating with day time are the mountain winds.

vallon de gelivation A small valley formed by the widening of bedrock joints by ice action rather than by normal fluvial processes.

valloni The drowned river valleys of a Dalmatian type coastline.

valve The shell or shells of certain organism, such as snails and clams.

Van't Hoff's rule The 'law' states the rate of increase of chemical reactions with rise in temperature. In general, the increase is 2–3 times for every 10°C rise. This rule helps to explain chemical weathering processes in tropical landscapes.

vaporisation The process by which a liquid is transformed into a gas.

vapour pressure The pressure exerted by the molecules of a given vapour. The maximum vapour pressure at any temperature occurs when the air is saturated.

varve A pair of thin sedimentary layers, one relatively coarse-grained and light-coloured and the other relatively fine-grained and dark-coloured, formed by deposition on a lake bottom. The grained layer is formed during spring run-off, and the fine-grained layer is formed during the winter when the surface of the lake is frozen.

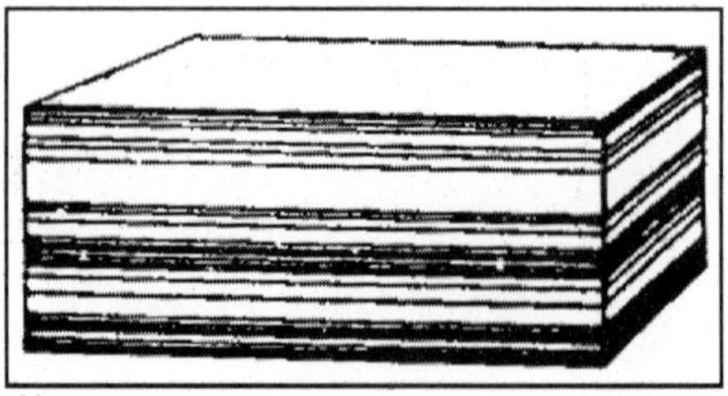

Varve

vascular system Plant tissue arranged into vessels that serves to conduct fluids from the roots to other parts of the plants.

veering wind The wind that changes direction in a clockwise sense—north to northeast to east, and so on.

vein A tabular rock body deposited in a fracture. Many ore minerals were deposited in veins where hot fluids flowed through fractures.

veld Grasslands, which are found in Southern Africa, specially in Northern Transvaal and Eastern Transvaal in the Republic of South Africa.

ventifact A pebble or cobble-shaped and polished by wind abrasion.

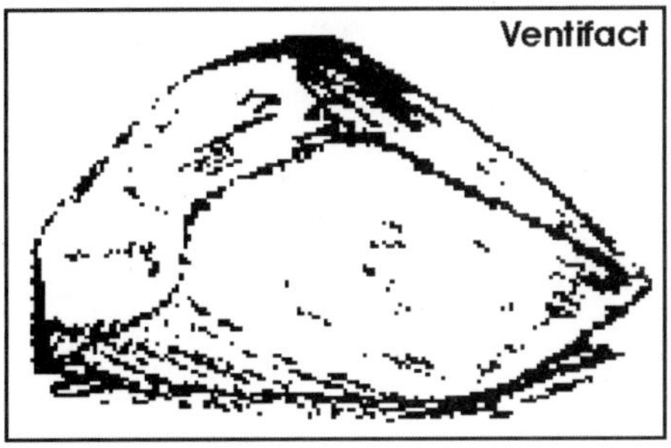

vermal equinox The equinox at which the sun approaches the northern hemisphere and passes directly over the equator. Occurs around March, 20.

vesicle A small hole formed in a volcanic rock by a gas bubble that becomes trapped as the lava solidifies.

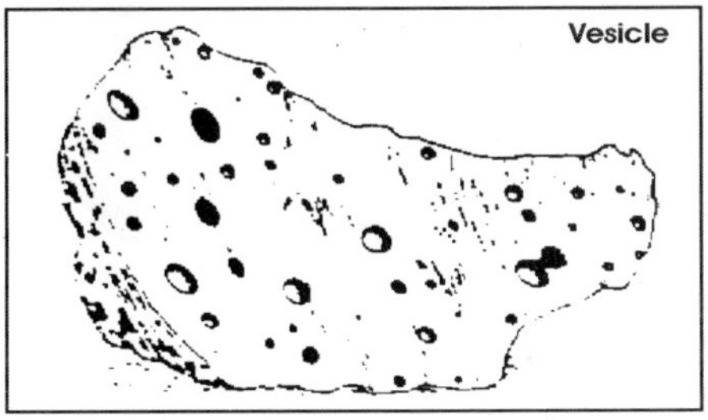

virga Precipitation that falls from a cloud but evaporates before reaching the ground.

viscosity The tendency of a body to resist flow because of internal friction due to molecular cohesion.

visibility It refers to the clarity of the atmosphere, which depends on the type of air mass and in part on the amount of human-induced pollution. It is lowest in conditions of fog, mist, haze or during a dust-storm.

volatile (1) Capable of being readily vaporised. (2) A substance that can readily be vaporised, such as water or carbon dioxide.

volcanic ash Dust-size particles ejected from a volcano.

volcanic bomb A hard fragment of lava that was liquid or plastic at the time of ejection and acquired its form and surface markings during flight through the air. Volcanic bombs range from a few millimetres to more than a metre in diameter.

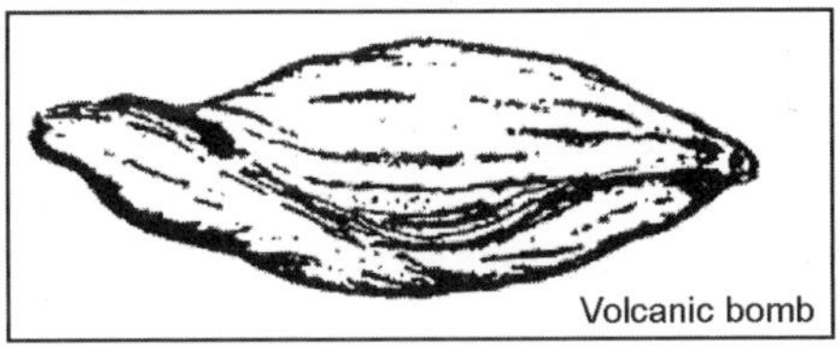

volcanic front The line in a volcanic ace system (parallel to a trench) along which volcanism abruptly begins.

volcanic neck The solidified magma that originally filled the vent or neck of an ancient volcano and has subsequently been exposed by erosion.

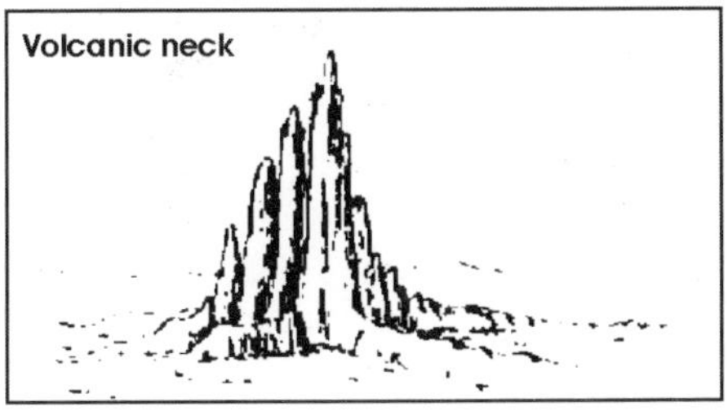

volcanic plug A hard mass of solidified lava in the vent of a volcano. In spite of erosion over a period of million years, the hard plug will survive as a small isolated hill in the crater.

Volcanic plug

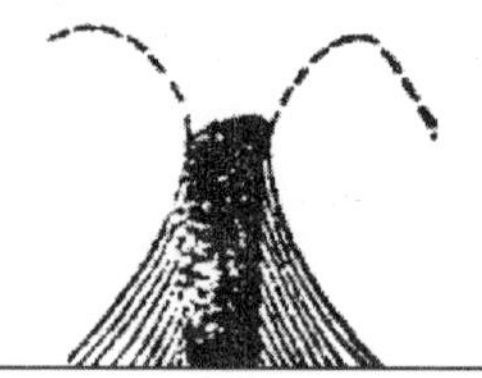

volcanic rock A rock formed on the surface of the earth as a result of igneous activity. They are mostly basic in chemical content and form basalt.

volcanism The process by which magma and gases are transferred from earth's interior to the surface.

volcano A conical hill built up by the ejection of igneous material from a vent. It is an opening in the crust out of which magma, ash and gases erupt. The shape of the volcano depends very much on the type of lava, its chemical composition (**acid lava** rich in silica and relatively viscous, forms steep-sided, dome-like volcanoes; **basic lava** containing little quartz, is more mobile, giving rise to more gently sloping volcanic cones) and subsequent erosional history of the volcano. Volcanoes are classified as **active**, **dormant** or **extinct**. Some volcanoes have built up large mountains as a result of many eruptions; for example, Mt. Etna, which has a circumference of over 120 km. There are over 600 active volcanoes in the world. The largest number can be found in circum-Pacific belt.

volcano coast The coast formed by volcanoes and lava flows built partly below and partly above sea level.

V-shaped valley A valley created by fluvial erosion. Which is characterised by evenly sloping sides and a V-shaped cross-profile. In cycle of erosion, the V-shaped valley has been equated with the early stages of youth.

Volcano

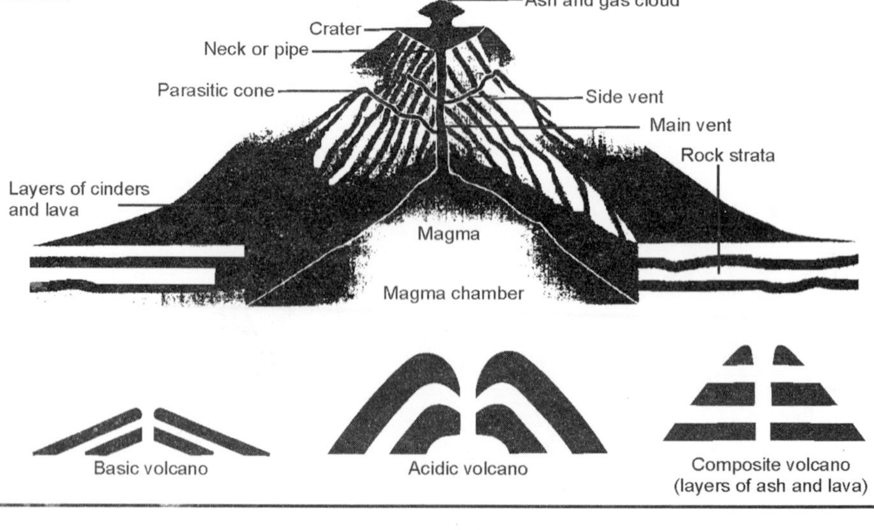

W

wadi An Arabic word, used for a narrow steep-sided valley found in deserts and semi-arid regions. These are dry valleys for much of the year, because rainfall is much less, when it does rain, it falls in heavy showers. As much of land is bare and without vegetation, surface run-off can be rapid. They will carry sand and small rock fragments, and so a great deal of erosion takes place. Wadis are common in Egypt, Libya and many other desert areas.

wall cloud An area of rotating clouds that extends beneath a severe thunderstorm and from which a funnel cloud may appear. It is also called **collar cloud.**

Wallace's realms A.R. Wallace has classified the world's fauna (zoogeographical regions) into six broad regions, known as Wallace realms. The realms are: (a) Palaearctic, (b) Ethiopian, (c) Oriental, (d) Australasian, (e) Nearctic, (f) Neotropical.

warm front The boundary between a warm and cold air mass, where the former is advancing and overriding the cold air, as at the leading edge of the warm sector of a depression. The gradient of the warm front is much less steep than that at the cold front; the relatively slow ascent of warm moist air at the front thus results in cooling and condensation over a wide area.

warm sector The region of warm air within a wave cyclone that lies between a retreating warm front and advancing cold front.

warm-core low A low pressure area that is warmer at its centre than its periphery. Tropical cyclones exhibit this temperature pattern as do some polar lows.

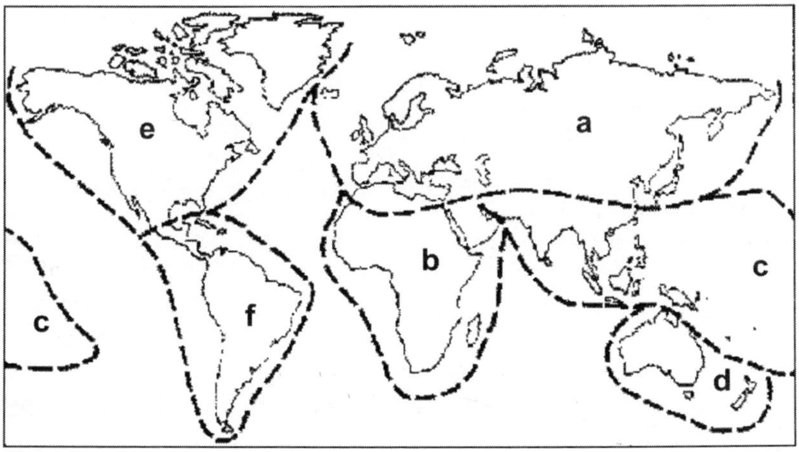

Wallace's realms

warm-core ring It is a circular eddy of warm water, surrounded by cold water and lying adjacent to a warm, poleward moving ocean current, such as Gulf stream.

wash A dry stream bed.

Wash

water balance The manner in which the precipitation received at a place is disposed of evapotranspiration, run-off and changes in the amounts of water stored within the soil in rocks. The water balance is calculated from the formula:

$$P = E + R \pm S$$

where P is precipitation, E is evapotranspiration, R is run-off and S represents changes in storage over the period of measurement. It is also referred as hydrological balance budget.

water equivalent The depth of water that would result from the melting of a snow sample. Typically about 10 cms of snow will melt to 1 cm of water, producing a water equivalent of 10 to 1.

water gap A valley through a ridge eroded by a river which continues to

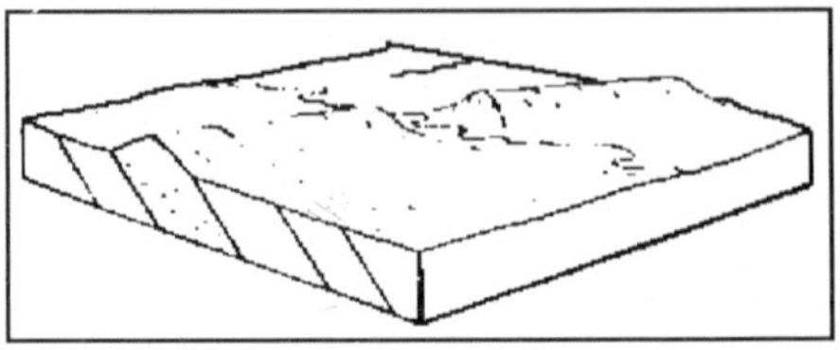

occupy the gap, e.g., Thames river cuts through the Chalk hills of the Berkshire Downs and Chilterns by way of Goring Gap.

water mass A large body of water with a distinctive set of properties, identifiable typically by its temperature and salinity.

water table The upper surface of the zone of saturation in rock, in other words the level below which the ground is saturated. Any hole in the ground will fills with water when the water table has been reached. This level often fluctuates with rainfall.

Water table

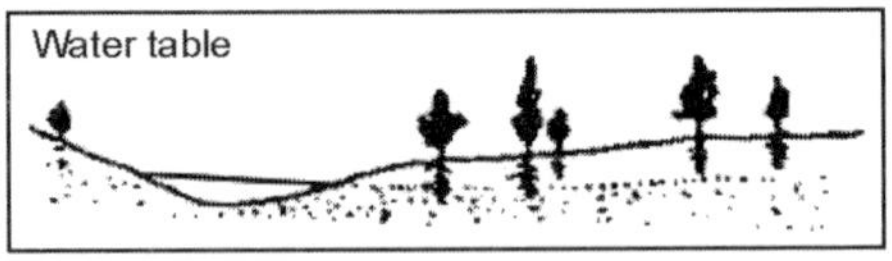

water vapour The existence of moisture in the atmosphere. It exists as an independent gas in variable quantities. The presence of water vapour is necessary for any form of precipitation and it also fulfils an important role in the earth's heat balance mechanism.

waterfall A vertical or near-vertical fall of water devolved where a river course is interrupted by a marked break of gradient. Waterfalls may be found at a band of more resistant rock at a knick point or where deposition has occurred. The famous Niagara Falls is having a drop of 55 m. Some waterfalls are the product of differential erosion; for example, along a Fault line soft rocks may be brought against hard rocks and will be rapidly eroded by fluvial activity, leading to the formation of a waterfall at the fault.

waterlogging The rise of a water table in alluvium to bring the zone of saturation into the root zone of plants.

watershed The boundary between two river systems. The watershed marks the divide between drainage basins and usually runs along the highest points of the interfluves.

waterspout It is just like marine tornado of shorter duration and lower intensity. It is due to local heating creating small-scale low pressure systems and their associated cumulonimbus clouds.

wave In water (specially oceans), the swell or forward movement caused by the friction of wind on the water surface. The oscillation is generated by the wind's pressure on the water surface and the wave magnitude is in proportion to the speed of the wind. Waves cause coastal erosion when they reach the shore. As the wave nears the shore it breaks, because of friction with sea-bed and carries debris up the beach as swash. Some waves are caused by submarine earthquakes as tsunami, which cause heavy destruction in coastal areas.

wave base The lower limit of wave transportation and erosion, equal to half the wave-length.

wave crest The highest part of a wave.

wave height The vertical distance between a wave crest and the preceding trough.

wave length The horizontal distance between similar points on two successive waves, measured perpendicular to the crest.

wave period The interval of time required for a wave crest to travel a distance equal to one wavelength; the interval of time required for two successive waves crests to pass a fixed point.

wave refraction The process by which a wave is bent or turned from its original direction. In sea waves, as a wave approaches a shore obliquely part of its reaches the shallow water near the shore while the rest is still advancing in deeper water. The part of the wave in the shallower water moves more slowly than the part in the deeper water.

wave steepness A ratio of waveheight to wavelength that measures the peakedness of a wave.

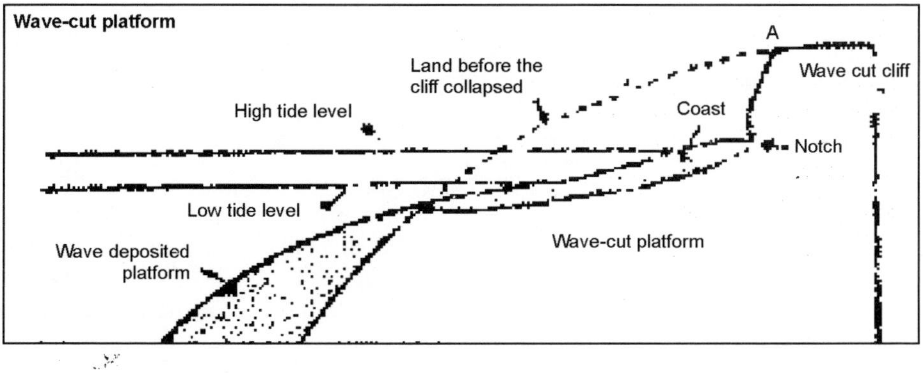

wave trough The lowest part of a wave, between successive crests.

wave-cut platform An outcrop of level or gently sloping rocks on the shore which have been scraped and smoothed by abrasion. Wave-cut platforms vary in width from a few metres to hundreds of metres, depending on rock-type and resistance and duration of marine erosion.

weather General atmospheric conditions at a particular time, vary from minutes to months. This include sunshine, temperature, rainfall, wind, cloud contribute to weather. In contrast to weather, climate is concerned with the long-term average conditions and behaviour of the atmosphere.

weather forecasting The science of predicting the future state of the atmosphere from very short period of less than one hour upto seven or ten days ahead. Modern science of weather forecasting has been revolutionised with the application of technology, e.g., satellites, radar, super computers, etc.

weather types Certain weather patterns categorised into similar groups. Used as an aid in weather prediction.

weather warning forecast indicting that hazardous weather is either imminent or actually occurring within the specified forecast area.

weather watch A forecast indicating that atmospheric conditions are favourable for hazardous weather to occur over a particular region during a specified time period.

weathering The process by which rocks are chemically altered or physically broken into fragments as a result of exposure to atmospheric agents and the pressures and temperatures at or near earth's surface, with little or no transportation of the loosened or altered materials. Weathering can be subdivided into **mechanical weathering**, **chemical weathering**, and **organic weathering**.

welded tuff A rock formed from particles of volcanic ash that were hot enough to become fused together.

westerlies The wind system which dominates the zones between latitudes 40° and 70°N and S of the equator. The flow of air is from the subtropical high pressure to the temperate zone of low pressure. The wind direction in N hemisphere is from SW and in the S hemisphere from the NW. Depressions are common in this wind system. The speed of westerlies reaching average high altitude speeds of 45–67 sec. In S hemisphere, these winds are stronger and known as **Roaring Forties**.

wet adiabatic lapse rate It is a reduced adiabatic lapse-rate when condensation is taking place in rising air, value ranges between 4 and 9°C per 1000 m.

wet spell Period of at least fifteen successive wet days.

wet-bulb thermometer An instrument for measuring temperature in which the bulb is enclosed in a moist muslin bag, thereby causing a loss of latent heat owing to evaporation. Thus, the temperature recorded by wet-bulb thermometer is lower than of dry-bulb, the difference being referred to as the wet-bulb depression. It helps in calculation of relative humidity.

wetlands The land areas of poor surface drainage, such as marshes and swamps. Maltby (1986) has defined wetland as "a collective term for ecosystems whose formation has been dominated by water, and whose processes and characteristics are largely controlled by water. A wetland is a place that has been wet enough for a long time to develop specially adopted vegetation and other organism".

whaleback (1) A term used for glacially-smoothed, elongated rock. (2) A very large longitudinal dune characterised by a flat top on which smaller branchans, also known as **seifduens.**

whirlpool A violent, circular, swirling movement in a water body.

whirlwind A rotating column of air on vortex, revolving around a local low pressure centre.

willy-willy This is a type of tropical cyclone occurred at the coastline of NW Australia.

wind The movement of air from one place to another. Air always moves from high pressure to low pressure. As the wind travels, it is diverted to its right in the N hemisphere and to its left in the S hemisphere. Winds are named by the direction from which they are blowing. The Beaufort scale is a traditional method of measuring wind speed.

wind direction The direction from which the wind is blowing.

wind energy Generated from wind by using windmills. It is one of the alternative sources of energy that could be used to generate electricity.

wind gap A gap in a ridge through which a stream, now abandoned as a result of stream piracy, once flowed.

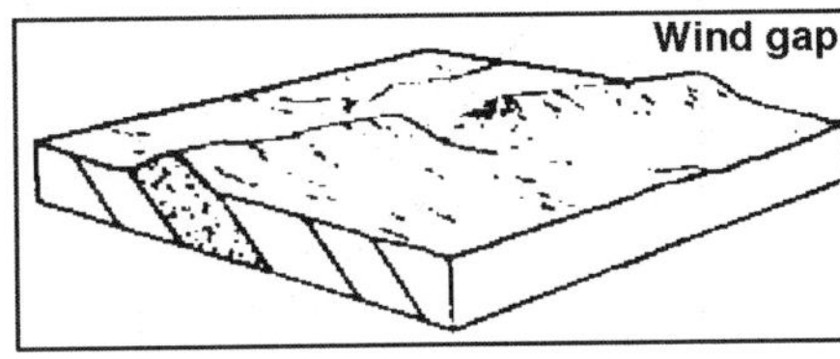

wind profiler A Doppler radar capable of measuring the turbulent eddies that move with the wind. Because of this, it is able to provide a vertical picture of wind speed and direction.

wind rose A diagram that shows the percentage of time that the wind blows from different directions at a given location over a given time. The percentage frequency of calms is usually indicated inside the circle and wind speed can be depicted by varying thickness of radiating lines.

wind shadow The area behind an obstacle where air movement is not capable of moving material.

wind shear Rate of change of wind speed or wind direction over a given distance.

wind vane An instrument used to indicate wind direction.

wind-chill factor The cooling effect of any combination of temperature and wind, expressed as the loss of body heat.

windward side The side of an object facing into the wind.

winter solstice The situation that occurred on 22 December in northern hemisphere and 21 June in the southern hemisphere.

winterbourne A stream which flows in winter but dries up in summer when the water table is lower. Such streams are found only in chalk areas in Southern England.

World Weather Watch (www) The programme planned by the World Meteorological Organisation as a coordinated attempt to give a global coverage of data relating to atmospheric phenomena.

wrinkle ridge A sinuous, irregular segmented ridge on the surface of a lunar mare, believed to be a result of compression of the lava.

X

xenocrysts Occurrence of crystals in igneous rocks. These are foreign to the rock body, because they have been introduced into the magma from an extraneous source.

xenolith This is an inclusion of pre-existing rock within an igneous rock. It may be a block of country rock, or a fragment of metamorphic rock.

xeralfs A suborder of the soil order **Alfisols**, includes Alfisols of the Mediterranean climate.

xerolls A suborder of the soil order Mollisols of the Mediterranean climate.

xerophyte A plant which is adopted to withstand seasons or perennial drought. Such plants are found in arid and semi-arid regions. Xerophytes are having long roots, thick barks, small glossy leaves and a capacity to store water when it becomes available. Cactus, acacia and baobab are examples of xerophilous species.

xerosere A plant succession developed in a dry environment.

X-ray diffraction In mineralogy, the process of identifying mineral structures by exposing crystals to a beam of X-ray and studying the resulting diffraction patterns.

Y

ya zoo stream A tributary stream that flows parallel to the main stream for a considerable distance before joining it. Such a tributary is forced to flow along the base of a natural levee formed by the main stream. It is named from River Yazoo, a tributary of the lower Mississippi.

yardang It is an elongated ridge carved by wind erosion. The ridges are parallel to the prevailing winds in arid regions with soft sediment at the surface.

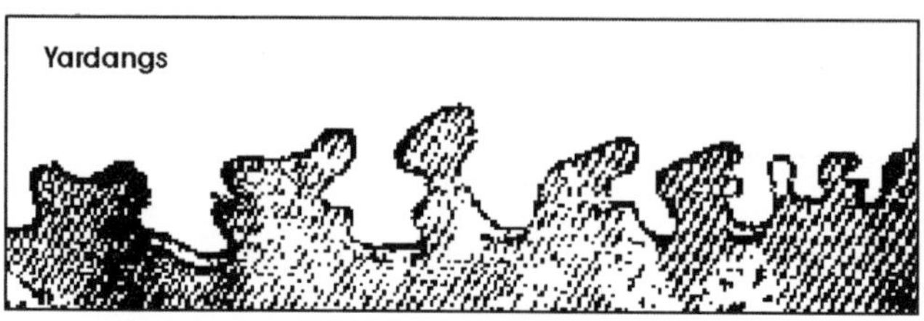

young fold mountains The fold mountains created by earth movements of the alpine orogeny of mid-Tertiary times. Himalayas, Andes and Alps are included in this category.

younger dryas The youngest of the five phases of the late-Glacial episode of the Devensiam stage in north-western Europe. This was a period of reduced temperatures and glacial advance. During this phase, areas like Mediterranean, Russia and North America experienced periglacial conditions.

youth (1) The first stage of the cycle of erosion, when rivers are engaged in active down-cutting and slope retreat is relatively less active. The river valley is narrow and V-shaped in cross-section. (2) In case of landscape, the landscape still covered with mountains; for example, the Himalayas or Alps.

Z

zenith For any location on earth, the point that is directly overhead to an observer. The zenith position of the sun is the one directly overhead.

zeugen This is an upstanding rock in a desert landscape, resulting from wind erosion in deserts. Zeugens stand upto 30 m in height, and are separated from each other by depressions. Zeugens are also known as **mushroom rock**.

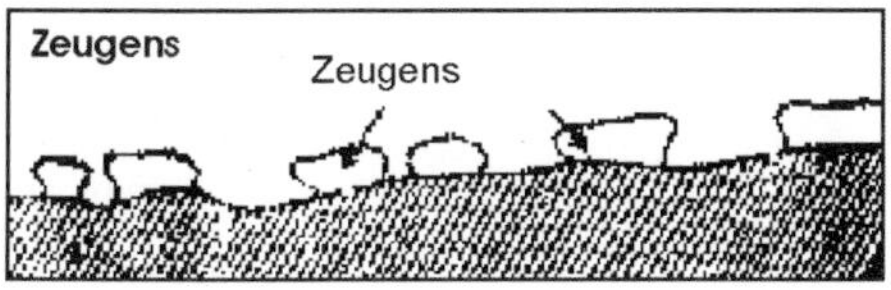

zonal circulation A reference, to geographically extensive atmosphere circulation reaching across different zones of latitude.

zonal soil Soils with well-developed horizons that reflect the climatic conditions of the region in which they are found.

zonal winds The winds associated with particular latitudinal zones, such as trade winds or westerlies.

zonation These are recognizable bands or zones of plant and animal communities in certain environments, such as within salt marshes and along rocky shorelines.

zone of eluviation The level, or zone, in a soil losing materials in the form of colloids and ions as part of soil formation; the zone of mineral removal.

zone of illuviation The level, or zone, in a soil where colloids and ions accumulate; the zone of mineral accumulation.

zoography The science of the distribution of animals, linking the subject matter of the discipline of zoology with the view point of geography.